Advance in Photoactive Nanomaterials

Advance in Photoactive Nanomaterials

Editor

Zhixing Gan

MDPI • Basel • Beijing • Wuhan • Barcelona • Belgrade • Manchester • Tokyo • Cluj • Tianjin

Editor
Zhixing Gan
School of Computer and
Electronic Information
Nanjing Normal University
Nanjing
China

Editorial Office
MDPI
St. Alban-Anlage 66
4052 Basel, Switzerland

This is a reprint of articles from the Special Issue published online in the open access journal *Nanomaterials* (ISSN 2079-4991) (available at: www.mdpi.com/journal/nanomaterials/special_issues/nano_photoactive).

For citation purposes, cite each article independently as indicated on the article page online and as indicated below:

LastName, A.A.; LastName, B.B.; LastName, C.C. Article Title. *Journal Name* **Year**, *Volume Number*, Page Range.

ISBN 978-3-0365-8251-1 (Hbk)
ISBN 978-3-0365-8250-4 (PDF)

Contents

About the Editor

Zhixing Gan

Zhixing Gan obtained his B.S. in Applied Physics in 2010 from Northeastern University, China, and then received his Ph.D. from Nanjing University in 2015. After that, he joined Nanjing Normal University as a Principal Investigator. His primary research interests include laser manipulation, photoluminescence, and the photothermal effect of emerging nanomaterials, such as carbon nanomaterials, graphene, 2D materials, and perovskites. He has co-authored 140 journal papers, receiving more than 3800 citations.

nanomaterials

MDPI

Editorial

Recent Developments in Photoluminescent, Photothermal and Photocatalytic Nanomaterials

Zhixing Gan

Center for Future Optoelectronic Functional Materials, School of Computer and Electronic Information/School of Artificial Intelligence, Nanjing Normal University, Nanjing 210023, China; zxgan@njnu.edu.cn

Photoactive nanomaterials exhibit myriad customized properties, including a photon converting ability, specific surface area, physicochemical stability, and chemical reactivity, making them appealing for a wide range of practical applications. Photoactive nanomaterials can convert photon to photon (photoluminescence) [1], heat (photothermal effect) [2], and separated charged carriers (photocatalytic reaction, photovoltaic effect) [3], which is very important for the manipulation of light and the utilization of solar energy. This Special Issue collates eleven papers, including nine research articles and two review articles, on recent research progress in emerging photoluminescent (PL), photothermal, and photocatalytic nanomaterials, and their applications in optoelectronics and energy conversion.

The photoluminescent materials investigated include crystalline tris(8-hydroxyquinoline) aluminum (Alq3) micro-rods (MRs), Fe^{2+}-doped $CsPb(Cl_xBr_{1-x})_3$ nanocrystals (NCs), organosilica films, and Tm^{3+}-doped SiO_2–HfO_2 [4–7]. Kim et al. fabricated hybrid Alq3/Ag structures using a self-assembly method under a mixed solution of protic and aprotic polar solvents [4]. An evident PL enhancement of approximately 26 times was obtained due to the localized surface plasmon resonance (LSPR) effects between crystalline Alq3 MRs and Ag nanowires. Rasadujjaman et al. revealed that the PL of organosilica films mainly originates from the carbon-containing components rather than the oxygen-deficient centers [5]. They showed that the PL intensity could be improved by increasing the porosity and internal surface area. Wu et al. prepared Fe^{2+}-doped $CsPb(Cl_xBr_{1-x})_3$ NCs using a hot injection method [6]. Since the defect density was reduced by the Fe^{2+} doping, the full width at half maximums declined, while PL quantum yields (QYs) and photostability increased. Zulfikri et al. deposited Tm^{3+}-doped SiO_2–HfO_2 in the form of nanofibers (NFs) and thin films (TFs) on a single substrate using the electrospinning and dip-coating methods, respectively [7]. Compared with single-layered NF and TF structures, the composites showed an about tenfold improvement in PL intensity. These articles reported the PL enhancement of different nanomaterials via various approaches, which will not only promote the applications of these PL materials, but also inspire the development of the PL modulation method.

Two review articles presented content related to photothermal nanomaterials [8,9]. Thermochromic smart windows that can actively regulate solar radiation according to the ambient temperature have important application prospects in the field of lower energy consumption buildings. Both photothermal effects and electrothermal effects can induce the thermochromic process. Thus, a large number of pioneering investigations have been conducted to change the optical performance of smart windows based on photothermal or electrothermal effects at room temperature. Recent advances in photo- and electro-driven thermochromic smart windows are summarized in these two review articles.

Another main contribution comes in the field of photocatalytic nanomaterials. Meng et al. fabricated black 3D-TiO_2 nanotube arrays on Ti meshes using a facile electrochemical reduction method [10]. The black 3D-TiO_2 nanotube arrays attained a maximal photocurrent density of 1.6 mA/cm^2 at 0.22 V vs. Ag/AgCl with a Faradic efficiency of 100%, resulting in an enhanced photoelectrochemical water-splitting performance.

Citation: Gan, Z. Recent Developments in Photoluminescent, Photothermal and Photocatalytic Nanomaterials. *Nanomaterials* **2023**, *13*, 1888. https://doi.org/10.3390/nano13121888

Received: 2 June 2023
Accepted: 15 June 2023
Published: 19 June 2023

Guan et al. prepared molybdenum-doped thin $ZnIn_2S_4$-containing S vacancies (Mo-doped Sv-$ZnIn_2S_4$) via a one-pot solvothermal method [11]. The cooperation of Mo doping and S vacancies enhanced light absorption and facilitated the separation of the electron-hole, while also providing active sites. Thus, the Mo-doped Sv-$ZnIn_2S_4$ exhibited a high hydrogen evolution rate of 5739 μmol g^{-1} h^{-1} with a high apparent quantum yield of 21.24% at 420 nm. Boaretti et al. explored the effects of combinations of different advanced oxidation processes on the degradation of formaldehyde at a low concentration [12]. They demonstrated that ultrasound treatment improved the kinetics, producing a final abatement of formaldehyde and breaking the bottleneck of photocatalysis. Santis et al. prepared 2D-$SnSe_2$ nanoflakes by using solvent-assisted sonication, and then embedded them in ordered mesoporous titania thin films [13]. The heterostructures formed between $SnSe_2$ and nanocrystalline anatase titania resulted in enhanced photocatalytic activity, which was used to remove fingerprints from the surface of smartphones, tablets, and computers. These articles reported highly efficient photocatalysts for water splitting, wastewater treatment, and anti-fingerprint purposes, which are significant for clean energy development and environmental protection.

In summary, photoactive nanomaterials, encompassing various applications, have attracted enormous research interest. This Special Issue mainly covers the applications of photoactive nanomaterials in photoluminescence, thermochromic smart windows, photocatalytic water splitting, photocatalytic degradation, and functional coatings. We hope that the contributions to this Special Issue enable readers to gain more insight into the recent advances in this area, and also provide helpful guidance for the future development of photoactive nanomaterials.

Conflicts of Interest: The authors declare no conflict of interest.

References

1. Gan, Z.; Shan, Y.; Chen, J.; Gui, Q.; Zhang, Q.; Nie, S.; Wu, X. The origins of the broadband photoluminescence from carbon nitrides and applications to white light emitting. *Nano Res.* **2016**, *9*, 1801–1812. [CrossRef]
2. Shan, X.; Zhao, A.; Lin, Y.; Hu, Y.; Di, Y.; Liu, C.; Gan, Z. Low-Cost, Scalable, and Reusable Photothermal Layers for Highly Efficient Solar Steam Generation and Versatile Energy Conversion. *Adv. Sustain. Syst.* **2020**, *4*, 1900153. [CrossRef]
3. Jia, R.; Gui, Q.; Sui, L.; Huang, Y.; Lu, H.; Dong, H.; Ma, S.; Gan, Z.; Dong, L.; Yu, L. Active sites provided by the surface autocatalytic effect and quantum confinement for stable and efficient photocatalytic hydrogen generation. *J. Mater. Chem. A* **2021**, *9*, 14768. [CrossRef]
4. Kim, M.; Kim, J.; Ju, S.; Kim, H.; Jung, I.; Jung, J.H.; Lee, G.S.; Hong, Y.K.; Park, D.H.; Lee, K.-T. Enhanced Photoluminescence of Crystalline Alq_3 Micro-Rods Hybridized with Silver Nanowires. *Nanomaterials* **2023**, *13*, 825. [CrossRef] [PubMed]
5. Wu, C.; Li, Y.; Xia, Z.; Ji, C.; Tang, Y.; Zhang, J.; Ma, C.; Gao, J. Enhancing Photoluminescence of $CsPb(Cl_xBr_{1-x})_3$ Perovskite Nanocrystals by Fe^{2+} Doping. *Nanomaterials* **2023**, *13*, 533. [CrossRef] [PubMed]
6. Rasadujjaman, M.; Zhang, J.; Spassky, D.A.; Naumov, S.; Vishnevskiy, A.S.; Vorotilov, K.A.; Yan, J.; Zhang, J.; Baklanov, M.R. UV-Excited Luminescence in Porous Organosilica Films with Various Organic Components. *Nanomaterials* **2023**, *13*, 1419. [CrossRef] [PubMed]
7. Zulfikri, N.I.Z.; Ibrahim, A.-B.M.A.; Mustaffa, N.A.; Yunus, R.M.; Kamil, S.A. Enhancing Photoluminescence Intensity and Spectral Bandwidth of Hybrid Nanofiber/Thin-Film Multilayer Tm^{3+}-Doped SiO_2–HfO_2. *Nanomaterials* **2022**, *12*, 3739. [CrossRef] [PubMed]
8. Zou, X.; Ji, H.; Zhao, Y.; Lu, M.; Tao, J.; Tang, P.; Liu, B.; Yu, X.; Mao, Y. Research Progress of Photo-/Electro-Driven Thermochromic Smart Windows. *Nanomaterials* **2021**, *11*, 3335. [CrossRef] [PubMed]
9. Zhao, Y.; Ji, H.; Lu, M.; Tao, J.; Ou, Y.; Wang, Y.; Chen, Y.; Huang, Y.; Wang, J.; Mao, Y. Thermochromic Smart Windows Assisted by Photothermal Nanomaterials. *Nanomaterials* **2022**, *12*, 3865. [CrossRef] [PubMed]
10. Meng, M.; Feng, Y.; Li, C.; Gan, Z.; Yuan, H.; Zhang, H. Black 3D-TiO_2 Nanotube Arrays on Ti Meshes for Boosted Photoelectrochemical Water Splitting. *Nanomaterials* **2022**, *12*, 1447. [CrossRef] [PubMed]
11. Guan, W.; Zhang, L.; Wang, P.; Wang, Y.; Wang, H.; Dong, X.; Meng, M.; Sui, L.; Gan, Z.; Dong, L.; et al. Highly Efficient Photocatalytic Hydrogen Evolution over Mo-Doped $ZnIn_2S_4$ with Sulfur Vacancies. *Nanomaterials* **2022**, *12*, 3980. [CrossRef] [PubMed]

12. Boaretti, C.; Roso, M.; Modesti, M.; Lorenzetti, A. Ultrasound-Promoted Abatement of Formaldehyde in Liquid Phase with Electrospun Nanostructured Membranes: The Synergy of Combined AOPs. *Nanomaterials* **2023**, *13*, 435. [CrossRef] [PubMed]
13. De Santis, J.; Paolucci, V.; Stagi, L.; Carboni, D.; Malfatti, L.; Cantalini, C.; Innocenzi, P. Bidimensional $SnSe_2$—Mesoporous Ordered Titania Heterostructures for Photocatalytically Activated Anti-Fingerprint Optically Transparent Layers. *Nanomaterials* **2023**, *13*, 1406. [CrossRef] [PubMed]

nanomaterials

MDPI

Article

Black 3D-TiO_2 Nanotube Arrays on Ti Meshes for Boosted Photoelectrochemical Water Splitting

Ming Meng [1,*], Yamin Feng [1], Chunyang Li [1], Zhixing Gan [2,*], Honglei Yuan [1] and Honghui Zhang [1]

[1] School of Physics and Telecommunication Engineering, Zhoukou Normal University, Zhoukou 466001, China; yadan205@126.com (Y.F.); lichunyang98@163.com (C.L.); yuanhenu@163.com (H.Y.); zhanghonghui4714@163.com (H.Z.)

[2] Key Laboratory of Optoelectronic Technology of Jiangsu Province, School of Physical Science and Technology, Nanjing Normal University, Nanjing 210023, China

* Correspondence: mengmingfly@163.com (M.M.); zxgan@njnu.edu.cn (Z.G.)

Abstract: Black 3D-TiO_2 nanotube arrays are successfully fabricated on the Ti meshes through a facile electrochemical reduction method. The optimized black 3D-TiO_2 nanotubes arrays yield a maximal photocurrent density of 1.6 mA/cm^2 at 0.22 V vs. Ag/AgCl with Faradic efficiency of 100%, which is about four times larger than that of the pristine 3D-TiO_2 NTAs (0.4 mA/cm^2). Such boosted PEC water splitting activity primarily originates from the introduction of the oxygen vacancies, which results in the bandgap shrinkage of the 3D-TiO_2 NTAs, boosting the utilization efficiency of visible light including the incident, reflected and/or refracted visible light captured by the 3D configuration. Moreover, the oxygen vacancies (Ti^{3+}) can work as electron donors, which leads to the enhanced electronic conductivity and upward shift of the Fermi energy level, and thereby facilitating the transfer and separation of the photogenerated charge carrier at the semiconductor-electrolyte interface. This work offers a new opportunity to promote the PEC water splitting activity of TiO_2-based photoelectrodes.

Keywords: 3D-TiO_2 nanotube arrays; electrochemical reduction; oxygen vacancies; photoelectrochemical water splitting

Citation: Meng, M.; Feng, Y.; Li, C.; Gan, Z.; Yuan, H.; Zhang, H. Black 3D-TiO_2 Nanotube Arrays on Ti Meshes for Boosted Photoelectrochemical Water Splitting. *Nanomaterials* **2022**, *12*, 1447. https://doi.org/10.3390/nano12091447

Academic Editor: Yuichi Negishi

Received: 28 February 2022
Accepted: 21 April 2022
Published: 24 April 2022

1. Introduction

Photoelectrochemical (PEC) water splitting technology capable of directly converting and storing the abundant solar energy into energy-dense hydrogen fuel has emerged as a promising strategy to alleviate the worsening energy crisis and environmental issues [1–9]. To achieve the practical application of this technology, the fabrication of stable and efficient photoelectrodes are desperately needed [10–17]. Three-dimensional TiO_2 nanotube arrays (3D-TiO_2 NTAs) formed on Ti mesh have been recognized as a competitive candidate in the design and fabrication of a photoanode for PEC water splitting owing to its larger internal and external surface areas, efficient charge separation and transportation features, and optimal adhesion with Ti mesh [18–28]. More importantly, the 3D-TiO_2 NTAs on Ti mesh exhibits significant improvement in the utilization efficiency of Ti source compared to the 2D-TiO_2 nanotube arrays formed on Ti foil [18,20,28]. Besides, the radial nature of 3D-TiO_2 NTAs endows it with capability of harvesting the incident, reflected and/or refracted ultraviolet and visible light from any direction surrounding the Ti wire, rendering a higher PEC water splitting activity to be achieved [18,20,29]. However, its PEC performances are still inhibited by the large bandgap (3.2 eV), which results in the photoexcited electron and hole not being produced by the visible light harvested by the 3D NTAs [30–37]. In addition, 3D-TiO_2 NTAs also suffer from poor electrical conductivity, and the bulk and surface recombination of photogenerated charge carriers, both of which are detrimental to the PEC water splitting activity [37–41]. Consequently, seeking an efficient strategy to boost the utilization of visible light and the electrical conductivity is vitally crucial.

Lately, the introduction of the oxygen vacancies (O-vacancies) has been demonstrated as an effective tactic to steer the optical and electronic characteristics of the metal oxide [41–44]. As illustrated by many research groups, the introduction of the O-vacancies can enable the Fermi energy level to shift toward the conduction band, which leads to the shrinkage of the bandgap, thus promoting the utilization efficiency of the visible light [31,41–44]. In addition, the presence of O-vacancies can also increase the electrical conductivity due to the high donor density, which facilitates the separation and transport of photogenerated charge carriers [31,38,39,44]. Accordingly, it is anticipated that rational introduction of the O-vacancies in 3D-TiO_2 NTAs may be a promising route to tackle the two abovementioned drawbacks. Unfortunately, the reported available strategies to produce O-vacancies generally involve the harsh experimental conditions or high-cost facilities, which are not suitable for the largescale practical application [31,38,39,44]. Hence, exploiting a simple and economical method to introduce O-vacancies into metal oxide still requires more endeavors.

Recently, an electrochemical reductive doping process has proved to be a simple and cost-effective route to introduce O-vacancies into the TiO_2 NTAs [44–48]. Under an external electric field, the Ti^{4+} is reduced to Ti^{3+}, which leads to the generation of O-vacancies. Three different reduction electrolytes have been utilized, including acidic (H_2SO_4), neutral (Na_2SO_4), and alkaline (KOH) aqueous solution [44–49]. It is found that the alkaline electrolytes are more favorable to the introduction of O-vacancies because of the occurrence of a gas-forming side reaction during reduction in acidic solution [47,49]. Nevertheless, the existing research mainly focused on electrochemical reduction in acidic and neutral aqueous solution. As such, electrochemical reduction in alkaline aqueous solution have not been comprehensively understood. For example, the fundamental questions are whether alkaline aqueous solution is general or just for KOH, which remains unclear so far.

Herein, black 3D-TiO_2 NTAs with substantial O-vacancies were prepared via a simple electrochemical reduction in NaOH solution, where the 3D-TiO_2 NTAs were reduced by cathodic polarization for 15 min. As expected, the optimally reduced 3D-TiO_2 NTAs generated a photocurrent density of 1.6 mA/cm^2 at 0.22 V vs. Ag/AgCl with Faradic efficiency of 100%, nearly four times higher than that of the pristine 3D-TiO_2 NTAs. Such boosted PEC water splitting activity primarily originates from the introduction of the O-vacancies, which results in bandgap shrinkage of the 3D-TiO_2 NTAs, boosting the utilization efficiency of visible light including the incident, reflected and/or refracted visible light captured by the 3D configuration. Moreover, the O-vacancies (Ti^{3+}) can work as electron donors, which leads to enhanced electronic conductivity and upward shift of the Fermi level, thereby facilitating the transfer and separation of the photogenerated charge carrier at the semiconductor-electrolyte interface. This work offers a new opportunity to promote the PEC water splitting activity of TiO_2-based photoelectrodes.

2. Materials and Methods

2.1. Preparation of the 3D-TiO_2 NTAs

The 3D-TiO_2 NTAs were fabricated by the electrochemical anodization of Ti meshes. Briefly, anodization was performed via a conventional two-electrode system, with clean Ti mesh (Alfa Aesar (China) Chemical Co. Ltd, Shanghang, China, 80-mesh) with size of 1.5 cm × 1 cm as the anode and Pt mesh as the cathode, respectively. The electrolytes solution was prepared by dissolving 0.3 wt% NH_4F and 2 vol% DI H_2O in ethylene glycol. The Ti mesh was anodized by 60 V for 1 h. After anodization, the as-prepared 3D-TiO_2 NTAs were thoroughly rinsed with ethanol and DI H_2O, respectively, and then were annealed in air at 400 °C for 2 h (denoted as pristine 3D-TiO_2 NTAs).

2.2. Electrochemical Reduction of the 3D-TiO_2 NTAs

The electrochemical reduction was conducted in a conventional three-electrode system. The as-prepared 3D-TiO_2 NTAs, Ag/AgCl (3 mol L-1 KCl-filled) and Pt mesh were employed as the working, reference, and counter electrode, respectively. NaOH aqueous solution (1 M, pH = 13.6) was utilized as electrolyte. The electrochemical reduction bias of −1.2, −1.3 and −1.4 (vs. Ag/AgCl) were used, and the corresponding photoelectrodes are denoted as ECR-3D-TiO_2 NTAs−x, where x = 1.2, 1.3 and 1.4 V. The time of the electrochemical reduction was 15 min.

2.3. Characterization

Morphologies, microstructures and crystal structures of the as-prepared samples were characterized by field-emission scanning electron microscopy (FE-SEM, S4800, Hitachi Ltd., Tokyo, Japan), field-emission transmission electron microscopy (FE-TEM, JEM-2100, JEOL Ltd., Tokyo, Japan) and X-ray powder diffractometry (XRD, Xpert, Philips, Amsterdam, The Netherlands). The diffuse reflectance spectra were measured by a VARIAN Cary5000 spectrophotometer (Varian, CA, USA). The X-ray photoelectron spectroscopy (XPS) data were collected by the PHI 5000 Versaprobe (Ulvac-Phi, Kanagawa, Japan).

2.4. Photoelectrochemical Measurements

The PEC tests were conducted in a three-electrode configuration connected to a CHI 660E electrochemical workstation (CH Instrument, Chenhua Ltd., Shanghai, China), with the pristine and ECR-3D-TiO_2 NTAs with an exposed area of 1 cm^2, Ag/AgCl (3 mol L^{-1} KCl-filled), and Pt mesh as the working, reference, and counter electrode, respectively. The supporting electrolyte was 1 M NaOH (pH = 13.6). The irradiation source was a 500 W Xe lamp (Solar 500, NBet Group Corp., Beijing, China) with calibrated intensity of 100 $mWcm^{-2}$. Moreover, a water filter was used between the lamp and electrochemical cell to remove solution heating from infrared light. An Ocean Optics oxygen sensor system equipped with a FOXY probe (NeoFox Phase Measurement System, Ocean optics, Orlando, FL, USA) was applied to determine the amount of evolved O_2. The experiment was carried out together with the stability tests. Before the O_2 measurement, the headspace of the anodic compartment was purged with high purity N_2 (99.9995%) for 1 h under vigorous stirring. PEC water splitting with O_2 sensing continued for 180 min at 0.22 V vs. Ag/AgCl, and the O_2 yield was quantified to calculate the Faradic efficiency. Electrochemical impedance spectroscopy was carried out to understand the charge transfer process between photoelectrodes/electrolyte interfaces. All the measurements were performed under the open circuit condition with the frequency ranging from 0.01 Hz to 100 kHz. Mott-Schottky plots were derived from impedance potential tests conducted at a frequency of 1 kHz in dark conditions.

3. Results

3.1. Morphological Characterization of the Pristine and ECR-3D-TiO_2 NTAs

The morphologies of the 3D-TiO_2 NTAs before and after electrochemical reduction were investigated by FE-SEM. The low-magnification overall FE-SEM image of the ECR-3D-TiO_2 NTAs−1.3 V displays that the diameter of a single Ti wires is about 0.12 mm and the percentage of the open area of Ti mesh is calculated to approximately 30%, suggesting the higher utilization efficiency of the Ti source (Figure 1a). Figure 1c,d are the magnified FE-SEM images of the area marked by the red ellipse in Figure 1b, which clearly exhibits that TiO_2 NTAs are radially grown outward around the Ti wires, leading to the formation of 3D-TiO_2 NTAs. This highly ordered structure can be described by the 3D representation in Figure S1. The top and cross-sectional view FE-SEM images show such ECR-3D-TiO_2 NTAs with an average diameter of approximately 150 nm, a wall thickness of about 10 nm, and a similar length of 6 μm (Figure 1c,d and Figure S2), which are identical to those of the pristine 3D-TiO_2 NTAs.

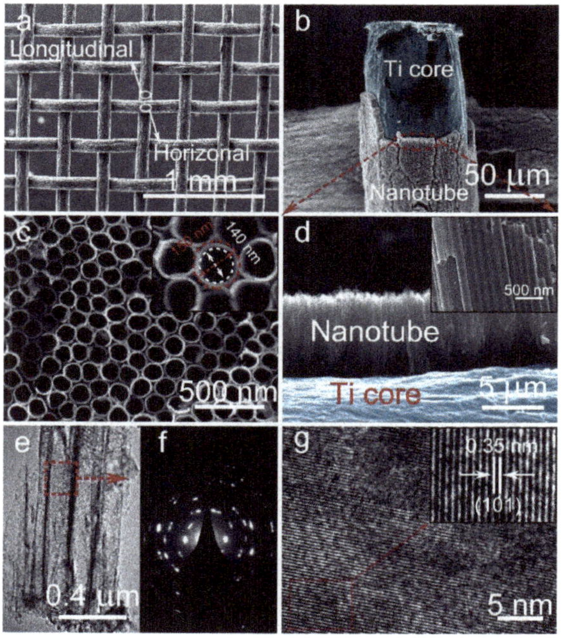

Figure 1. (**a**,**b**) Low-magnification FE-SEM images of the ECR-3D-TiO_2 NTAs−1.3 V. (**c**,**d**) Corresponding top and cross-sectional view FE-SEM images; Insets: magnified FE-SEM images. (**e**) Low-magnification FE-TEM of the ECR-3D-TiO_2 NTAs−1.3 V. (**f**,**g**) Selected area electron diffraction (SAED) pattern and HR-TEM image of the area highlighted by the red dashed box in (**e**). Insets: Inverse fast Fourier transform filtered TEM image recorded from the area bounded by the red dashed box in (**g**).

The effect of the electrochemical reduction on the morphologies and microstructures of 3D-TiO_2 NTAs were further investigated by FE-TEM. From the low-magnification FE-SEM images, all the products possess a tightly packed tubular nanostructures with a mean external diameter of 150 mm, which is consistent with the FE-SEM results (Figure 1e and Figure S3a,d,g). The selected electron diffraction patterns display very similar diffraction patterns, which demonstrate the polycrystalline structures of the 3D-TiO_2 NTAs before and after electrochemical reduction (Figure 1f and Figure S3b,e,h). In addition, the well-resolved lattice spacing of 0.305 nm are observed in all the products (Figure 1g and Figure S3c,f,i), which corresponds to the {101} plane of anatase TiO_2 [38,39]. The phase transition of the 3D-TiO_2 NTAs induced by electrochemical reduction were analyzed by XRD. As shown in Figure S4, all the diffraction peaks match well with crystal structure of the anatase TiO_2 (JCPDS 21-1272) and metal Ti [38,39]. No other phase is detected, suggesting no change in the lattice structures after electrochemical reduction. The above FE-SEM, FE-TEM and XRD results imply that electrochemical reduction does not destroy the morphology, microstructures or phase of the 3D-TiO_2 NTAs.

3.2. Optical Absorption Properties of the Pristine and ECR-3D-TiO_2 NTAs

We have investigated the UV-vis reflectance spectra of the ECR-3D-TiO_2 NTAs as a function of external bias applied in the electrochemical reduction and then compared with that of the pristine 3D-TiO_2 NTAs. Clearly, the pronounced absorption can be clearly observed in the UV region (<390 nm) of all the products, which can be attributed to the intrinsic band-to-band absorption of TiO_2 [38,39,44]. Compared with the pristine 3D-TiO_2 NTAs, the visible light absorption (400–800 nm) is significantly enhanced after electrochemical reduction. As the applied bias changes from −1.2 to −1.4 V, the visible light absorption increases gradually, which are further verified by the color variation of the ECR-3D-TiO_2 NTAs. This implies that the ECR-3D-TiO_2 NTAs may respond to the visible light region (Figure 2a). Moreover, the bandgaps of the pristine 3D-TiO_2 NTAs and ECR-3D-TiO_2 NTAs−1.2, −1.3 and −1.4 V, estimated from the intercept of the tangents to the curves of $(\alpha h\upsilon)^2$ vs. photon energy by assuming TiO_2 as a direct semiconductor, are about 3.09, 2.95, 2.65 and 2.63, respectively (Figure 2b). These results suggest that the electrochemical reduction not only promote the visible light absorption, but also reduce the bandgap of the 3D-TiO_2 NTAs, which can be ascribed to the presence of the defect state in the bandgap of TiO_2 created by the O-vacancies. The boosted visible light absorption and bandgap shrinkage means that visible light trapped by the 3D configuration can excite electron-hole pairs and thus effectively improve the PEC water splitting activity of the 3D-TiO_2 NTAs.

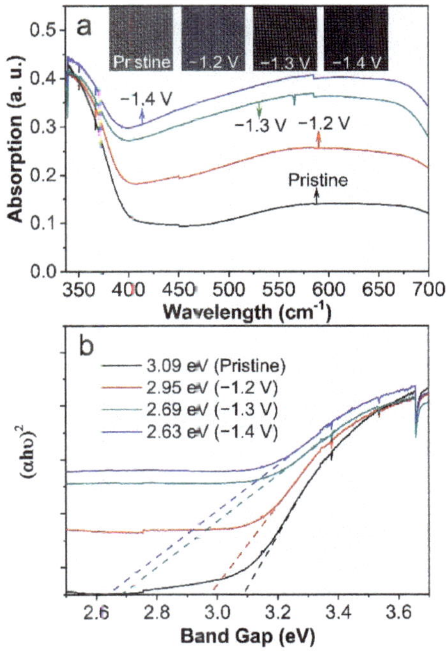

Figure 2. (**a**) UV-vis reflectance spectra and photographs (inset) of the pristine 3D-TiO_2 NTAs and ECR-3D-TiO_2 NTAs electrochemically reduced under the different applied bias −1.2, −1.3 and −1.4 V. (**b**) Corresponding curves of the transformed Kubelka–Munk function vs. the energy of light.

3.3. Surface Oxidation State of the Pristine and ECR-3D-TiO_2 NTAs

To solidify the presence of O-vacancies in the ECR-3D-TiO_2 NTAs, the chemical composition and surface oxidation states of the pristine 3D-TiO_2 NTAs and ECR-3D-TiO_2 NTAs were further examined by XPS. Only Ti, O and C signals are observed in the survey spectra of all the products, which reveals that electrochemical reduction does not introduce other impurities (Figure S5a). For the pristine 3D-TiO_2 NTAs, the Ti 2p core level spectrum has two peaks centered at 458.3 and 464.1 eV, which are typical for the Ti $2p_{3/2}$ and $2p_{1/2}$ peaks of Ti^{4+} in TiO_2 (Figure 3a) [39,43,50]. After the electrochemical reduction, the Ti $2p_{3/2}$ and $2p_{1/2}$ peaks shift to the low binding energy of 457.9 and 463.7 eV, illustrating the different bonding environment of the Ti atom. By subtracting the normalized Ti 2p spectra of the ECR-3D-TiO_2 NTAs−1.3 V with that of the pristine 3D-TiO_2 NTAs, two extra peaks at 457.7 and 463.3 eV were observed, which were indexed to the Ti $2p_{3/2}$ and $2p_{1/2}$ peaks of Ti^{3+} [39,43,50]. This indicates that O-vacancies are introduced in the ECR-3D-TiO_2 NTAs−1.3 V. In addition, the O1s spectra of the ECR-3D-TiO_2 NTAs−1.3 V was also different from that of that of the pristine 3D TiO_2 NTAs. In the O1s spectra, the main peak located 529.7 eV is the characteristic peak reported for lattice oxygen of TiO_2, while other peaks centered at 531.4 eV can be associated with oxygen species absorbed at O-vacancies [39]. As displayed in Figure 3b and Figure S5b, the peaks of area of 531.4 eV of ECR-3D-TiO_2 NTAs increase gradually with electrochemical reduction bias reducing from −1.2 V to −1.4 V, which suggests that the amount of the O-vacancies increases with the deceasing electrochemical reduction bias. This is why the visible light absorption increases gradually with the electrochemical reduction bias reducing from −1.2 to −1.4 V.

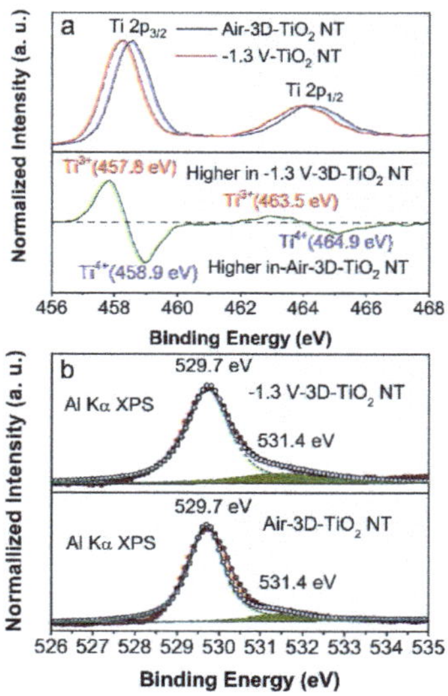

Figure 3. (**a**) Normalized Ti 2p XPS spectra of the pristine 3D-TiO_2 NTAs and ECR-3D-TiO_2 NTAs−1.3 V, and their difference spectrum (ECR-3D-TiO_2 NTAs−1.3 V minus pristine 3D-TiO_2 NTAs). (**b**) Normalized O1s XPS spectra of the pristine 3D-TiO_2 NTAs and ECR-3D-TiO_2 NTAs−1.3 V. The red circles represent the experimental XPS data. The blue circles are the fitting of the experimental data and can be divided into two peaks displayed by the green dashed lines.

3.4. PEC Water Splitting Activity of the Pristine and ECR-3D-TiO_2 NTAs

The influence of the electrochemical reduction bias on the PEC water splitting activity of 3D-TiO_2 NTAs were also studied, and the results are shown in Figure 4. All the 3D-TiO_2 NTAs-based photoelectrodes display negligible dark currents in comparison with their respective photocurrents, suggesting no occurrence of the electrocatalytic water splitting. Under irradiation, the photocurrent densities of the ECR-3D-TiO_2 NTAs increase steeply and are distinctly larger than that of the pristine 3D-TiO_2 NTAs in the whole potential window from −0.9 to 0.6 V vs. Ag/AgCl, which reveals that the electrochemical reduction can significantly promote the PEC performance of the 3D-TiO_2 NTAs. Figure 4b compares the transient photocurrent responses of the pristine and ECR-3D-TiO_2 NTAs measured at 0.22 V vs. Ag/AgCl. It can be seen that all the 3D-TiO_2 NTAs-based photoelectrodes show excellent sensitivity to the light irradiation. There is a steep rise in current density from almost zero in dark conditions to a stable value upon illumination. In addition, the ECR-3D-TiO_2 NTAs−1.3 V generate a maximal photocurrent density of 1.6 mA/cm^2, which is about four times larger than that of the pristine 3D-TiO_2 NTAs (0.4 mA/cm^2). This photocurrent density value is superior or comparable to the previously reported values on self-doping TiO_2 NTAs formed on Ti foil (Table S1) [39,43,46–48,51]. This means that the optimal electrochemical reduction bias is −1.3 V, which can be attributed to the two-faced effect of the O-vacancies on the PEC water splitting performance, and will be discussed thoroughly in the following text.

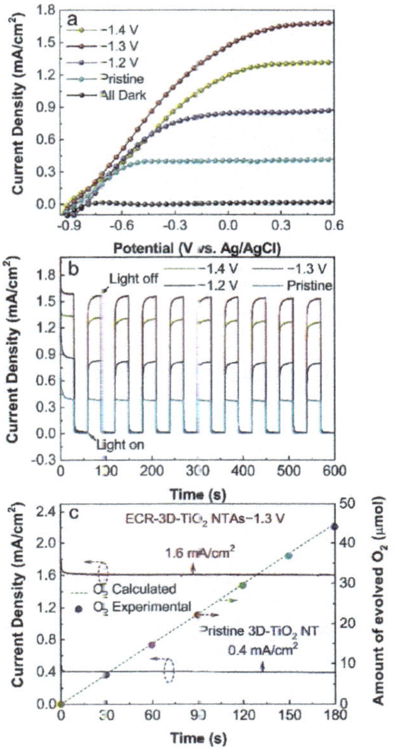

Figure 4. (**a**) Current density vs. voltage (J-V) plots of the pristine 3D-TiO_2 NTAs and ECR-3D-TiO_2 NTAs electrochemically reduced under the different applied bias −1.2, −1.3 and −1.4 V. (**b**) Corresponding transient photocurrent responses measured at 0.22 V vs. Ag/AgCl. (**c**) Photocurrent vs. time (*J-t*) curves of the pristine and ECR-3D-TiO_2 NTAs−1.3 V obtained at 0.22 V vs. Ag/AgCl. The dashed line and colorful circles are the amount of the evolved O_2 calculated theoretically and detected experimentally of the ECR-3D-TiO_2 NTAs−1.3 V, respectively.

The structural and chemical stability is a critical parameter for a photoelectrode during the PEC water splitting. To assess this property, the photocurrent density vs. time (*J-t*) curves of the pristine and ECR-3D-TiO_2 NTAs−1.3 V are obtained at 0.22 V vs. Ag/AgCl under continuous illumination (Figure 4c). No sign of decrease in photocurrent densities for the pristine and ECR-3D-TiO_2 NTAs-1.3 V are detected during the entirely measured 180 min. To further identify whether the observed photocurrents derive from the water splitting reaction, the amount of oxygen evolved from the ECR-3D-TiO_2 NTAs−1.3 V was determined by a fluorescence sensor. The amount of evolved oxygen increases linearly with test time with unity Faradic efficiency. Figure S6 presents the FE-SEM image and XRD pattern of the ECR-3D-TiO_2 NTAs−1.3 V after continuous PEC water splitting for 180 min, which prove that the surface morphology and crystal phase of the ECR-3D-TiO_2 NTAs-1.3 V remains intact. These results sufficiently confirm that excellent stability of the ECR-3D-TiO_2 NTAs−1.3 V, which is suitable for the potential long-term PEC water splitting application.

To investigate the effect of the electrochemical reduction on the electronic characteristics of 3D-TiO_2 NTAs, electrochemical impedance spectra (EIS) measurements were performed and the Nyquist plots are shown in Figure 5a, where the scatter points are the original experimental data, and the solid lines are the fitted curves utilizing the equivalent circuit mode in the inset of Figure 5a. It can be clearly seen that the equivalent circuit model fits well with the two samples. In this equivalent circuit model, R_s corresponds to the overall series resistance of the circuit, and R_{ct} represents the charge transfer resistance [47,52]. As depicted in Figure 5a, the ECR-3D-TiO_2 NTAs−1.3 V has a smaller semicircle diameter than the pristine 3D-TiO_2 NTAs under illumination, suggesting the smaller charge transfer resistance of the ECR-3D-TiO_2 NTAs−1.3 V. The charge transfer resistance can be obtained by fitting the Nyquist plots with the equivalent circuit model. As expected, the charge transfer resistance R_{ct} of the ECR-3D-TiO_2 NTAs−1.3 V is reduced from 440.45 to 133.08 Ω, which indicates a more effective separation of the photogenerated electron and hole and/or a faster interfacial charge transfer of the ECR-3D-TiO_2 NTAs−1.3 V. Moreover, the electrochemical active surface areas of the pristine 3D-TiO_2 NTAs and ECR-3D-TiO_2 NTAs−1.3 V are estimated from the capacitive region of cyclic voltammograms (CV). The data shown in Figure S7 reveal that the electrochemically active area of the ECR-3D-TiO_2 NTAs−1.3 V is only 1.05 times than that of the pristine -3D-TiO_2 NTAs, indicating that both samples have comparable electrochemically active areas. In addition, the slope of the Mott–Schottky plot collected from the ECR-3D TiO_2 NTAs−1.3 V is much smaller than that of the pristine 3D-TiO_2 NTAs, which suggest an improvement of donor densities (Figure 5b). The donor densities were estimated from the slopes of Mott–Schottky plots using the following equation:

$$N_D = -(\frac{2}{e_0 \varepsilon \varepsilon_0})[\frac{d(1/C^2)}{d(U_s)}]^{-1} \tag{1}$$

where $e_0 = -1.6 \times 10^{-19}$, $\varepsilon_0 = 8.86 \times 10^{-12}$ and $\varepsilon = 48$ for the anatase TiO_2. The calculated donor densities of the pristine and ECR-3D-TiO_2 NTAs−1.3 V are about 1.03×10^{19} and 1.46×10^{21} cm^{-3}, respectively. The increased the donor density can be attributed to the generation of the O-vacancies that works as electron donors. The increased donor density can effectively boost the transport property of the photogenerated charge carrier, which are of benefit to enhance the PEC water splitting activity. Moreover, the increased donor densities can also shift the Fermi level of the TiO_2 toward the conduction band, which facilitate the charge separation at the semiconductor–electrolyte interface.

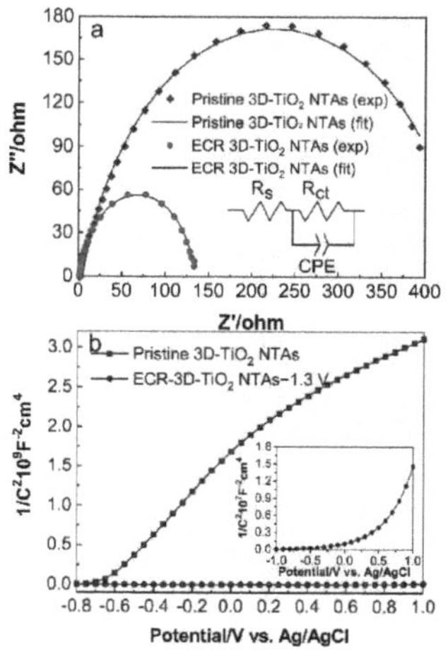

Figure 5. (**a**) Electrochemical impedance spectra of the pristine 3D-TiO_2 NTAs and ECR- 3D-TiO_2 NTAs−1.3 V under illumination, and (**b**) Mott–Schottky curves of the pristine 3D-TiO_2 NTAs and ECR- 3D-TiO_2 NTAs−1.3 V tested at a frequency of 1 kHz in dark conditions.

4. Discussion

Based on the above experimental results, the boosted photoelectrochemical water splitting performance of ECR-3D-TiO_2 NTAs can be ascribed to the introduction of the

O-vacancies. Firstly, PEC water splitting performance of the photoelectrode largely depend on its capability of effectively absorbing visible light. In the present case, the presence of O-vacancies results in the generation of a new defect energy level near the conduction band, which lead to the bandgap shrinkage, hence being favorable for the visible light harvesting. More importantly, the incident, reflected and/or refracted visible light captured by the 3D configuration is also absorbed by defect energy level near CB created by oxygen vacancy. Secondly, the introduction of the O-vacancies (Ti^{3+}) in ECR-3D-TiO_2 NTAs generally work as electron donors, which leads to the enhanced electronic conductivity and upward shift of the Fermi energy level, thereby facilitating the transfer and separation of photogenerated charge carrier at the semiconductor–electrolyte interface. Nevertheless, the excess O-vacancies may be the recombination centers for photogenerated carriers, hence limiting the generation of photocurrent [53,54]. Therefore, the optimized amount of the O-vacancies is essential to the PEC water splitting performance. The XPS result illustrates that the amount of the O-vacancies increases with deceasing electrochemical reduction bias (Figure 3 and Figure S5). Consequently, it can be included that the ECR-3D-TiO_2 NTAs−1.4 V may possess excess amount of the O-vacancies (Ti^{3+}), which lead to the recombination of photogenerated carriers before reaching the TiO_2/electrolyte interface. Accordingly, the optimal electrochemical reduction bias is −1.3 V from the perspective of PEC water splitting activity.

5. Conclusions

In conclusion, black 3D-TiO_2 NTAs have been successfully fabricated via an electrochemical reduction and employed as a photoanode for PEC water splitting. The introduction of the O-vacancies results in bandgap shrinkage, which can effectively boost the utilization efficiency of visible light including the incident, reflected and/or refracted visible light captured by the 3D configuration. Moreover, the O-vacancies (Ti^{3+}) can work as electron donors, which leads to the enhanced electronic conductivity and upward shift of the Fermi energy level, thereby facilitating the transfer and separation of photogenerated charge carrier at the semiconductor–electrolyte interface. Benefiting from the oxygen vacancy, the optimized photocurrent density of ECR-3D-TiO_2 NTAs under white light illumination generated the photocurrent density of 1.6 mA/cm^2 at 0.22 V vs. Ag/AgCl, which is superior or comparable to the previously reported values on self-doping TiO_2 NTAs formed on Ti foil.

Supplementary Materials: The following supporting information can be downloaded at: https://www.mdpi.com/article/10.3390/nano12091447/s1, Figure S1: 3D schematic diagram of the 3D-TiO_2 NTAs, which clearly exhibits the growth of TiO_2 NTAs on Ti mesh in a radially outward direction.; Figure S2: FE-SEM image of the pristine 3D-TiO_2 NTAs, ECR-3D-TiO_2 NTAs−1.2 V and ECR-3D-TiO_2 NTAs−1.4 V; Figure S3: FE-TEM image of the pristine 3D-TiO_2 NTAs, ECR-3D-TiO_2 NTAs−1.2 V and ECR-3D-TiO_2 NTAs-1.4 V; Figure S4: XRD patterns of the pristine 3D-TiO_2 NTAs and ECR-3D-TiO_2 NTAs electrochemically reduced under different applied bias −1.2, −1.3 and −1.4 V; Figure S5: (a) Survey spectrum of the pristine and ECR-3D-TiO_2 NTAs electrochemically reduced under the different applied bias −1.2, −1.3 and −1.4 V. (b) O1s XPS spectra of pristine and ECR-3D-TiO_2 NTAs electrochemically reduced under the different applied bias −1.2, −1.3 and −1.4 V; Figure S6: (a) FE-SEM image and (b) XRD pattern of the ECR-3D TiO_2 NTAs−1.3 V after undergoing the PEC water splitting reaction for 180 min. The results obviously show that the morphology of the ECR-3D-TiO_2 NTAs−1.3 V maintained intact and without observed structural degradation. Figure S7: Cyclic voltammetry (CV) for (a) ECR-3D TiO_2 NTAs-1.3 V (b) Pristine-3D-TiO_2 NTAs under different scan rates. (c) Relative electrochemical surface areas of the ECR-3D-TiO_2 NTAs−1.3 V and Pristine-3D-TiO_2 NTAs photoanodes: linear relationship between the capacitive current and scan rate. Table S1: Comparison of the PEC performance for the self-doping TiO_2 NTAs on formed on Ti foil [39,43,46–48,51].

Author Contributions: Conceptualization, M.M. and Z.G.; methodology, Y.F.; investigation, C.L.; data curation, H.Y. writing—original draft preparation, H.Z. All authors have read and agreed to the published version of the manuscript.

Funding: This research was funded by the Natural Science Foundation of Henan Province (222300420597), the National Natural Science Foundation of China (51702379, 12104523), the Science and Technology Project of Henan Province (222102240100, 212102210536, 222102230006, 222102230029, 22B120006), and the Taishan Scholars Program of Shandong Province (tsqn201909117).

Institutional Review Board Statement: Not applicable.

Informed Consent Statement: Not applicable.

Data Availability Statement: Data are available from the authors on request.

Acknowledgments: We would like to thank Lei Wang (Nanjing Normal University) for carrying out the FE-SEM and FE-TEM measurements.

Conflicts of Interest: The authors declare no conflict of interest.

References

1. Fu, H.C.; Varadhan, P.; Lin, C.H.; He, J.H. Spontaneous solar water splitting with decoupling of light absorption and electrocatalysis using silicon back-buried junction. *Nat. Commun.* **2020**, *11*, 3930. [CrossRef] [PubMed]
2. Huang, D.W.; Li, L.T.; Wang, K.; Li, Y.; Feng, K.; Jiang, F. Wittichenite semiconductor of Cu_3BiS_3 films for efficient hydrogen evolution from solar driven photoelectrochemical water splitting. *Nat. Commun.* **2021**, *12*, 3795. [CrossRef]
3. Nandal, V.; Pihosh, Y.; Higashi, T.; Minegishi, T.; Yamada, T.; Seki, K.; Sugiyama, M.; Domen, K. Probing fundamental losses in nanostructured Ta_3N_5 photoanodes: Design principles for efficient water oxidation. *Energy Environ. Sci.* **2021**, *14*, 4038–4047. [CrossRef]
4. Lu, Y.; Yang, Y.L.; Fan, X.Y.; Li, Y.Q.; Zhou, D.H.; Cai, B.; Wang, L.Y.; Fan, K.; Zhang, K. Boosting charge transport in $BiVO_4$ photoanode for solar water oxidation. *Adv. Mater.* **2022**, *34*, 2108178. [CrossRef] [PubMed]
5. Nellist, M.R.; Laskowski, F.A.L.; Qiu, J.J.; Hajibabaei, H.; Sivula, K.; Hamann, T.W.; Boettcher, S.W. Potential-sensing electrochemical atomic force microscopy for in operando analysis of water splitting catalysts and interfaces. *Nat. Mater.* **2020**, *19*, 69–76. [CrossRef]
6. Li, Y.; Mei, Q.; Liu, Z.J.; Hu, X.S.; Zhou, Z.H.; Huang, J.W.; Bai, B.; Liu, H.; Ding, F.; Wang, Q.Z. Fluorine-doped iron oxyhydroxide cocatalyst: Promotion on the WO_3 photoanode conducted photoelectrochemical water splitting. *Appl. Catal. B Environ.* **2022**, *304*, 120995. [CrossRef]
7. Narangari, P.R.; Narangari, R.; Butson, J.D.; Tan, H.H.; Jagadish, C.; Karuturi, S. Surface-tailored InP nanowires via self-assembled Au nanodots for efficient and stable photoelectrochemical hydrogen evolution. *Nano Lett.* **2021**, *21*, 6967–6974. [CrossRef]
8. Zhang, B.B.; Huang, X.J.; Zhang, Y.; Lu, G.X.; Chou, L.J.; Bi, Y.P. Unveiling the activity and stability origin of $BiVO_4$ photoanodes with FeNi oxyhydroxides for oxygen evolution. *Angew. Chem. Int. Ed.* **2020**, *59*, 18990–18995. [CrossRef]
9. Ye, S.; Shi, W.W.; Liu, Y.; Li, D.F.; Yin, H.; Chi, H.B.; Luo, Y.L.; Ta, N.; Fan, F.T.; Wang, X.L.; et al. Unassisted photoelectrochemical cell with multimediator modulation for solar water splitting exceeding 4% solar-to-hydrogen efficiency. *J. Am. Chem. Soc.* **2021**, *143*, 12499–12508. [CrossRef]
10. Yang, Y.; Niu, S.W.; Han, D.D.; Liu, T.Y.; Wang, G.M.; Li, Y. Progress in developing metal oxide nanomaterials for photoelectrochemical water splitting. *Adv. Energy Mater.* **2017**, *7*, 1700555. [CrossRef]
11. Wang, W.R.; Guo, B.D.; Dai, H.T.; Zhao, C.; Xie, G.C.; Ma, R.P.; Akram, M.Z.; Shan, H.Y.; Cai, C.Z.; Fang, Z.Y.; et al. Improving the water oxidation efficiency with a light-induced electric field in nanograting photoanodes. *Nano Lett.* **2019**, *19*, 6133–6139. [CrossRef] [PubMed]
12. Wei, T.C.; Zhu, Y.N.; Gu, Z.N.; An, X.Q.; Liu, L.M.; Wu, Y.X.; Liu, H.J.; Tang, J.W.; Qu, J.H. Multi-electric field modulation for photocatalytic oxygen evolution: Enhanced charge separation by coupling O-vacancies with faceted heterostructures. *Nano Energy* **2018**, *51*, 764–773. [CrossRef]
13. Samuel, E.; Joshi, B.; Kim, M.W.; Swihart, M.T.; Yoon, S.S. Morphology engineering of photoelectrodes for efficient photoelectrochemical water splitting. *Nano Energy* **2020**, *72*, 104648. [CrossRef]
14. Zhang, X.M.; Zhai, P.L.; Zhang, Y.X.; Wu, Y.Z.; Wang, C.; Ran, L.; Gao, J.F.; Li, Z.W.; Zhang, B.; Fan, Z.Z.; et al. Engineering single-atomic NiN_4O sites on semiconductor photoanodes for high-performance photoelectrochemical water splitting. *J. Am. Chem. Soc.* **2021**, *143*, 20657–20669. [CrossRef]
15. Qiu, Y.C.; Liu, W.; Chen, W.; Chen, W.; Zhou, G.M.; Hsu, P.C.; Zhang, R.F.; Liang, Z.; Fan, S.S.; Zhang, Y.G.; et al. Efficient solar-driven water splitting by nanocone $BiVO_4$-perovskite tandem cells. *Sci. Adv.* **2016**, *2*, e1501764. [CrossRef]
16. Yang, Q.; Du, J.Y.; Nie, X.Q.; Yang, D.M.; Bian, L.; Yang, L.; Dong, F.Q.; He, H.C.; Zhou, Y.; Yang, H.M. Magnetic field-assisted photoelectrochemical water splitting: The photoelectrodes gave weaker nonradiative recombination of carrier. *ACS Catal.* **2021**, *11*, 1242–1247. [CrossRef]
17. Hu, Y.X.; Pan, Y.Y.; Wang, Z.L.; Lin, T.G.; Gao, Y.Y.; Luo, B.; Hu, H.; Fan, F.T.; Liu, G.; Wang, L.Z. Lattice distortion induced internal electric field in TiO_2 photoelectrode for efficient charge separation and transfer. *Nat. Commun.* **2020**, *11*, 2129. [CrossRef]
18. Liu, Z.Y.; Zhang, Q.Q.; Zhao, T.Y.; Zhai, J.; Jiang, L. 3D vertical arrays of TiO_2 nanotubes on Ti meshes: Efficient photoanodes for water photoelectrolysis. *J. Mater. Chem.* **2011**, *21*, 10354–10358. [CrossRef]

19. Kołodziej, J.K.; Chudecka, A.; Sulka, G.D. 3D nanoporous titania formed by anodization as a promising photoelectrode material. *J. Electroanal. Chem.* **2018**, *823*, 221–233. [CrossRef]
20. Liao, J.J.; Lin, S.W.; Zhang, L.; Pan, N.Q.; Cao, X.K.; Li, J.B. Photocatalytic degradation of methyl orange using a TiO_2/Ti mesh electrode with 3D nanotube arrays. *ACS Appl. Mater. Interfaces* **2012**, *4*, 171–177. [CrossRef]
21. Bao, R.Y.; Zhao, Y.; Ma, F.F.; Wu, J.H.; Xia, J.X.; Li, H. 3D-TNAs composite electrodes with enhanced visible-light photoelectrocatalytic performance and stability. *J. Phys. Chem. Solids* **2022**, *161*, 110435. [CrossRef]
22. Liu, Z.Y.; Wang, Q.Y.; Cao, D.D.; Wang, Y.J.; Jin, R.C.; Gao, S.M. Vertical grown BiOI nanosheets on TiO_2 NTs/Ti meshes toward enhanced photocatalytic performances. *J. Alloys Compd.* **2020**, *820*, 153109. [CrossRef]
23. Liu, Z.Y.; Song, Y.D.; Wang, Q.Y.; Jia, Y.; Tan, X.Y; Du, X.X.; Gao, S.M. Solvothermal fabrication and construction of highly photoelectrocatalytic TiO_2 NTs/Bi_2MoO_6 heterojunction based on titanium mesh. *J. Colloid. Interf. Sci.* **2019**, *556*, 92–101. [CrossRef] [PubMed]
24. Li, T.T.; Wang, Z.H.; Liu, C.C.; Tang, C.M.; Wang, X.K.; Ding, G.S.; Ding, Y.C.; Yang, L.X. TiO_2 nanotubes/Ag/MoS_2 meshy photoelectrode with excellent photoelectrocatalytic degradation activity for tetracycline hydrochloride. *Nanomaterials* **2018**, *8*, 666. [CrossRef] [PubMed]
25. Jia, Y.; Liu, P.B.; Wang, Q.Y.; Wu, Y.; Cao, D.D.; Qiao, Q.A. Construction of Bi_2S_3-BiOBr nanosheets on TiO_2 NTA as the effective photocatalysts: Pollutant removal, photoelectric conversion and hydrogen generation. *J. Colloid Interf. Sci.* **2021**, *585*, 459–469. [CrossRef] [PubMed]
26. Bao, R.Y.; Chen, C.; Xia, J.X.; Chen, H.Y.; Li, H. Controlled synthesis and enhanced photoelectrocatalytic activity of a 3D TiO_2 nanotube array/TiO_2 nanoparticle heterojunction using a combined dielectrophoresis/sol-gel method. *J. Mater. Chem. C* **2019**, *7*, 4981–4987. [CrossRef]
27. Yang, X.C.; Chen, C. Cu_2O sensitized flexible 3D-TiO_2 nanotube arrays for enhancing visible photo-electrochemical performance. *RSC Adv.* **2016**, *6*, 70978–70983. [CrossRef]
28. Smith, Y.R.; Subramanian, V. Heterostructural composites of TiO_2 mesh-TiO_2 nanoparticles photosensitized with CdS: A new flexible photoanode for solar cells. *J. Phys. Chem. C* **2011**, *115*, 8376–8385. [CrossRef]
29. Kar, A.; Smith, Y.R.; Subramanian, V. Improved photocatalytic degradation of textile dye using titanium dioxide nanotubes formed over titanium wires. *Environ. Sci. Technol.* **2009**, *43*, 3260–3265. [CrossRef]
30. Foo, C.; Li, Y.Y.; Lebedev, K.; Chen, T.Y.; Day, S.; Tang, C.; Tsang, S.C.E. Characterisation of oxygen defects and nitrogen impurities in TiO_2 photocatalysts using variable-temperature X-ray powder diffraction. *Nat. Commun.* **2021**, *12*, 661. [CrossRef]
31. Gao, J.Q.; Xue, J.B.; Jia, S.F.; Shen, Q.Q.; Zhang, X.C.; Jia, H.S.; Liu, X.G.; Li, Q.; Wu, Y.C. Self-doping surface oxygen vacancy-induced lattice strains for enhancing visible light-driven photocatalytic H_2 evolution over black TiO_2. *ACS Appl. Mater. Interfaces* **2021**, *13*, 18758–18771. [CrossRef] [PubMed]
32. Cheng, X.; Dong, G.J.; Zhang, Y.J.; Feng, C.C.; Bi, Y.P. Dual-bonding interactions between MnO_2 cocatalyst and TiO_2 photoanodes for efficient solar water splitting. *Appl. Catal. B Environ.* **2020**, *267*, 118723. [CrossRef]
33. Paidi, V.K.; Lee, B.H.; Ahn, D.; Kim, K.J.; Kim, Y.; Hyeon, T.; Hyeon, T.; Lee, K.S. Oxygen-vacancy-driven orbital reconstruction at the surface of TiO_2 core-shell Nanostructures. *Nano Lett.* **2021**, *21*, 7953–7959. [CrossRef] [PubMed]
34. Meng, M.; Qin, W.; Li, C.Y.; Xu, K.; Xu, L.Y.; Li, J.; Ma, L.; Liu, K.L.; Li, J.T.; Qin, N.; et al. Synergistic effect of photonic crystals and oxygen vacancies on photoelectrochemical water splitting of TiO_2 nanotube. *J. Nanoelectron. Optoelectron.* **2020**, *15*, 226–230. [CrossRef]
35. Liu, Q.H.; He, J.F.; Yao, T.; Sun, Z.H.; Cheng, W.R.; He, S.; Xie, Y.; Peng, Y.H.; Cheng, H.; Sun, Y.F.; et al. Aligned Fe_2TiO_5-containing nanotube arrays with low onset potential for visible-light water oxidation. *Nat. Commun.* **2014**, *5*, 5122. [CrossRef] [PubMed]
36. Liu, X.Y.; Zhu, G.L.; Wang, X.; Yuan, X.T.; Lin, T.Q.; Huang, F.Q. Progress in black titania: A new material for advanced photocatalysis. *Adv. Energy Mater.* **2016**, *6*, 1600452. [CrossRef]
37. Wei, N.; Liu, Y.; Feng, M.; Lia, Z.X.; Chen, S.G.; Zheng, Y.B.; Wang, D.A. Controllable TiO_2 core-shell phase heterojunction for efficient photoelectrochemical water splitting under solar light. *Appl. Catal. B Environ.* **2019**, *244*, 519–528. [CrossRef]
38. Cui, H.L.; Zhao, W.; Yang, C.Y.; Yin, H.; Lin, T.Q.; Shan, Y.F.; Xie, Y.; Gua, H.; Huang, F.Q. Black TiO_2 nanotube arrays for high-efficiency photoelectrochemical water-splitting. *J. Mater. Chem. A* **2014**, *2*, 8612–8616. [CrossRef]
39. Meng, M.; Zhou, S.H.; Yang, L.; Gan, Z.X.; Liu, K.L.; Tian, F.S.; Zhu, Y.; Li, C.Y.; Liu, W.F.; Yuan, H.L.; et al. Hydrogenated TiO_2 nanotube photonic crystals for enhanced photoelectrochemical water splitting. *Nanotechnology* **2018**, *29*, 155401. [CrossRef]
40. Li, Z.H.; Zhou, C.; Hua, J.H.; Hong, X.F.; Sun, C.L.; Li, H.W.; Xu, X.; Mai, L.Q. Engineering O-vacancies in a polysulfde-blocking layer with enhanced catalytic ability. *Adv. Mater.* **2020**, *32*, 1907444. [CrossRef]
41. Lei, F.C.; Sun, Y.F.; Liu, K.T.; Gao, S.; Liang, L.; Pan, B.C.; Xie, Y. O-vacancies confined in ultrathin indium oxide porous sheets for promoted visible-light water splitting. *J. Am. Chem Soc.* **2014**, *136*, 6826–6829. [CrossRef] [PubMed]
42. Lin, T.Q.; Yang, C.Y.; Wang, Z.; Yin, H.; Lu, X.J.; Huang, F.Q.; Lin, J.H.; Xie, X.M.; Jiang, M.H. Effective nonmetal incorporation in black titania with enhanced solar energy utilization. *Energy Environ. Sci.* **2014**, *7*, 967–972. [CrossRef]
43. Kang, Q.; Cao, J.Y.; Zhang, Y.J.; Liu, L.Q.; Xu, H.; Ye J.H. Reduced TiO_2 nanotube arrays for photoelectrochemical water splitting. *J. Mater. Chem. A* **2013**, *1*, 5766–5774. [CrossRef]
44. Wang, G.M.; Yang, Y.; Ling, Y.C.; Wang, H.Y.; Lu, X.H.; Pu, Y.C.; Zhang, J.Z.; Tong, Y.X.; Li, Y. An electrochemical method to enhance the performance of metal oxides for photoelectrochemical water oxidation. *J. Mater. Chem.* **2016**, *4*, 2849–2855. [CrossRef]

45. Chang, X.; Thind, S.S.; Chen, A.C. Electrocatalytic enhancement of salicylic acid oxidation at electrochemically reduced TiO_2 nanotubes. *ACS Catal.* **2014**, *4*, 2616–2622. [CrossRef]
46. Zhang, Z.H.; Hedhili, M.N.; Zhu, H.B.; Wang, P. Electrochemical reduction induced self-doping of Ti3+ for efficient water splitting performance on TiO_2 based photoelectrodes. *Phys. Chem. Chem. Phys.* **2013**, *15*, 15637–15644. [CrossRef] [PubMed]
47. Song, J.N.; Zheng, M.J.; Yuan, X.L.; Li, Q.; Wang, F.Z.; Ma, L.G.; You, Y.X.; Liu, S.H.; Liu, P.J.; Jiang, D.K.; et al. Electrochemically induced Ti^{3+} self-doping of TiO_2 nanotube arrays for improved photoelectrochemical water splitting. *J. Mater. Sci.* **2017**, *52*, 6976–6986. [CrossRef]
48. Xu, C.; Song, Y.; Lu, L.F.; Cheng, C.W.; Liu, D.F.; Fang, X.H.; Chen, X.Y.; Zhu, X.F.; Li, D.D. Electrochemically hydrogenated TiO_2 nanotubes with improved photoelectrochemical water splitting performance. *Nanoscale Res. Lett.* **2013**, *8*, 391. [CrossRef]
49. Close, T.; Tulsyan, G.; Diaz, C.A.; Weinstein, S.J.; Richter, C. Reversible oxygen scavenging at room temperature using electrochemically reduced titanium oxide nanotubes. *Nat. Nanotechnol.* **2015**, *10*, 418–422. [CrossRef]
50. Lu, X.H.; Wang, G.M.; Zhai, T.; Yu, M.H.; Gan, J.Y.; Tong, Y.X.; Li, Y. Hydrogenated TiO_2 nanotube arrays for supercapacitors. *Nano Lett.* **2012**, *12*, 1690–1696. [CrossRef]
51. Li, Z.Z.; Xin, Y.M.; Wu, W.L.; Fu, B.H.; Zhang, Z.H. Phosphorus cation doping: A new strategy for boosting photoelec trochemical performance on TiO_2 nanotube photonic crystals. *ACS Appl. Mater. Interfaces* **2016**, *8*, 30972–30979. [CrossRef] [PubMed]
52. Cheng, X.; Zhang, Y.J.; Bi, Y. Spatial dual-electric fields for highly enhanced the solar water splitting of TiO_2 nanotube arrays. *Nano Energy* **2019**, *57*, 542–548. [CrossRef]
53. Gan, J.Y.; Lu, X.H.; Wu, J.S.; Xie, S.L.; Zhai, T.; Yu, M.H.; Zhang, Z.S.; Mao, Y.C.; Wang, S.C.; Shen, Y.; et al. O-vacancies promoting photoelectrochemical performance of In_2O_3 nanocubes. *Sci. Rep.* **2013**, *3*, 1021. [CrossRef] [PubMed]
54. Meng, M.; Yang, L.; Wu, X.L.; Gan, Z.X.; Pan, W.Y.; Liu, K.L.; Li, C.Y.; Qin, N.; Li, J. Boosted photoelectrochemical performance of In_2O_3 nanowires via modulating O-vacancies on crystal facets. *J. Alloys Compd.* **2020**, *845*, 156311. [CrossRef]

nanomaterials MDPI

Article

Highly Efficient Photocatalytic Hydrogen Evolution over Mo-Doped $ZnIn_2S_4$ with Sulfur Vacancies

Wei Guan [1], Lin Zhang [1], Peng Wang [1], Ying Wang [1], Haoyu Wang [1], Xingchen Dong [1], Ming Meng [2], Lina Sui [1], Zhixing Gan [1,3,*], Lifeng Dong [1,*] and Liyan Yu [1,*]

1 College of Materials Science and Engineering, Qingdao University of Science and Technology, Qingdao 266042, China
2 School of Physics and Telecommunication Engineering, Zhoukou Normal University, Zhoukou 466001, China
3 Center for Future Optoelectronic Functional Materials, School of Computer and Electronic Information/School of Artificial Intelligence, Nanjing Normal University, Nanjing 210023, China
* Correspondence: zxgan@njnu.edu.cn (Z.G.); donglifeng@qust.edu.cn (L.D.); liyanyu@qust.edu.cn (L.Y.)

Abstract: The introduction of impure atoms or crystal defects is a promising strategy for enhancing the photocatalytic activity of semiconductors. However, the synergy of these two effects in 2D atomic layers remains unexplored. In this case, the preparation of molybdenum-doped thin $ZnIn_2S_4$-containing S vacancies (Mo-doped Sv-$ZnIn_2S_4$) is conducted using a one-pot solvothermal method. The coordination of Mo doping and S vacancies not only enhances visible light absorption and facilitates the separation of photogenerated carriers but also provides many active sites for photocatalytic reactions. Meanwhile, the Mo-S bonds play function as high-speed channels to rapidly transfer carriers to the active sites, which can directly promote hydrogen evolution. Consequently, Sv-$ZnIn_2S_4$ with an optimized amount of Mo doping exhibits a high hydrogen evolution rate of 5739 μmol g^{-1} h^{-1} with a corresponding apparent quantum yield (AQY) of 21.24% at 420 nm, which is approximately 5.4 times higher than the original $ZnIn_2S_4$. This work provides a new strategy for the development of highly efficient and sustainable 2D atomic photocatalysts for hydrogen evolution.

Keywords: photocatalytic hydrogen production; element doping; atomic vacancies

Citation: Guan, W.; Zhang, L.; Wang, P.; Wang, Y.; Wang, H.; Dong, X.; Meng, M.; Sui, L.; Gan, Z.; Dong, L.; et al. Highly Efficient Photocatalytic Hydrogen Evolution over Mo-Doped $ZnIn_2S_4$ with Sulfur Vacancies. *Nanomaterials* **2022**, *12*, 3980. https://doi.org/10.3390/nano12223980

Academic Editor: Diego Cazorla-Amorós

Received: 15 October 2022
Accepted: 8 November 2022
Published: 11 November 2022

1. Introduction

With the acceleration of global population growth and social development, environmental problems and energy deprivation emerged as two major issues that need to be addressed. Therefore, the search for alternative energy sources is of great significance to the development of society. H_2 is considered as an ideal alternative energy source due to its clean, renewable and transportable advantages [1–4]. However, the conventional hydrogen production process has high energy consumption, low efficiency and serious environmental pollution, which seriously restrict the development of hydrogen. In contrast, photocatalytic fracking is a cleaner and is a more sustainable way of converting sustained solar energy into chemical energy. Its large-scale application will help alleviate problems such as the greenhouse effect and environmental pollution [5]. However, the low photocatalytic efficiency still remains a great challenge for practical applications. Therefore, in-depth research on hydrogen production by the photocatalytic cracking of water is of great significance [6–10].

In recent years, 2D semiconductors, such as C_3N_4, TiO_2, CdS and MoS_2, attracted much attention [11–13]. The two-dimensional matter shows a unique confinement of electrons in ultra-thin layers, resulting in superior optical and electronic properties. Two-dimensional ultrathin materials with suitable band gap structure have shown great potential in achieving efficient photocatalysis due to their unique structure and electronic properties. [14]. $ZnIn_2S_4$ is a classical trimetallic sulfide semiconductor with a tunable band gap of 2.06 eV to 2.86 eV. Due to its unique crystal growth mechanism and interatomic interactions, it is easy to form

2D layered structures. Currently, controllable morphologies include nanotubes [15], nanoribbons [16] and nanoflowers, and they are assembled from 2D thin nanosheets [17–19]. In addition, $ZnIn_2S_4$ has the advantages of being low in toxicity and possesses good photostability. However, similarly to most 2D semiconductors, $ZnIn_2S_4$ still has some problems in practical applications, such as the rapid recombination of light-generated electron-hole pairs, short carrier lifetime and low light absorption capacity.

To improve the photocatalytic activity of $ZnIn_2S_4$, a series of modifications have been carried out, including element doping, defect engineering, cocatalyst loading, morphology tuning and heterostructure construction [20–24]. Among them, element doping and defect engineering are considered the most effective means to enhance photocatalytic activities [25–31]. On the one hand, the concentration and energy distribution of carriers near the conduction band edge can be adjusted by doping the semiconductor with ions to introduce donor/acceptor energy levels, thus improving the electronic transition behavior [32–37]. For example, Huang et al. reported Mo-doped $ZnIn_2S_4$ flower-like hollow microspheres for efficient photocatalytic hydrogen evolution, and the results showed that the hydrogen evolution activity of Mo-doped $ZnIn_2S_4$ was 9 times higher (4.62 mmol g^{-1} h^{-1}) than the pristine $ZnIn_2S_4$ (0.465 mmol g^{-1} h^{-1}) [38]. They found that the doping of Mo into the $ZnIn_2S_4$ crystal lattices can introduce a doping energy level within the band gap. Therefore, the electronic conversion from the valence band to doping energy levels or from doping energies to the conduction band can effectively expand the light absorption range and improve the electronic conversion efficiency of solar energy. In addition, Mo doping can form Mo-S bonds, which accelerate the photogenerated carrier transport and separation, thus facilitating the transfer of light-excited carriers to the adsorbed molecules of reactants for photocatalytic reactions. A similar improvement in photocatalytic activity was also observed in Ni-doped $ZnIn_2S_4$ [39]. On the other hand, surface defects can lead to trapped photogenerated carriers, tuning their spatial distribution and prolonging their lifetime. Previous reports validated that S vacancies (Sv) enhances the absorption of visible light and adds to the photogenerated electric charge density, thereby enhancing photocatalytic activities [40]. For example, Yang et al. reported an Sv-containing half-cell $ZnIn_2S_4$, which exhibited a 7.8 times enhancement of photocatalytic hydrogen generation performance under exposure to visible light compared to pristine ZIS. Similarly, improved photocatalytic activities are also observed in $ZnIn_2S_4$ with In vacancies [41] and Zn vacancies [42]. Although metal doping and atomic vacancies have been employed separately to improve the photocatalytic action of $ZnIn_2S_4$, the coordination of these two effects is still unexplored.

Herein, a simple one-pot soluble heat method was developed to dope Mo into $ZnIn_2S_4$ nanoflakes containing Sv. An excess of thioacetamide (TAA) was added to the reaction. The adsorption of thioacetamide on the surface of the crystal hindered the growth of the crystal, leading to the formation of vacancy structure [43–45]. Meanwhile, Na_2MoO_4 was added to incorporate Mo atoms into the lattice of $ZnIn_2S_4$. The coordination of Mo doping and sulfur vacancies not only enhances the light absorption and the separation of photogenerated carriers but also provides a large number of active sites for photocatalytic reactions [46,47]. Therefore, the as-prepared Mo-doped Sv-$ZnIn_2S_4$ exhibited improved photocatalytic performance and stability. The photocatalytic hydrogen evolution rate of Sv-$ZnIn_2S_4$ with optimal Mo doping amounts reached 5.74 mmol g^{-1} h^{-1} under visible light irradiation, which is 5.4 times that of pure $ZnIn_2S_4$ and 2.6 times that of Sv-$ZnIn_2S_4$.

2. Results

As shown in Figure 1, ZIS, Sv-ZIS and Mo-Sv-ZIS were obtained by a simple single-pot soluble heat method. The number of sulfur vacancies was controlled by dosage of TAA. The amount of TAA during the synthesis was used to inhibit the growth of ZIS primary crystals, thus introducing Sv in ZIS [6]. The surface morphologies of the obtained ZIS, Sv-ZIS and Mo-Sv-ZIS are described by a field emission transmission electron microscopy (FE-TEM). As shown in Figure 2a,d, the basic morphology of the pristine ZIS is made up of a great number of two-dimensional hexagonal nanoflakes, which facilitates the acquisition of a large contact area to expose more reactive sites for the photocatalytic interaction. As shown in Figure 2b, the introduction of sulfur vacancies hardly changes the morphology. Sv-ZISs are also irregular flakes. The HRTEM image in Figure 2e shows that the crystal spacing between the planes of Sv-ZIS is 0.32 nm, which belongs to the (102) plane of $ZnIn_2S_4$. As shown in Figure 2c, after Mo doping, Mo-Sv-ZIS still maintains thin nanosheet structures. Due to the thinness of the nanosheets, there are different degrees of curling and bending (Figure S1). The HRTEM image shown in Figure 2f suggests that the interplanar crystal spacing of Mo-Sv-ZIS is 0.32 nm, which is similar to that of Sv-ZIS, indicating minor lattice distortions induced by Mo doping. The elemental distribution in the Mo-Sv-ZIS prototype is further studied by high-angle annular dark-field scanning transmission electron microscopy (HAADF-STEM) (Figure 2g) Figure 2h–k show the distribution of the corresponding elements (Zn, In, S and Mo) for Mo-SV-ZIS, and Figure S2 shows the corresponding energy dispersive spectra (EDS), showing the existence of Zn, In, S and Mo components. The uniform distribution of Mo indicates that it is doped successfully into the lattice of ZIS.

ZIS, Sv-ZIS and Mo-Sv-ZIS are investigated further by XPS for chemical states and apparent chemical compositions. Figure S3 displays the full spectrum of ZIS, Sv-ZIS and Mo-Sv-ZIS containing the typical peaks of Zn 2*p*, In 3*d*, S 2*p* and Mo 3*d*, which is consistent with the EDS' results. Figure 3a shows that the Zn 2*p* characteristic peaks of ZIS are located at 1022.05 eV and 1045.06 eV, which belong to $2p_{3/2}$ and $2p_{1/2}$, respectively. In comparison, the Zn 2*p* XPS peak positions of Sv-ZIS undergo negative shifts of 0.28 eV and 0.26 eV, separately indicating a decrease in the coordination number of Zn atoms due to the presence of sulfur vacancies [5]. The doping of Mo shifted the peak positions of Mo-Sv-ZIS positively by 0.1 eV and 0.08 eV, which means that the Zn atoms returned to the higher binding energy region after the doping of Mo. As shown in Figure 3b, the In 3*d* characteristic peaks of ZIS are located at 445.14 eV and 452.68 eV, which are assigned to In $3d_{5/2}$ and $3d_{3/2}$, respectively. The In 3*d* peaks of Mo-Sv-ZIS and Sv-ZIS are essentially the same, with negative shifts of 0.19 eV and 0.15 eV, respectively. After the addition of Mo, the XPS peak position of In is almost unchanged, while the binding energy of Zn varies more than that of In, indicating that Mo replaces the position of Zn rather than that of In. The radius of Zn atom is 1.39 Å, and the radius of Mo atom is 1.40 Å, which are very close. Therefore, it is feasible to replace the Zn atom with the Mo atom. As shown in Figure 3c, the S 2*p* characteristic peaks of ZIS are located at 161.88 eV and 163.18 eV, which belong to $2p_{3/2}$ and $2p_{1/2}$, respectively. On the other hand, the S 2*p* XPS bands of Sv-ZIS underwent negative shifts of 0.09 eV and 0.18 eV, respectively, proving the existence of S vacancies in ZIS. After doping with Mo atoms, S $2p_{3/2}$ and S $2p_{1/2}$ negatively shift to 0.13 eV and 0.14 eV, respectively. The XPS results show that Mo doping leads to a further reduction in the coordination number of S in addition to the unpaired electrons brought by Mo. These two effects make the Mo-Sv-ZIS have a high density of unpaired electrons, which are active sites for photocatalytic reactions [48]. In addition, the atomic contents are calculated from the XPS peak areas. As shown in Table S1, the atomic ratios of Zn/In/S in ZIS, Sv-ZIS and Mo-Sv-ZIS are 1/1.6/3.5, 1/1.25/2.6 and 1/1.3/2.61, respectively. The lower S content in Sv-ZIS and Mo-Sv-ZIS supports the existence of a large number of S vacancies. In addition, the elemental content was tested by inductively coupled plasma mass spectrometry (ICP-MS) (Table S2). Moreover, the sub-band located at 227.5 eV due to the formation of Mo-S bonds is observed in the Mo 3*d* XPS spectrum [49], which further demonstrates the successful synthesis of Mo-doped Sv-ZIS (Figure 3d). The XPS O 1*s* peak of metal-O bonding typically is located at about 530 eV [50,51]. Herein, the O 1*s* peak positions of ZIS and Sv-ZIS are 532.36 eV and 532.17 eV, respectively (Figure S4), suggesting that there is no metal-O bonding.

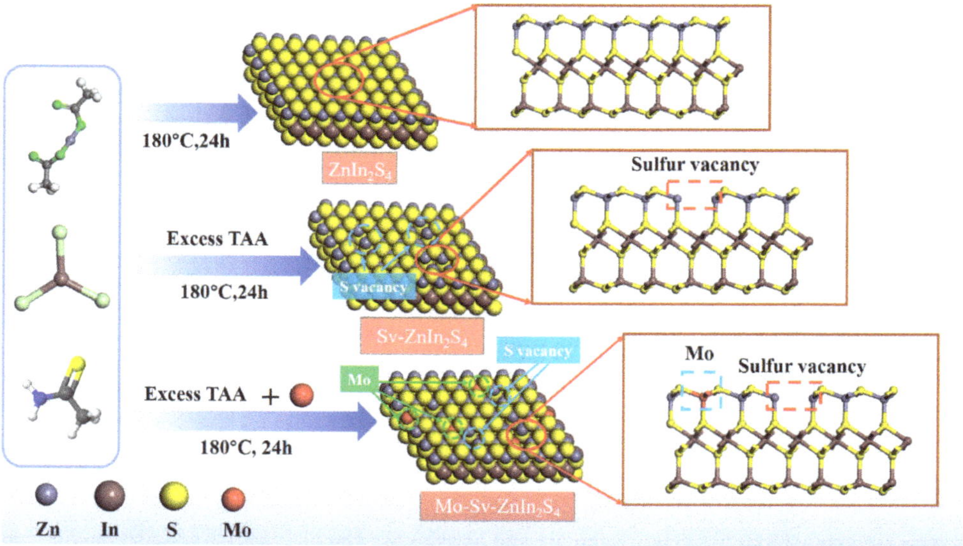

Figure 1. Schematic diagram of the synthetic routes for ZIS, Sv-ZIS and Mo-Sv-ZIS. The dotted red squares highlight the Sv. (The squares show the differences between the crystals).

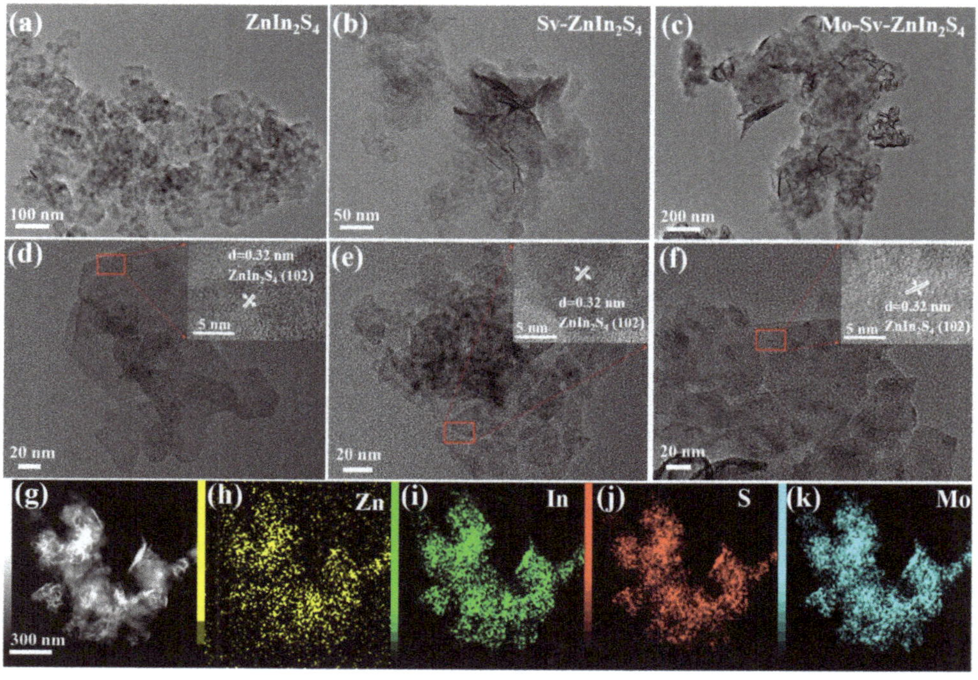

Figure 2. (**a**,**d**) TEM (**a**) and HRTEM (**d**) images of the ZIS. (**b**,**e**) TEM (**b**) and HRTEM (**e**) images of the Sv-ZIS. (**c**,**f**) TEM (**c**) and HRTEM (**f**) images of the 1.5% Mo-Sv-ZIS. (**g**) HAADF-STEM image of the 1.5% Mo-Sv-ZIS. (**h**–**k**) EDS element mappings of Zn (**h**), In (**i**), S (**j**) and Mo (**k**) in 1.5% Mo-Sv-ZIS. (Note: The area inside the red square is the area magnified by HRTEM).

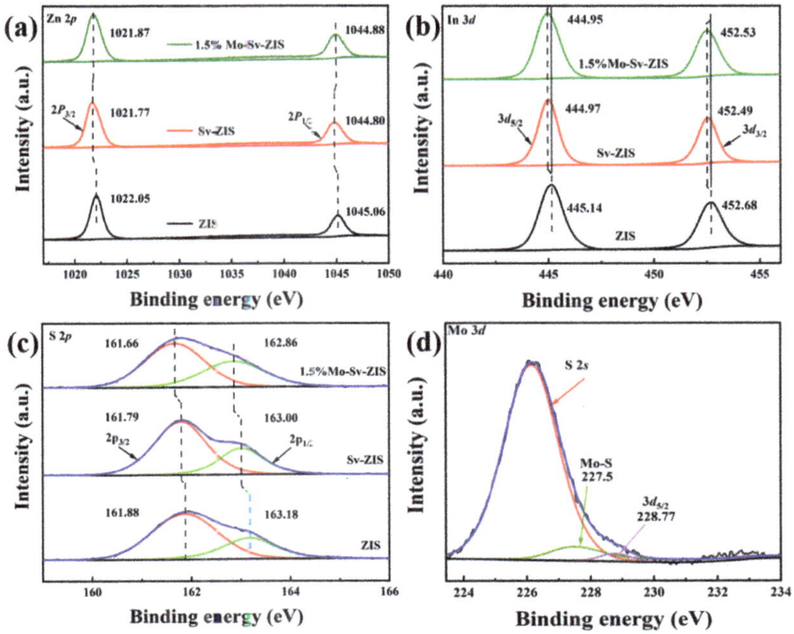

Figure 3. (**a**–**d**) High-resolution XPS spectra of Zn 2*p* (**a**), In 3*d* (**b**), S 2*p* (**c**) and Mo 3*d* (**d**) of ZIS, Sv-ZIS and 1.5% Mo-Sv-ZIS.

The crystal structures of ZIS, Sv-ZIS and Mo-Sv-ZIS are investigated by powder X-ray diffraction (XRD). In Figure 4a, the XRD motif of ZIS is the same as that of $ZnIn_2S_4$ in hexagonal form (JCFDS file card No. 72-0773) without impurity peaks. The characteristic peaks at 21.6°, 27.6°, 30.5°, 39.8°, 47.1°, 52.2° and 55.6° belong to diffractions of (006), (102), (104), (108), (110), (116) and (202) crystal planes of $ZnIn_2S_4$, respectively. Among them, the diffraction peaks of (104), (108) and (116) crystal planes become weaker in the XRD of Sv-ZIS due to the presence of sulfur vacancies. The diffraction peaks of Mo-Sv-ZIS are similar to that of Sv-ZIS, which can prove the existence of sulfur vacancies in Mo-Sv-ZIS [39,41]. The XRD of Mo-ZIS is also measured for comparison. Interestingly, a diffraction peak at 15.5° is observed in the XRD of Mo-ZIS, which corresponds to the (002) planes of MoS_2. Moreover, the diffraction of (104) planes at 30.4° is noticeable. When an appropriate amount of TAA is added, no additional S vacancies form in ZIS. The doped Mo atoms tend to bond with S atoms to form MoS_2, which is not conducive to the full incorporation of Mo into the lattice of ZIS. In other words, the sulfur vacancies induced by the excess TAA can facilitate the doping of Mo into ZIS lattices. As illustrated in Figure S5, with an increase in Mo doping, the (006) and (110) crystal planes of Mo-Sv-ZIS shifted to high angles, demonstrating that the Mo atoms are included in the crystal lattice of ZIS [52,53]. The intensity of the individual diffraction peak diminishes with the increase in Mo doping. In addition, no significant peaks of diffraction of Mo species are found for the highly doped 5% Mo-Sv-ZIS. These results indicate that the Mo elements are uniformly incorporated into the lattice of ZIS and have not damaged the ZIS crystal structure. The number of unpaired electrons in ZIS, Sv-ZIS and Mo-Sv-ZIS are evaluated by electron paramagnetic resonance (EPR). As shown in Figure 4b, the original ZIS does not show any discernible signals. In contrast, Sv-ZIS shows an EPR indication for a g value of 1.998, confirming the existence of the S-vacancy. In addition, the EPR signal of Mo-Sv-ZIS is more intense, indicating that the increased number of unpaired electrons in Mo-doped Sv-ZIS is due to the simultaneous presence of undercoordinated Mo and S vacancies [54]. Therefore, the EPR results suggest that the density of active sites for photocatalytic reactions increased by Mo doping and S vacancies.

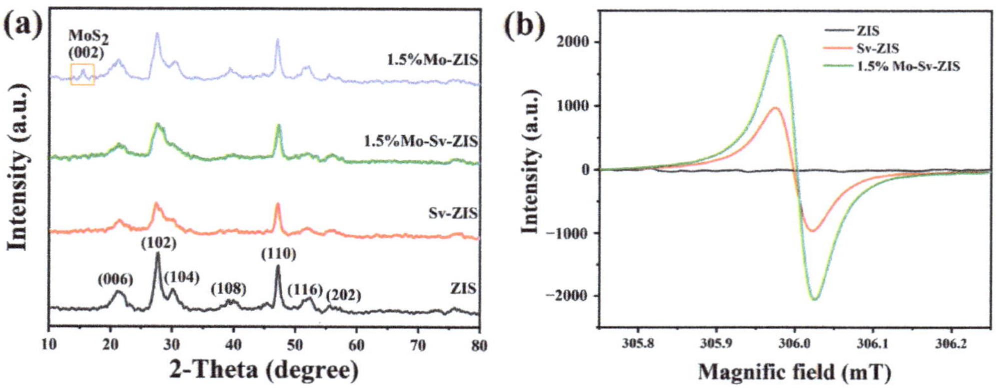

Figure 4. (**a**) XRD patterns of ZIS, Sv-ZIS, 1.5% Mo-ZIS and 1.5% Mo-Sv-ZIS. (**b**) The EPR Spectra of ZIS, Sv-ZIS and 1.5% Mo-Sv-ZIS.

Furthermore, as shown in Figure S6, the analysis of the functional groups on the surface of the composites is performed by FTIR. The observed peaks at 1633 and 1394 cm^{-1} are composed of water and hypdroxyl moieties adsorbed on the surface [55], implying the ready attraction of free H_2O molecules for continuous H_2O dissociation and the facilitation of the reaction kinetics. Meanwhile, Raman spectroscopy is also performed to obtain more details about the crystalline structure. As shown in Figure S7, in the case of ZIS, the four Raman peaks are located around 253, 308, 339 and 370 cm^{-1}, which are assigned to the longitudinal optical mode (LO_1), transverse optical mode (TO_2), longitudinal optical mode (LO_2) and A1g mode of crystal ZIS, separately. For the Mo-Sv-ZIS, all peaks are relatively weak, and the peaks located at 308 and 339 cm^{-1} are barely observable because of the doping influence and the inferior degree of crystallinity. A new band at 405 cm^{-1}, as shown in the Raman spectrum of Mo-Sv-ZIS, comes from the coupling of Mo-S stretching vibrations and A1g modes in Mo-Sv-ZIS, further confirming the existence of Mo-S bonds [5].

The strong light absorption is beneficial to the generation of photogenerated carriers, which directly contribute to the improvement of photocatalytic performance. To investigate the optical absorption ability, UV-Vis spectra in the range of 300 to 800 nm are measured. As illustrated in Figure 5a, the primary light absorption of pristine ZIS and Sv-ZIS occurs in the wavelength range of 300–550 nm. With the introduction of S vacancies, the optical absorbance of Sv-ZIS is slightly enhanced compared to that of ZIS. After Mo doping, the optical absorption of Mo-Sv-ZIS further increased. With the increase in Mo content, the absorption in the wavelength range of 500 to 800 nm is significantly enhanced, and the color of the sample gradually deepens (inset in Figure 5a). Meanwhile, the absorption edge sightly redshifts. ZIS is a type of semiconductor with a direct band gap that can be computed from the following formula: $ahv^2 = A\ (hv\text{- } Eg)$. As shown in Figure 5b, the bandgaps of ZIS, Sv-ZIS and Mo-Sv-ZIS are 2.53 eV, 2.48 eV and 2.45 eV, respectively. Ultraviolet photoelectron spectra (UPS) are tested to determine the valence band position (E_{VB}). As shown in Figure 5c, the valence band potentials ($E_{VB,\ XPS}$) of ZIS, Sv-ZIS and Mo-Sv-ZIS are 1.38 eV, 1.58 eV and 1.36 eV, respectively. The E_{VB} of the corresponding standard hydrogen electrodes relative to ZIS, Sv-ZIS and Mo-Sv-ZIS are then obtained based on the following equation: $E_{VB,\ NHE} = \varphi + E_{VB,\ XPS} - 4.44$ eV; here, φ is the power functional of the apparatus (5.1 eV). Therefore, for ZIS, Sv-ZIS and Mo-Sv-ZIS, the $E_{VB,\ NHE}$ are 2.04 eV, 2.24 eV and 2.02 eV, respectively [56].

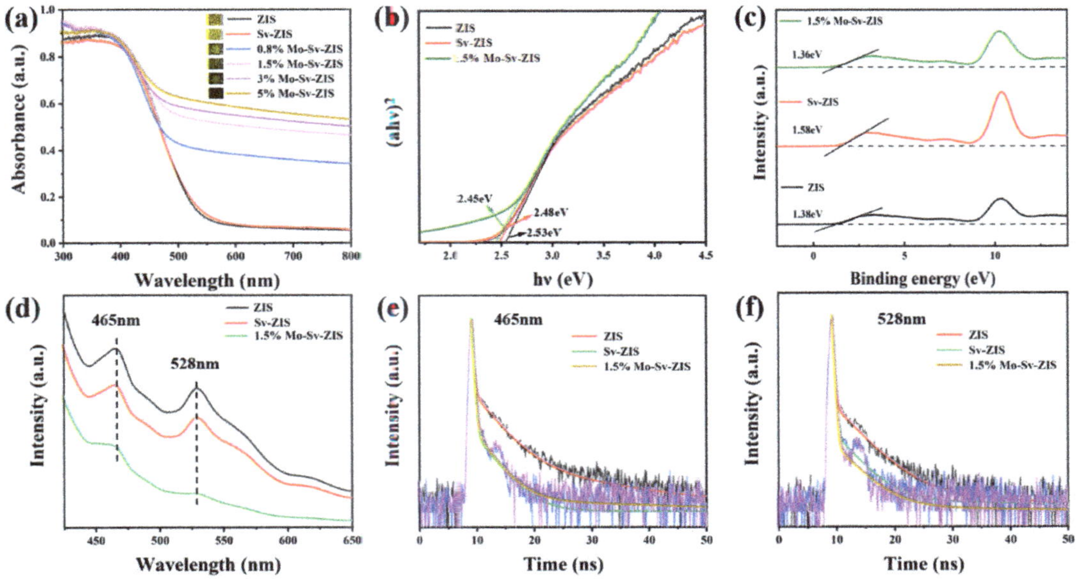

Figure 5. (**a**) UV–vis absorption spectra of ZIS, Sv-ZIS and x% Mo-Sv-ZIS (x = 0.8, 1.5, 3, 5). (**b**) $(\alpha h\nu)^2$ versus hν curves. (**c**) UPS spectra of the ZIS, Sv-ZIS and 1.5% Mo-Sv-ZIS. (**d**) PL spectra of ZIS, Sv-ZIS and 1.5% Mo-Sv-ZIS at 350 nm excitation. (**e**) Time-resolved PL spectra of ZIS, Sv-ZIS and 1.5% Mo-Sv-ZIS monitored at 465 nm. (**f**) Time-resolved PL spectra of ZIS, Sv-ZIS and 1.5% Mo-Sv-ZIS monitored at 528 nm.

The dynamics of excited carriers is studied through photoluminescence (PL) with time-resolved PL (TRPL). Two emission peaks at 465 nm and 528 nm are observed in the PL spectra of all samples, as indicated in Figure 5d. Among these three samples, ZIS has the highest PL intensity, which suggests an effective recombination of photogenerated carriers, greatly limiting the photocatalytic activity. However, the PL intensity of Sv-ZIS is obviously reduced, which suggests that the sulfur vacancy can act as a carrier trap and promote the separation of photocarriers [56]. It is clear that the PL intensity of the Mo-doped Sv-ZIS further decreases, indicating that photogenerated carriers can be also rapidly captured by the undercoordinated Mo, thereby significantly inhibiting the recombination of photogenerated electron-hole pairs. Time-resolved PL is tested at 465 nm and 528 nm (Figure 5e,f). The average PL lifetime (τ_A) is calculated according to the following equation [48]:

$$\tau_A = \frac{A_1\tau_1{}^2 + A_2\tau_2{}^2 + A_3\tau_3{}^2}{A_1\tau_1 + A_2\tau_2 + A_3\tau_3}$$

where τ_1, τ_2 and τ_3 are the PL lifetimes, and A_1, A_2 and A_3 are the corresponding amplitudes obtained by tri-exponential fittings. The fitting details are listed in Tables S3 and S4. The average PL lifetimes of the pristine ZIS are 5.33 ns and 3.31 ns for 465 nm emission and 528 nm emission, respectively. For the Mo-Sv-ZIS, the average PL lifetimes of 465 nm and 528 nm emissions shorten to 0.08 ns and 0.8 ns. The ultra-short PL lifetimes are in good agreement with the rapid capture of photogenerated carriers by the doped Mo and sulphur vacancies.

To investigate the role of Mo doping and S vacancies on photogenerated carrier migration, photoelectrochemical tests are carried out by using ZIS, Sv-ZIS, Mo-ZIS and Mo-Sv-ZIS as photoelectrodes. As shown in Figure 6a, the EIS radius of Sv-ZIS is smaller than the original ZIS, indicating that S vacancies can reduce the interfacial transfer resistance (detailed parameters are listed in Table S5). In addition, the EIS radius of Mo-Sv-ZIS is less than Sv-ZIS and ZIS due to the presence of both doped Mo and S vacancies [41,56]. The EIS spectra of Mo-Sv-ZIS with different Mo doping levels are further investigated. As shown in Figure 6b, the EIS semicircle becomes smaller as the Mo doping level increases. However, when the Mo doping concentration is as high as 5%, the radius of the EIS semicircle increases, implying that Mo doping that is too high is not conducive to lowering the surface charge's transport resistance. As illustrated in Figure 6c, all tested samples exhibit evident photo-responses. As expected, Mo-Sv-ZIS exhibits the highest photocurrent density, which is four times higher than that of the pristine ZIS. Therefore, we demonstrate that the coordination of Mo doping and sulfur vacancies can significantly improve the photo-response of ZIS due to optical absorption enhancement, accelerated separation and the transport of carriers. The high photocatalytic activity of Mo-Sv-ZIS is expected in combination with the dense active sites.

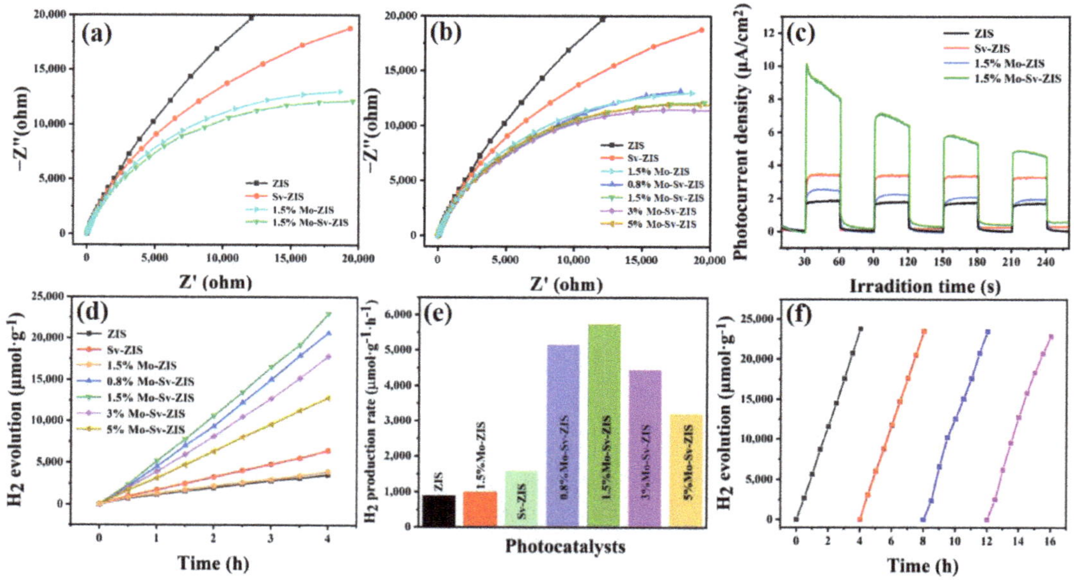

Figure 6. (**a**) The Nyquist plot displays the EIS of ZIS, Sv-ZIS, 1.5% Mo-ZIS and 1.5% Mo-Sv-ZIS. (**b**) EIS of all the samples. (**c**) Transient photocurrent responses of ZIS, Sv-ZIS, 1.5% Mo-ZIS and 1.5% Mo-Sv-ZIS electrodes under visible-light irradiation. (**d**) Time-dependent hydrogen evolution by photocatalysis in visible irradiation (>420 nm). (**e**) Photocatalytic H_2 evolution rates of ZIS, Sv-ZIS, 1.5% Mo-ZIS and Mo-Sv-ZIS with different Mo contents. (**f**) Hydrogen evolution cycle 16 h test on 1.5% Mo-Sv-ZIS.

The photocatalytic activities are assessed by hydrogen evolution in visible light ($\lambda > 420$ nm) irradiation. As shown in Figure 6d,e, the original ZIS shows poor hydrogen evolution activity of about 887 μmol g^{-1} h^{-1}. Mo-ZIS and Sv-ZIS exhibit slightly higher hydrogen evolution activities of 983 μmol g^{-1} h^{-1} and 1607 μmol g^{-1} h^{-1}, respectively. Among all the samples, the 1.5% Mo-Sv-ZIS shows the highest photocatalytic hydrogen evolution rate of 5739 μmolg^{-1} h^{-1}, which is 5.4 time more than the original ZIS, 2.5 times better than that of Sv-ZIS, and 4.8 times higher than that of 1.5% Mo-ZIS. AQY value of 1.5% Mo-Sv-ZIS at 420 nm shows an excellent value of 21.24%. As summarized in Table S6, photocatalytic hydrogen evolution activities over 1.5% Mo-Sv-ZIS outcompete most reported photocatalysts based on $ZnIn_2S_4$. Moreover, Mo-Sv-ZIS has stable and sustainable photocatalytic activities. As shown in Figure 6f, the hydrogen evolution rate over 1.5% Mo-Sv-ZIS did not show a significant decrease after four periods. To verify the stability of the structure, XRD and XPS of the Mo-Sv-ZIS are tested after continuous photocatalytic hydrogen evolution for 16 h. As shown in Figures S8 and S9, both XRD and

XPS results show negligible change after the long-term photocatalytic reaction, verifying the excellent stability of the Mo-Sv-ZIS.

A scheme to explain the excellent photocatalytic activity of the Mo-Sv-ZIS is shown in Figure 7. The presence of S-vacancies and doped Mo atoms in $ZnIn_2S_4$ contribute to the change in electronic structure, which leads to a shift in the charged potential. In spite of the change in the position of both VB and CB, their CB potentials are still higher than the H^+/H_2 reduction potential and are still capable of photocatalytic hydrogen evolution. To gain more insight into the change of band positions, the density of states (DOS) of ZIS, Sv-ZIS and Mo-Sv-ZIS is calculated by using the Castep module of Materials Studio software (Figure S10). Under visible light irradiation, photogenerated electrons respond with H^+ to form H_2 ($H^+ + e^- \rightarrow H_2$), while holes react with TEOA to form $TEOA^+$ ($TEOA + h^+ \rightarrow TEOA^+$). First, Mo doping significantly improves visible light absorption. Second, S vacancy and undercoordinated Mo can capture photogenerated carriers to inhibit electron-hole recombination. Third, the introduction of S vacancies and the doping of Mo atoms allow the ZIS system to have a lower coordination number, providing more reaction sites for photocatalytic reactions. Last but not the least, the formation of Mo-S bonds accelerates the charge transfer and reduce the internal resistance. As a consequence, Mo-Sv-ZIS exhibits an extremely high photocatalytic hydrogen evolution activity.

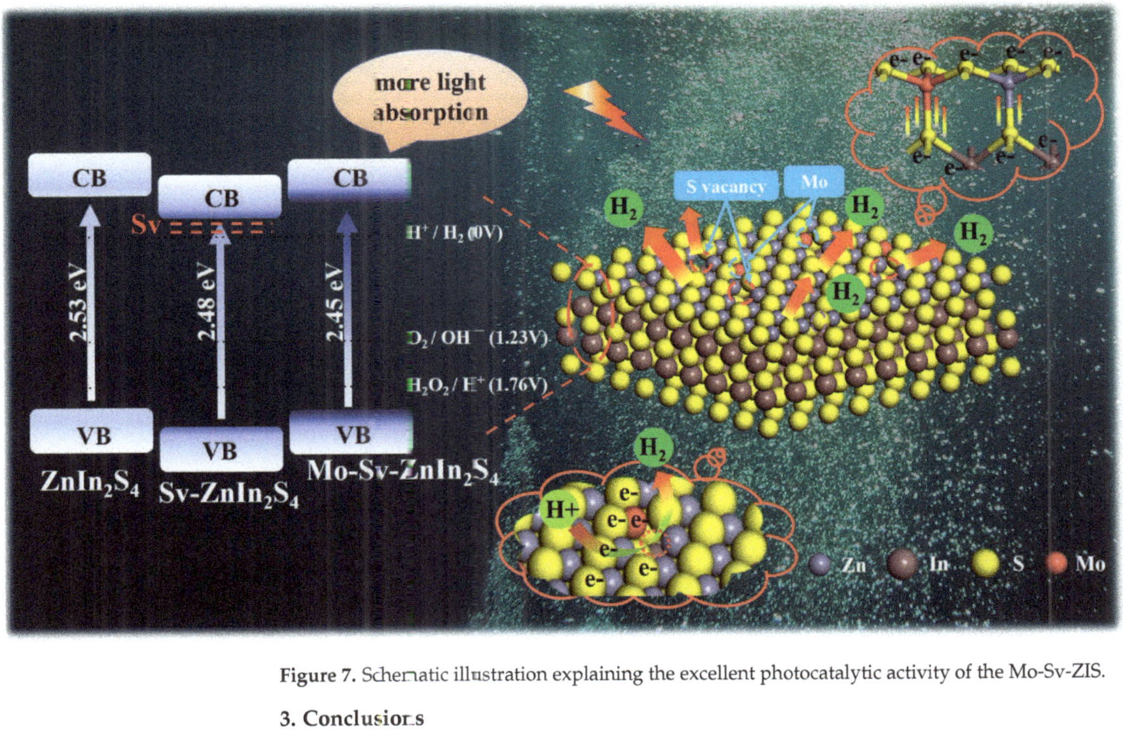

Figure 7. Schematic illustration explaining the excellent photocatalytic activity of the Mo-Sv-ZIS.

3. Conclusions

In summary, we develop a facile one-pot solvothermal method to synthesize Mo-doped 2D $ZnIn_2S_4$ nanoflake with S vacancies. Compared to pristine $ZnIn_2S_4$, the incorporation of Mo doping and S vacancies into the crystal structure modulates the electronic structure and photo-response of $ZnIn_2S_4$, resulting in a higher visible light absorption, faster carrier transfer rate and lower internal resistance. Meanwhile, the doping of Mo atoms with S vacancies reduces the coordination number of $ZnIn_2S_4$, gaining more active sites and, thus, accelerating the catalytic reaction. Thus, the hydrogen evolution rate of 1.5% Mo-Sv-ZIS reaches 5739 $\mu mol\ g^{-1}\ h^{-1}$ under visible light irradiation, which is 5.4 and 2.5 times higher than that of pristine ZIS and Sv-ZIS, respectively. The corresponding AQY at 420 nm reaches 21.24%. Therefore, we demonstrate that the synergy between elemental doping and surface defects is an effective strategy to enhance the evolution of photocatalytic hydrogen in two-dimensional semiconductors, providing an additional perspective on the evolution of photocatalytic materials.

Supplementary Materials: The following supporting information can be downloaded at: https://www.mdpi.com/article/10.3390/nano12223980/s1, Experimental section; Characterizations; Photoelectrochemical measurements; Photocatalytic hydrogen production; Calculation methods; Figure S1: (a–d) TEM images of 1.5% Mo-Sv-ZIS; Figure S2: EDS of the 1.5% Mo-Sv-ZIS shows the presence of Zn, In, S and Mo; Figure S3: Full survey XPS spectra of ZIS, Sv-ZIS and 1.5% Mo-Sv-ZIS; Figure S4: High-resolution XPS spectra of O 1s of ZIS (a) and Sv-ZIS (b); Figure S5: XRD patterns of ZIS, Sv-ZIS and 1.5%Mo-Sv-ZIS with different contents of Mo; Figure S6: FT-IR spectra of ZIS, Sv-ZIS and 1.5% Mo-Sv-ZIS; Figure S7: Raman spectra of ZIS and 1.5% Mo-Sv-ZIS; Figure S8: XRD patterns of the as- prepared 1.5% Mo-Sv-ZIS and 1.5% Mo-Sv-ZIS after photocatalytic test; Figure S9: XPS spectra of 1.5% Mo-Sv-ZIS after photocatalytic test. (a) Full survey XPS spectrum, (b) Zn 2*p*, (c) In 3*d*, (d) S 2*p* and (e) Mo 3*d*; Figure S10 The partial density of sates of (a) total orbit of ZIS and Sv-ZIS, (b) total orbit of Sv-ZIS and Mo-Sv-ZIS, (c) S atomic orbital of ZIS and Sv-ZIS, (d) Mo atomic orbital of Mo-Sv-ZIS; Table S1: Elemental compositions of ZIS Sv-ZIS and 1.5%Mo-Sv-ZIS according to XPS; Table S2 Element content of ZIS, Sv-ZIS and Mo-Sv-ZIS tested by ICP-MS; Table S3: Exponential decay-fitting parameters for time-resolved PL lifetime of ZIS, Sv-ZIS and 1.5% Mo-Sv-ZIS (465 nm); Table S4: Exponential decay-fitting parameters for time-resolved PL lifetime of ZIS, Sv-ZIS and 1.5% Mo-Sv-ZIS (528 nm); Table S5 Impedance values of ZIS, Sv-ZIS and 1.5%Mo-Sv-ZIS obtained by fitting on the EIS results; Table S6: Comparison of the 1.5% Mo-Sv-ZIS with other $ZnIn_2S_4$ related photocatalysts. References [38–40,57–67] are cited in the supplementary materials.

Author Contributions: Conceptualization, L.Y., Z.G. and L.D.; methodology, W.G and M.M.; software, L.Z.; validation, W.G., P.W. and Y.W.; formal analysis, H.W.; investigation, X.D.; resources, L.S.; data curation, W.G.; writing—original draft preparation, W.G.; writing—review and editing, Z.G.; visualization, W.G.; supervision, L.Y.; project administration, L.D.; funding acquisition, L.Y., Z.G. and L.D. All authors have read and agreed to the published version of the manuscript.

Funding: This work was financially backed by the Natural Science Foundation of Shandong Province (ZR2021YQ32) and the National Natural Science Foundation of China (21776147, 21606140); the Taishan Scholar Project of Shandong Province (tsqn201909117); the Qingdao Municipal Science and Technology Bureau, China (19-6-1-91-nsh); and the International Science and Technology Cooperation Program of China (2014DFA60150).

Data Availability Statement: Not applicable.

Conflicts of Interest: The authors declare no conflict of interest. The funders had no role in the design of the study; in the collection, analyses or interpretation of data; in the writing of the manuscript; or in the decision to publish the results.

References

1. Wang, Y.; Liu, L.; Ma, T.; Zhang, Y.; Huang, H. 2D Graphitic Carbon Nitride for Energy Conversion and Storage. *Adv. Funct. Mater.* **2021**, *31*, 1807013. [CrossRef]
2. Zhang, Y.; Wang, T.; Zheng, B.; Shi, J.; Cai, C.; Mao, L.; Cheng, C.; Zong, S.; Guo, X.; Chen, Q. EDTA-dominated hollow tube-like porous graphitic carbon nitride towards enhanced photocatalytic hydrogen evolution. *J. Colloid Interface Sci.* **2022**, *619*, 289–297. [CrossRef] [PubMed]
3. Wang, W.; Bai, X.; Ci, Q.; Du, L.; Ren, X.; Phillips, D.L. Near-Field Drives Long-Lived Shallow Trapping of Polymeric C_3N_4 for Efficient Photocatalytic Hydrogen Evolution. *Adv. Funct. Mater.* **2021**, *31*, 2103978. [CrossRef]
4. Jia, R.; Gui, Q.; Sui, L.; Huang, Y.; Lu, H.; Dong, H.; Ma, S.; Gan, Z.; Dong, L.; Yu, L. Active sites provided by the surface autocatalytic effect and quantum confinement for stable and efficient photocatalytic hydrogen generation. *J. Mater. Chem. A* **2021**, *9*, 14768–14774. [CrossRef]
5. Wang, X.; Wang, X.; Huang, J.; Li, S.; Meng, A.; Li, Z. Interfacial chemical bond and internal electric field modulated Z-scheme Sv-$ZnIn_2S_4$/$MoSe_2$ photocatalyst for efficient hydrogen evolution. *Nat. Commun.* **2021**, *12*, 4112. [CrossRef]
6. Xie, J.; Zhang, H.; Li, S.; Wang, R.; Sun, X.; Zhou, M.; Zhou, J.; Lou, X.W.; Xie, Y. Defect-rich MoS_2 ultrathin nanosheets with additional active edge sites for enhanced electrocatalytic hydrogen evolution. *Adv. Mater.* **2013**, *25*, 5807–5813. [CrossRef]

7. Du, C.; Yan, B.; Yang, G. Self-integrated effects of 2D $ZnIn_2S_4$ and amorphous Mo_2C nanoparticles composite for promoting solar hydrogen generation. *Nano Energy* **2020**, *76*, 105031. [CrossRef]
8. Xiao, F.-X.; Miao, J.; Liu, B. Layer-by-layer self-assembly of CdS quantum dots/graphene nanosheets hybrid films for photoelectrochemical and photocatalytic applications. *J. Am. Chem. Soc.* **2014**, *136*, 1559–1569. [CrossRef]
9. Maeda, K.; Higashi, M.; Lu, D.; Abe, R.; Domen, K. Efficient nonsacrificial water splitting through two-step photoexcitation by visible light using a modified oxynitride as a hydrogen evolution photocatalyst. *J. Am. Chem. Soc.* **2010**, *132*, 5858–5868. [CrossRef]
10. Maeda, K.; Takata, T.; Hara, M.; Saito, N.; Inoue, Y; Kobayashi, H.; Domen, K. GaN: ZnO solid solution as a photocatalyst for visible-light-driven overall water splitting. *J. Am. Chem. Soc.* **2005**, *127*, 8286–8287. [CrossRef]
11. Kim, Y.; Coy, E.; Kim, H.; Mrówczyński, R.; Torruella, P.; Jeong, D.-W.; Choi, K.S.; Jang, J.H.; Song, M.Y.; Jang, D.-J. Efficient photocatalytic production of hydrogen by exploiting the polydopamine-semiconductor interface. *Appl. Catal. B Environ.* **2021**, *280*, 119423. [CrossRef]
12. Wang, S.; Wang, Y.; Zhang, S.L.; Zang, S.Q.; Lou, X.W. Supporting ultrathin $ZnIn_2S_4$ nanosheets on Co/N-Doped graphitic carbon nanocages for efficient photocatalytic H_2 generation. *Adv. Mater.* **2019**, *31*, 1903404. [CrossRef] [PubMed]
13. Kageshima, Y.; Shiga, S.; Ode, T.; Takagi, F.; Shiiba, H.; Htay, M.T.; Hashimoto, Y.; Teshima, K.; Domen, K.; Nishikiori, H. Photocatalytic and Photoelectrochemical Hydrogen Evolution from Water over $Cu_2Sn_xGe_{1-x}S_3$ Particles. *J. Am. Chem. Soc.* **2021**, *143*, 5698–5708. [CrossRef] [PubMed]
14. Peng, Y.; Geng, M.; Yu, J.; Zhang, Y.; Tian, F.; Guo, Y.n.; Zhang, D.; Yang, X.; Li, Z.; Li, Z.; et al. Vacancy-induced 2H@1T MoS_2 phase-incorporation on $ZnIn_2S_4$ for boosting photocatalytic hydrogen evolution. *Appl. Catal. B Environ.* **2021**, *298*, 120570. [CrossRef]
15. Gou, X.; Cheng, F.; Shi, Y.; Zhang, L.; Peng, S.; Chen, J.; Shen, P. Shape-controlled synthesis of ternary chalcogenide $ZnIn_2S_4$ and $CuIn(S, Se)_2$ nano-/microstructures via facile solution route. *J. Am. Chem. Soc.* **2006**, *128*, 7222–7229. [CrossRef]
16. Shi, L.; Yin, P.; Dai, Y. Synthesis and photocatalytic performance of $ZnIn_2S_4$ nanotubes and nanowires. *Langmuir* **2013**, *29*, 12818–12822. [CrossRef]
17. Wang, J.; Chen, Y.; Zhou, W.; Tian, G.; Xiao, Y.; Fu, H.; Fu, H. Cubic quantum dot/hexagonal microsphere $ZnIn_2S_4$ heterophase junctions for exceptional visible-light-driven photocatalytic H_2 evolution. *J. Mater. Chem. A* **2017**, *5*, 8451–8460. [CrossRef]
18. Zhang, Q.; Zhang, J.; Zhang, L.; Cao, M.; Yang, F.; Dai, W.-L. Facile construction of flower-like black phosphorus nanosheet@ $ZnIn_2S_4$ composite with highly efficient catalytic performance in hydrogen production. *Appl. Surf. Sci.* **2020**, *504*, 144366. [CrossRef]
19. Gao, Z.; Chen, K.; Wang, L.; Bai, B.; Liu, H.; Wang, Q. Aminated flower-like $ZnIn_2S_4$ coupled with benzoic acid modified g-C_3N_4 nanosheets via covalent bonds for ameliorated photocatalytic hydrogen generation. *Appl. Catal. B Environ.* **2020**, *268*, 118462. [CrossRef]
20. Xiao, Y.; Peng, Z.; Zhang, W.; Jiang, Y.; Ni, L. Self-assembly of Ag_2O quantum dots on the surface of $ZnIn_2S_4$ nanosheets to fabricate pn heterojunctions with wonderful bifunctional photocatalytic performance. *Appl. Surf. Sci.* **2019**, *494*, 519–531. [CrossRef]
21. Yang, W.; Zhang, L.; Xie, J.; Zhang, X.; Liu, Q.; Yao, T.; Wei, S.; Zhang, Q.; Xie, Y. Enhanced photoexcited carrier separation in oxygen-doped $ZnIn_2S_4$ nanosheets for hydrogen evolution. *Angew. Chem. Int. Ed.* **2016**, *55*, 6716–6720. [CrossRef] [PubMed]
22. Shi, X.; Mao, L.; Yang, P.; Zheng, H.; Fujitsuka, M.; Zhang, J.; Majima, T. Ultrathin $ZnIn_2S_4$ nanosheets with active (110) facet exposure and efficient charge separation for cocatalyst free photocatalytic hydrogen evolution. *Appl. Catal. B Environ.* **2020**, *265*, 118616. [CrossRef]
23. Sun, Y.; Gao, S.; Lei, F.; Xie, Y. Atomically-thin two-dimensional sheets for understanding active sites in catalysis. *Chem. Soc. Rev.* **2015**, *44*, 623–636. [CrossRef] [PubMed]
24. Gao, D.; Wu, X.; Wang, P.; Yu, H.; Zhu, B.; Fan, J.; Yu, J. Selenium-enriched amorphous $NiSe_{1+x}$ nanoclusters as a highly efficient cocatalyst for photocatalytic H_2 evolution. *Chem. Eng. J.* **2021**, *408*, 127230. [CrossRef]
25. Zhang, S.; Zhang, Z.; Si, Y.; Li, B.; Deng, F.; Yang, L.; Liu, X.; Dai, W.; Luo, S. Gradient Hydrogen Migration Modulated with Self-Adapting S Vacancy in Copper-Doped $ZnIn_2S_4$ Nanosheet for Photocatalytic Hydrogen Evolution. *ACS Nano* **2021**, *15*, 15238–15248. [CrossRef]
26. Pan, J.; Zhang, G.; Guan, Z.; Zhao, Q.; Li, G.; Yang, J.; Li, Q.; Zou, Z. Anchoring Ni single atoms on sulfur-vacancy-enriched $ZnIn_2S_4$ nanosheets for boosting photocatalytic hydrogen evolution. *J. Energy Chem.* **2021**, *58*, 408–414. [CrossRef]
27. Wang, H.; Thangamuthu, M.; Wu, Z.; Yang, J.; Yuan, H.; Tang, J. Self-assembled sulphur doped carbon nitride for photocatalytic water reforming of methanol. *Chem. Eng. J.* **2022**, *445*, 136790. [CrossRef]
28. Zhou, T.; Li, T.; Hou, J.; Wang, Y.; Hu, B.; Sun, D.; Wu, Y.; Jiang, W.; Che, G.; Liu, C. Tailoring boron doped intramolecular donor–acceptor integrated carbon nitride skeleton with propelling photocatalytic activity and mechanism insight. *Chem. Eng. J.* **2022**, *445*, 136643. [CrossRef]
29. Liu, Y.; Zhou, Y.; Zhou, X.; Jin, X.; Li, B.; Liu, J.; Chen, G. Cu doped SnS_2 nanostructure induced sulfur vacancy towards boosted photocatalytic hydrogen evolution. *Chem. Eng. J.* **2021**, *407*, 127180. [CrossRef]
30. Li, W.; Wang, F.; Liu, X.-y.; Dang, Y.-y.; Li, J.-y.; Ma, T.-h.; Wang, C.-y. Promoting body carriers migration of CdS nanocatalyst by N-doping for improved hydrogen production under simulated sunlight irradiation. *Appl. Catal. B Environ.* **2022**, *313*, 121470. [CrossRef]

31. Wang, Y.; Zhang, Y.; Zhao, S.; Huang, Z.; Chen, W.; Zhou, Y.; Lv, X.; Yuan, S. Bio-template synthesis of Mo-doped polymer carbon nitride for photocatalytic hydrogen evolution. *Appl. Catal. B Environ.* **2019**, *248*, 44–53. [CrossRef]
32. Zhang, N.; Jalil, A.; Wu, D.; Chen, S.; Liu, Y.; Gao, C.; Ye, W.; Qi, Z.; Ju, H.; Wang, C. Refining defect states in $W_{18}O_{49}$ by Mo doping: A strategy for tuning N_2 activation towards solar-driven nitrogen fixation. *J. Am. Chem. Soc.* **2018**, *140*, 9434–9443. [CrossRef] [PubMed]
33. Wang, H.; Zhang, L.; Wang, K.; Sun, X.; Wang, W. Enhanced photocatalytic CO_2 reduction to methane over $WO_3{\cdot}0.33\ H_2O$ via Mo doping. *Appl. Catal. B Environ.* **2019**, *243*, 771–779. [CrossRef]
34. Wang, Y.; Xu, Y.; Wang, Y.; Qin, H.; Li, X.; Zuo, Y.; Kang, S.; Cui, L. Synthesis of Mo-doped graphitic carbon nitride catalysts and their photocatalytic activity in the reduction of CO_2 with H_2O. *Catal. Commun.* **2016**, *74*, 75–79. [CrossRef]
35. Etogo, A.; Liu, R.; Ren, J.; Qi, L.; Zheng, C.; Ning, J.; Zhong, Y.; Hu, Y. Facile one-pot solvothermal preparation of Mo-doped Bi_2WO_6 biscuit-like microstructures for visible-light-driven photocatalytic water oxidation. *J. Mater. Chem. A* **2016**, *4*, 13242–13250. [CrossRef]
36. Xie, P.; Yang, F.; Li, R.; Ai, C.; Lin, C.; Lin, S. Improving hydrogen evolution activity of perovskite $BaTiO_3$ with Mo doping: Experiments and first-principles analysis. *Int. J. Hydrog. Energy* **2019**, *44*, 11695–11704. [CrossRef]
37. Zhang, R.; Li, P.; Wang, F.; Ye, L.; Gaur, A.; Huang, Z.; Zhao, Z.; Bai, Y.; Zhou, Y. Atomically dispersed Mo atoms on amorphous g-C_3N_4 promotes visible-light absorption and charge carriers transfer. *Appl. Catal. B Environ.* **2019**, *250*, 273–279. [CrossRef]
38. Xing, F.; Liu, Q.; Huang, C. Mo-Doped $ZnIn_2S_4$ Flower-Like Hollow Microspheres for Improved Visible Light-Driven Hydrogen Evolution. *Sol. RRL* **2019**, *4*, 1900483. [CrossRef]
39. Qiu, B.; Huang, P.; Lian, C.; Ma, Y.; Xing, M.; Liu, H.; Zhang, J. Realization of all-in-one hydrogen-evolving photocatalysts via selective atomic substitution. *Appl. Catal. B Environ.* **2021**, *298*, 120518. [CrossRef]
40. Du, C.; Zhang, Q.; Lin, Z.; Yan, B.; Xia, C.; Yang, G. Half-unit-cell $ZnIn_2S_4$ monolayer with sulfur vacancies for photocatalytic hydrogen evolution. *Appl. Catal. B Environ.* **2019**, *248*, 193–201. [CrossRef]
41. Luan, Q.; Xue, X.; Li, R.; Gu, L.; Dong, W.; Zhou, D.; Wang, X.; Li, B.; Wang, G.; Hou, C. Boosting photocatalytic hydrogen evolution: Orbital redistribution of ultrathin $ZnIn_2S_4$ nanosheets via atomic defects. *Appl. Catal. B Environ.* **2022**, *305*, 121007. [CrossRef]
42. Jiao, X.; Chen, Z.; Li, X.; Sun, Y.; Gao, S.; Yan, W.; Wang, C.; Zhang, Q.; Lin, Y.; Luo, Y.; et al. Defect-Mediated Electron-Hole Separation in One-Unit-Cell $ZnIn_2S_4$ Layers for Boosted Solar-Driven CO_2 Reduction. *J. Am. Chem. Soc.* **2017**, *139*, 7586–7594. [CrossRef] [PubMed]
43. Hou, L.; Wu, Z.; Jin, C.; Li, W.; Wei, Q.; Chen, Y.; Wang, T. Flower-Like Dual-Defective Z-Scheme Heterojunction g-C_3N_4/$ZnIn_2S_4$ High-Efficiency Photocatalytic Hydrogen Evolution and Degradation of Mixed Pollutants. *Nanomaterials* **2021**, *11*, 2483. [CrossRef] [PubMed]
44. Jia, T.; Liu, M.; Zheng, C.; Long, F.; Min, Z.; Fu, F.; Yu, D.; Li, J.; Lee, J.H.; Kim, N.H. One-Pot Hydrothermal Synthesis of La-Doped $ZnIn_2S_4$ Microspheres with Improved Visible-Light Photocatalytic Performance. *Nanomaterials* **2020**, *10*, 2026. [CrossRef] [PubMed]
45. Pan, X.; Shang, C.; Chen, Z.; Jin, M.; Zhang, Y.; Zhang, Z.; Wang, X.; Zhou, G. Enhanced Photocatalytic H_2 Evolution over $ZnIn_2S_4$ Flower-Like Microspheres Doped with Black Phosphorus Quantum Dots. *Nanomaterials* **2019**, *9*, 1266. [CrossRef]
46. Zhang, S.; Liu, X.; Liu, C.; Luo, S.; Wang, L.; Cai, T.; Zeng, Y.; Yuan, J.; Dong, W.; Pei, Y.; et al. MoS_2 Quantum Dot Growth Induced by S Vacancies in a $ZnIn_2S_4$ Monolayer: Atomic-Level Heterostructure for Photocatalytic Hydrogen Production. *ACS Nano* **2018**, *12*, 751–758. [CrossRef]
47. Turner, N.; Single, A. Determination of peak positions and areas from wide-scan XPS spectra. *Surf. Interface Anal.* **1990**, *15*, 215–222. [CrossRef]
48. Duan, Y.; Xue, J.; Dai, J.; Wei, Y.; Wu, C.; Chang, S.-H.; Ma, J. Interface engineering of ZnO/In_2O_3 Z-scheme heterojunction with yolk-shell structure for efficient photocatalytic hydrogen evolution. *Appl. Surf. Sci.* **2022**, *592*, 153306. [CrossRef]
49. Zhang, Y.; Qiu, J.; Zhu, B.; Fedin, M.V.; Cheng, B.; Yu, J.; Zhang, L. ZnO/COF S-scheme heterojunction for improved photocatalytic H_2O_2 production performance. *Chem. Eng. J.* **2022**, *444*, 136584. [CrossRef]
50. Wang, P.; Shen, Z.; Xia, Y.; Wang, H.; Zheng, L.; Xi, W.; Zhan, S. Atomic Insights for Optimum and Excess Doping in Photocatalysis: A Case Study of Few-Layer Cu- $ZnIn_2S_4$. *Adv. Funct. Mater.* **2019**, *29*, 1807013. [CrossRef]
51. Zhong, X.; Sun, Y.; Chen, X.; Zhuang, G.; Li, X.; Wang, J.-G. Mo Doping Induced More Active Sites in Urchin-Like $W_{18}O_{49}$Nanostructure with Remarkably Enhanced Performance for Hydrogen Evolution Reaction. *Adv. Funct. Mater.* **2016**, *26*, 5778–5786. [CrossRef]
52. Huang, X.; Gu, W.; Ma, Y.; Liu, D.; Ding, N.; Zhou, L.; Lei, J.; Wang, L.; Zhang, J. Recent advances of doped graphite carbon nitride for photocatalytic reduction of CO_2: A review. *Res. Chem. Intermed.* **2020**, *46*, 5133–5164. [CrossRef]
53. Ye, L.; Fu, J.; Xu, Z.; Yuan, R.; Li, Z. Facile one-pot solvothermal method to synthesize sheet-on-sheet reduced graphene oxide (RGO)/$ZnIn_2S_4$ nanocomposites with superior photocatalytic performance. *ACS Appl. Mater. Interfaces* **2014**, *6*, 3483–3490. [CrossRef] [PubMed]
54. Li, X.; Kang, B.; Dong, F.; Zhang, Z.; Luo, X.; Han, L.; Huang, J.; Feng, Z.; Chen, Z.; Xu, J.; et al. Enhanced photocatalytic degradation and H_2/H_2O_2 production performance of S-pCN/$WO_{2.72}$ S-scheme heterojunction with appropriate surface oxygen vacancies. *Nano Energy* **2021**, *81*, 105671. [CrossRef]

55. Wang, B.; Wang, X.; Lu, L.; Zhou, C.; Xin, Z.; Wang, J.; Ke, X.-k.; Sheng, G.; Yan, S.; Zou, Z. Oxygen-Vacancy-Activated CO_2 Splitting over Amorphous Oxide Semiconductor Photocatalyst. *ACS Catal.* **2017**, *8*, 516–525. [CrossRef]
56. Jia, R.; Lu, H.; Wang, C.; Guan, W.; Dong, H.; Pang, B.; Sui, L.; Gan, Z.; Dong, L.; Yu, L. Construction of 2D-layered quantum dots/2D-nanosheets heterostructures with compact interfaces for highly efficient photocatalytic hydrogen evolution. *J. Colloid Interface Sci.* **2022**, *608*, 284–293. [CrossRef]
57. Xue, J.; Liu, H.; Zeng, S.; Feng, Y.; Zhang, Y.; Zhu, Y.; Cheng, M.; Zhang, H.; Shi, L.; Zhang, G. Bifunctional Cobalt-Doped $ZnIn_2S_4$ Hierarchical Nanotubes Endow Noble-Metal Cocatalyst-Free Photocatalytic H_2 Production Coupled with Benzyl Alcohol Oxidation. *Solar RRL* **2022**, *6*, 2101042. [CrossRef]
58. Zhan, X.; Zheng, Y.; Li, B.; Fang, Z.; Yang, H.; Zhang, H.; Xu, L.; Shao, G.; Hou, H.; Yang, W. Rationally designed $Ta_3N_5/ZnIn_2S_4$ 1D/2D heterojunctions for boosting Visible-Light-driven hydrogen evolution. *Chem. Eng. J.* **2022**, *431*, 134053. [CrossRef]
59. Su, H.; Lou, H.; Zhao, Z.; Zhou, L.; Pang, Y.; Xie, H.; Rao, C.; Yang, D.; Qiu, X. In-situ Mo doped $ZnIn_2S_4$ wrapped MoO_3 S-scheme heterojunction via Mo-S bonds to enhance photocatalytic HER. *Chem. Eng. J.* **2022**, *430*, 132770. [CrossRef]
60. Bhavani, P.; Praveen Kumar, D.; Hussain, M.; Aminabhavi, T.M.; Park, Y.-K. Eco-friendly rice husk derived biochar as a highly efficient noble Metal-Free cocatalyst for high production of H_2 using solar light irradiation. *Chem. Eng. J.* **2022**, *434*, 134743. [CrossRef]
61. Kong, D.; Hu, X.; Geng, J.; Zhao, Y.; Fan, D.; Lu, Y.; Geng, W.; Zhang, D.; Liu, J.; Li, H.; et al. Growing $ZnIn_2S_4$ nanosheets on $FeWO_4$ flowers with p-n heterojunction structure for efficient photocatalytic H_2 production. *Appl. Surf. Sci.* **2022**, *591*, 153256. [CrossRef]
62. She, P.; Qin, J.S.; Sheng, J.; Qi, Y.; Rui, H.; Zhang, W.; Ge, X.; Lu, G.; Song, X.; Rao, H. Dual-Functional Photocatalysis for Cooperative Hydrogen Evolution and Benzylamine Oxidation Coupling over Sandwiched-Like Pd@TiO_2@ $ZnIn_2S_4$ Nanobox. *Small* **2022**, *18*, e2105114. [CrossRef] [PubMed]
63. Su, H.; Rao, C.; Zhou, L.; Pang, Y.; Lou, H.; Yang, D.; Qiu, X. Mo-Doped/Ni-supported $ZnIn_2S_4$-wrapped $NiMoO_4$ S-scheme heterojunction photocatalytic reforming of lignin into hydrogen. *Green Chem.* **2022**, *24*, 2027–2035. [CrossRef]
64. Dai, M.; He, Z.; Zhang, P.; Li, X.; Wang, S. $ZnWO_4$- $ZnIn_2S_4$ S-scheme heterojunction for enhanced photocatalytic H_2 evolution. *J. Mater. Sci. Technol.* **2022**, *122*, 231–242. [CrossRef]
65. Cao, M.; Yang, F.; Zhang, Q.; Zhang, J.; Zhang, L.; Li, L.; Wang, X.; Dai, W.-L. Facile construction of highly efficient MOF-based Pd@UiO-66-NH_2@ $ZnIn_2S_4$ flower-like nanocomposites for visible-light-driven photocatalytic hydrogen production. *J. Mater. Sci. Technol.* **2021**, *76*, 189–199. [CrossRef]
66. Dhingra, S.; Sharma, M.; Krishnan, V.; Nagaraja, C.M. Design of noble metal-free $NiTiO_3/ZnIn_2S_4$ heterojunction photocatalyst for efficient visible-light-assisted production of H_2 and selective synthesis of 2,5-Bis(hydroxymethyl)furan. *J. Colloid Interface Sci.* **2022**, *615*, 346–356. [CrossRef]
67. Liu, L.; Liu, J.; Yang, W.; Wan, J.; Fu, F.; Wang, D. Constructing a Z-scheme $ZnIn_2S_4$-S/CNTs/RP nanocomposite with modulated energy band alignment for enhanced photocatalytic hydrogen evolution. *J. Colloid Interface Sci.* **2022**, *608*, 482–492. [CrossRef]

nanomaterials

MDPI

Article

Ultrasound-Promoted Abatement of Formaldehyde in Liquid Phase with Electrospun Nanostructured Membranes: The Synergy of Combined AOPs

Carlo Boaretti *, Martina Roso, Michele Modesti and Alessandra Lorenzetti *

Department of Industrial Engineering, University of Padova, Via Marzolo 9, 35131 Padova, Italy
* Correspondence: carlo.boaretti@unipd.it (C.B.); alessandra.lorenzetti@unipd.it (A.L.);
Tel.: +39-049-827-5544 (C.B.); +39-049-827-5556 (A.L.)

Abstract: The present work investigates the effect of ultrasounds in the performance of combined advanced oxidation processes (AOPs) on the degradation of formaldehyde (HCHO)-polluted aqueous solutions for potential application in wastewater treatment. Different heterogeneous nanostructured catalysts based on TiO_2 and $FeSO_4$ for photocatalysis and the Fenton process were employed after electrospray deposition on electrospun nanofibrous membranes. Such systems were tested, without the use of any added hydrogen peroxide, by varying the combinations among the selected AOPs in a batch reactor configuration. The results show that, in the absence of a Fenton reaction, ultrasounds provided a significantly increased formaldehyde photocatalytic abatement, probably by increasing the concentration of active species through a different set of reactions while providing a favorable mass transfer regime by the cavitational effect. Due to the faster kinetics of the photo–Fenton process, thanks to its partial homogeneous nature, such a beneficial effect is more limited for the sono–photo–Fenton configuration. On the other hand, the employment of a sono–photocatalytic–Fenton process revealed a synergic effect that provided the best results, reducing the formaldehyde concentration to less than 99% after 240 min. Further analysis showed that, due to a mutual influence, only a tailored $TiO_2/FeSO_4$ ratio on the membranes was able to display the best performance.

Keywords: formaldehyde; titanium dioxide; photocatalysis; ultrasound; Fenton; AOPs; synergy; nanofibers

Citation: Boaretti, C.; Roso, M.; Modesti, M.; Lorenzetti, A. Ultrasound-Promoted Abatement of Formaldehyde in Liquid Phase with Electrospun Nanostructured Membranes: The Synergy of Combined AOPs. *Nanomaterials* **2023**, *13*, 435. https://doi.org/10.3390/nano13030435

Academic Editor: Yu Dong

Received: 10 December 2022
Revised: 18 January 2023
Accepted: 18 January 2023
Published: 20 January 2023

1. Introduction

Formaldehyde (HCHO) is the simplest chemical compound of the aldehyde family, but at the same time, it is one of the most produced chemicals all over the world, with a global market that reached about 22.6 million tons in 2020 [1]. Such high production volumes are justified by its widespread employment as a building block for the synthesis of some of the most important synthetic thermosetting resins (melamine–formaldehyde, urea–formaldehyde, and phenol–formaldehyde) and for the production of important chemicals such as methylene diphenyl diisocyanate for polyurethane production and functional polyalcohols, such as pentaerythritol and 1,4-butanediol. Therefore, its common use in the chemical industry in large quantities, combined with its high toxicity, has led to its classification as a toxic industrial chemical of high hazard [2] due to its potential involvement in terrorist events [3]. Chemical production plants can produce wastewater with up to 10^4 mg/L of formaldehyde [4], which is difficult to treat, even in the range of 10^2 mg/L using conventional biological processes due to the adverse effect on the cellular structure of microorganisms [5,6]. Such evidence requires the employment of alternative degradation strategies to induce formaldehyde decomposition [7–12] and limit its adverse effects on human health and living organisms.

Advanced oxidation processes (AOPs) is a collective term embracing a wide set of homogeneous and/or heterogeneous abatement techniques that rely on the generation of reactive oxidant species to convert different types of recalcitrant pollutants, such as pharmaceutical, dyes, pesticides, and so on, into a more biodegradable form or up to their complete mineralization [13]. To such end, AOPs involve the use of strong oxidizing agents, a specific catalyst or photocatalyst, and different energy vectors, such as ultraviolet (UV) light irradiation, ultrasound (US), or electrical sources, under relatively mild operating conditions. Despite the great debate around the specific species involved in the degradation mechanism, it is generally considered that the hydroxyl radical plays a pivotal role thanks to its high oxidizing potential and non-selective nature, which is useful for the treatment of different types of organics [14]. Therefore, the proper exploitation of such techniques is based on the maximization of the production of such species, combined with relatively high reaction rates and possible complete or at least partial mineralization of the targeted pollutant into non-toxic compounds. From this point of view, the use of single AOPs can suffer from many possible limitations that still hinder their applicability at an industrial scale [15]. For example, photocatalysis with suspended particles is known to suffer from mass transfer limitations, high electron–hole recombination rates, and possible particle agglomeration [16]; the Fenton process is affected by a slow rate of Fe^{3+} reduction to Fe^{2+}, high dependence on the pH, the generation of sludge, and the consumption of expensive hydrogen peroxide (H_2O_2) [17,18] while ultrasound can provide very limited pollutant degradation, representing an expensive technology from an economic point of view [19]. However, while an exhaustive solution for a real breakthrough is yet to come, it has been recognized that a possible compromise to solve some of these issues can be found in the proper coupling of these AOPs as a possible way to also find synergic effects [20–25].

In such a context, the use of heterogeneous systems based on membranes has been recognized as a promising approach for wastewater treatment [26]. In this sense, electrospun nanofibrous membranes are considered an interesting solution in water filtration due to an ideal combination of different properties in terms of mechanical resistance, high surface area, porosity, and wettability [27]. Despite this, the main application of these types of membranes is still related to filtration, the adsorption of contaminants and heavy metals, and the photocatalytic degradation of organic compounds [28–31]. A less studied solution is the use of these membranes with a combination of different AOPs. For this purpose, among the different polymeric substrates that can be employed, PAN has been indicated as a viable support for both photocatalysts [32,33] and photo–Fenton [34,35] catalysts thanks to its high UV stability and propensity for surface functionalization.

Many studies dealing with different combinations of AOPs for the degradation of formaldehyde aqueous solution are present in the literature, often focusing on the analysis of the effect of different process parameters. Kajitvichyanukul et al. [36] compared different AOPs based on photocatalysis and Fenton reactions for the abatement of highly concentrated formalin solution. They observed that the photo–Fenton system had the best performance both in terms of formaldehyde abatement and complete mineralization rate. In a second study [37], the same authors investigated the photo–Fenton degradation of formalin in lab-scale experiments with UV light by analyzing the effects of different parameters on the oxidation rates of both formaldehyde and methanol. They observed that formaldehyde removal was less affected by pH, although a pH of 2.6 was necessary to provide the best efficiency, while generally an increase in both the concentrations of Fe^{3+} ions and H_2O_2 was able to increase the degradation rate. From their analysis, it was possible to conclude that methanol was found to be more difficult to oxidize when in competition with formaldehyde in the aqueous stream to be treated. Sekiguchi et al. [38] analyzed the degradation of model aldehydes such as formaldehyde and benzaldehyde to assess the possible synergistic effect between different techniques of advanced oxidation, such as ultrasounds, UV photocatalysis, and sono–photocatalysis. The results of the study highlighted that the sono–photocatalytic process was able to display the best performance with a first-order degradation rate and a synergic effect on both of the two analyzed pollutants. Liang et al. [39,40] analyzed the kinetics of formaldehyde degradation using UV–Fenton oxidation, identifying a suitable model for describing the evolution of formaldehyde concentration as a function of the concentrations of Fe^{2+} and H_2O_2 in the photochemical reactor. In their conclusions, they highlighted that the main intermediate product of oxidation was formic acid, whose further oxidation was the main rate-limiting step of complete mineralization. Guimaraes et al. [41] compared the degradation efficiency of formaldehyde with the employment of different abatement techniques. They observed that formaldehyde was not possible to degrade by photo-oxidation alone, while peroxidation had a limited effect due to difficulty with the degradation of the formic acid generated as an intermediate oxidation product. In terms of performance, the degradation rates for both UV/H_2O_2 and photo–Fenton were similar, while the latter was found to produce a faster decrease in the dissolved organic concentration for highly concentrated formaldehyde solutions. Tymoyrimoghadam et al. [42] degraded formaldehyde in aqueous solutions with the use of a $UV/S_2O_3{}^{2-}$ process in a photoreactor operated in rotary mode. In their experiments, a certain dependence on pH was noted, which favored a higher removal efficiency under acidic conditions, while specific persulfate-to-formaldehyde ratios were identified to obtain the best results. The specific effects of anionic species on the removal efficiency were also discussed. Deniz et al. [43] analyzed formaldehyde degradation using H_2O_2 as an oxidant species under UV and fluorescent light irradiation. They noticed a higher dependence of the performance on pH by using fluorescent light, but the best performance was identified for conditions of a neutral pH with a combination of UV light and a suitable mass ratio between formaldehyde and H_2O_2. Recently, Lai et al. [44] explored an in situ degradation of formaldehyde by the electro–Fenton process using nitrogen-activated carbon to promote both H_2O_2 and formaldehyde adsorption. Through nitrogen doping, they enhanced the degradation of formaldehyde by promoting Fe^{2+} regeneration to improve the production of $^{\bullet}OH$ at the cathode and achieve a higher probability of interaction with the adsorbed HCHO. With this solution, they were able to reach a 49% increase in degradation compared to the unmodified cathode.

The traditional approach in many of these studies has been to improve the degradation rate of formaldehyde removal but without a specific combination of techniques to achieve both a synergic effect and a partial overcoming of the limitations of each single AOP adopted. In this work, we present the use of such a combined strategy, never reported before, to degrade low concentrations of formaldehyde (90 ppm) in aqueous solutions. Starting from the conclusions of our previous work [45], we adopted a membrane substrate to immobilize the catalytic system on a porous support to avoid its suspension, boosting the production of radical species using both the Fenton reaction and UV light without any added H_2O_2, along with the introduction of ultrasound irradiation. More specifically, we investigated the effect of the latter as it can provide improvements both in terms of mass transfer conditions and degradation of the selected pollutant. The main aim was to analyze if an additive or synergic effect can arise on both the formaldehyde degradation rate and the final conversion by a specific combination of processes involving the use of ultrasound. The optimal catalytic system composition was also considered, with reference to the best-performing combination of AOPs involved.

2. Materials and Methods

2.1. Materials

Polyacrylonitrile (PAN, Mw = 150,000 g/mol, Sigma Aldrich, St. Louis, MO, USA) was used as a nanofibrous support for catalyst deposition and electrospinning, using *N,N*-dimethylformamide (DMF, Sigma Aldrich, St. Louis, MO, USA) as a solvent. TiO_2 powder Aeroxide® P90 (d = 14 nm, BET = 90 m^2/g, 90:10 anatase/rutile ratio, crystal size = 11 nm [46]) was supplied by Evonik (Essen, Germany) and used for the photocatalytic tests, while ferrous sulfate heptahydrate ($FeSO_4{\cdot}7H_2O$) was supplied by Sigma Aldrich (St. Louis, MO, USA) and employed as a catalyst for all the tests involving the Fenton reaction. The catalysts' immobilization on the surfaces of PAN nanofibers was realized by electrospraying the corresponding dispersions, using ethanol (Sigma Aldrich, St. Louis, MO, USA) as a solvent and Dynasylan® 4144 (Evonik, Essen, Germany) as a dispersing agent. A 90 ppm formaldehyde aqueous solution was employed for all the abatement tests and realized by the dilution of formalin (37% *w*/*w* of formaldehyde and 10% *w*/*w* of methanol), supplied by VWR Chemicals (Radnor, PA, USA), with MilliQ water. Whenever the Fenton reaction was involved, the pH of such a solution was adjusted to a value of 3 with the use of a 0.2 N sulfuric acid solution (H_2SO_4 at 98% *w*/*w*, Sigma Aldrich, St. Louis, MO, USA), as such a process is known to be highly favored under these conditions [47]. For the quantification of the formaldehyde concentration, a specific derivatization method was employed using o-(2,3,4,5,6-pentafluorobenzyl)hydroxylamine hydrochloride (PFBHA-HCl) purchased from Alfa Aesar (Haverhill, MA, USA) and 1,2-dibromopropane at 97% *w*/*w* as internal standards (Sigma Aldrich, St. Louis, MO, USA) and using hexane (Sigma Aldrich, St. Louis, MO, USA) for the extraction process.

2.2. Methods

Thermogravimetric analysis (TGA) was the technique employed for the quantification of the amount of photocatalyst deposited on the surface of the electrospun composite membranes. The analysis was carried out with an SDT-Q600 apparatus (TA Instruments, New Castle, DE, USA) equipped with alumina pans, employing a 100 cc/min airflow and a 20 °C/min heating rate from room temperature up to 900 °C. The quantification of the residual formaldehyde concentration was carried out according to the U.S. Environmental Protection Agency (EPA) method 556 [48] using a derivatization procedure with PFBHA-HCl on samples withdrawn from the aqueous solution employed during the degradation tests. The oxime derivative formed was extracted from water with hexane and processed with an acidic wash step before being analyzed by gas chromatography–mass spectroscopy (gas chromatograph TRACE 1300® directly coupled with a single quadrupole mass spectrometer ISQ QD®, both from Thermo Scientific, Waltham, MA, USA). The column for the chromatographic separation was a non-polar DB5 capillary column (0.25 mm i.d., 30 m length, supplied by Agilent, Santa Clara, CA, USA). The oxime derivative ($m/z = 225$) and 1,2 dibromopropane ($m/z = 121$), used as internal standards, were identified as target analytes and employed for the quantification of the formaldehyde concentration using a calibration curve derived from solutions at different but known formaldehyde concentrations. For each data point, an average value based on three replicates was employed to analyze the trends of the different tests.

2.3. Nanostructured Membranes Preparation

A 5%wt solution of PAN in DMF was prepared and then electrospun on a rotating drum collector. After that, a suitable amount of fibers was collected, the electrospinning was stopped, and electrospraying of the proper catalytic system suspension was performed. The optimized conditions for all the deposition processes are shown in Table 1, according to a previous study [45], and identified in order to avoid excessive catalyst agglomeration with membrane clogging and to obtain tailored amounts on the membrane surface.

Table 1. Deposition parameters for membrane production.

System	Flow Rate (mL/h)	Voltage (kV)	Electrodes Distance (cm)	Relative Humidity (%)	Deposition Time (min)
PAN support	2	23	20–25	30–40	120
Catalyst deposition	2	23	15	20–30	variable

The electrospray deposition of the catalytic systems was realized with 5% *w*/*w* concentrated suspensions in ethanol. Before deposition, the catalysts dispersions were preliminarily sonicated in an ice-cold bath for 40 min at a 40% amplitude using a 500 W ultrasonic probe (Vibra-cell VC505® of Sonics & Materials, Newtown, CT, USA), alternating 1 min of sonication with 1 min of non-sonication to avoid solvent evaporation. Subsequently, Dynasylan® 4144 was added at a 1% *w*/*w* concentration based on the ethanol content, followed by a further sonication of the resultant dispersion for 15 min. For the following analysis and discussion of the results, the membranes realized are named PAN_TiO_2, PAN_Fe, and PAN_TiO_2_Fe, according to the fact that the deposited catalytic systems were TiO_2, $FeSO_4$, and TiO_2 + $FeSO_4$, respectively.

2.4. Set-Up for the Formaldehyde Degradation Tests

The set-up for the degradation tests is reported in detail in our previous work [45]. Briefly, about 70 mL of solution containing approximately 90 ppm of formaldehyde was placed in a Petri dish along with a 50 cm^2 nanostructured membrane of PAN fibers covered by the electrosprayed particles of the specific catalytic system. The container with the formaldehyde solution was then suspended inside an ultrasonic bath (Branson Bransonic 1510, Emerson Electric, Ferguson, MI, USA) with a consumption power of 70 W and a working frequency of about 42 kHz. For the tests involving the use of UV radiation, a 16 W UVC lamp (UV Stylo E16, Light Progress, Anghiari, Italy), with λ_{max} = 254 nm, was placed on top of the ultrasonic bath and at a distance of 5 cm from the Petri dish, using a parallel configuration with respect to the surface of the membrane. During all catalytic experiments, the temperature was controlled by a continuous circulation of the water contained in the sonication bath.

3. Results and Discussion

3.1. Membrane Analysis

Thermogravimetric analysis was carried out to quantify the amounts of the different catalytic systems deposited by electrospraying in order to compare the results of the abatement tests among the different membrane systems as a function of the catalyst weight. In order to assess the relative quantities of the catalysts employed in the case of the mixed composition (TiO_2 + $FeSO_4$), a first analysis was carried out in terms of the typical degradation profiles of pure materials. As can be seen in Figure 1a, PAN thermo-oxidative degradation, with a specific heating rate employed (20 °C/min) for the analysis, is a two-step process that can be identified by the mass-loss intervals of the polymer as a function of temperature [49]. The material mass was stable up to 250 °C, with an onset degradation temperature of about 300 °C, above which the decomposition of the material can be reasonably ascribed to a combination of reactions involving dehydrogenation, cyclization, and oxidation. The latter two are recognized to occur simultaneously, generating an unstable and partially hydrogenated cyclic structure, while oxidation, despite preceding cyclization, is characterized by a slower rate and becomes predominant in the second degradation stage, which involves PAN cyclization products [50,51]. Due to the high heating rate adopted in the test, the material was not stabilized, as in the case of carbon fiber pretreatment [52], and therefore, above 400 °C, the temperature was able to promote chain scission and the generation of volatile products up to the complete mass loss [53]. TiO_2 is a stable material with a negligible mass loss (around 2%) at low temperatures due to the evaporation of adsorbed water, which can be considered not relevant for the membranes developed due to the relatively low amount of TiO_2 deposited. On the other hand, the thermal degradation of $FeSO_4$ is known to proceed with two different and subsequent degradation steps, divided into dehydration and decomposition [54,55]. In the first phase, crystallization water is lost upon heating at relatively low temperatures, producing anhydrous $FeSO_4$ according to the following Equations (1)–(3) [56,57]:

$$FeSO_4{\cdot}7H_2O \xrightarrow{40\text{–}90\ °C} FeSO_4{\cdot}4H_2O + 3H_2O \quad (1)$$

$$FeSO_4{\cdot}4H_2O \xrightarrow{140\text{–}200\ °C} FeSO_4{\cdot}H_2O + 3H_2O \quad (2)$$

$$FeSO_4{\cdot}H_2O \xrightarrow{270\text{–}350\ °C} FeSO_4{+}H_2O \quad (3)$$

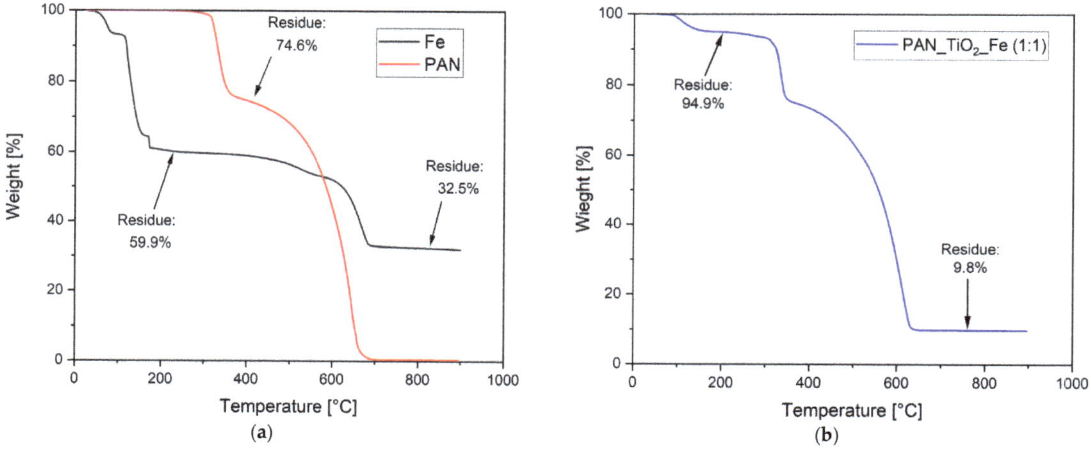

Figure 1. Weight residues of (**a**) PAN, $FeSO_4{\cdot}7H_2O$ and (**b**) PAN_TiO_2_Fe (1:1) under air atmosphere.

Above 500 °C, $FeSO_4$ starts to decompose by releasing SO_2 up to 700 °C, and it is converted in Fe_2O_3 according to Equations (4) and (5) [58]:

$$6\ FeSO_4 \xrightarrow{525\text{–}650\ ^\circ C} Fe_2(SO_4)_3 + 2\ Fe_2O_3 + 3\ SO_2 \tag{4}$$

$$Fe_2(SO_4)_3 \xrightarrow{625\text{–}710\ ^\circ C} Fe_2O_3 + 3\ SO_2 + \frac{3}{2}O_2 \tag{5}$$

As can be seen in the thermogram of pure $FeSO_4$ in Figure 1a, there were three distinct but partially overlapping mass losses up to 200 °C that can be associated with the dehydration reactions. Subsequently, the mass of the sample remained stable up to 400 °C, when the decomposition reaction occurred, according to the reactions suggested in the previous equations.

From the stoichiometry of the dehydration and decomposition reactions, it was possible to quantify the amount of volatile products released from the thermo-oxidative process and the theoretical residue, which was approximately 28.7%. Such a residue was slightly different from the final result, reported in Figure 1a, of about 32.5%. This could be explained as a partial loss of crystallization water during storage as the first dehydration step starts at a very low temperature. Indeed, according to the total mass loss at 800 °C, the true water content of the sample corresponds approximately to a $FeSO_4{\cdot}5H_2O$ composition. This conclusion is in accordance with the fact that the first mass-loss step is around 6.8%, and it is close to the theoretical loss of 7.4% for the reaction $FeSO_4{\cdot}5H_2O \rightarrow FeSO_4{\cdot}4H_2O + H_2O$, while the total dehydration (40.1% mass loss) is in reasonable agreement with the theoretical value of 37%. Due to this evidence, the values obtained from the TGA were taken as a reference for the evaluation of the relative content of iron sulfate during electrospray deposition.

The introduction of $FeSO_4$ in the membranes altered the degradation profile between the first and second mass-loss steps as Fenton reagents can act as hydroperoxide-decomposing agents during the formation of organic radicals [59–61], promoting a faster degradation of PAN and, therefore, a lower decomposition temperature. However, by comparing the TGA curves of the catalysts with that of the composite membrane PAN_TiO_2_Fe (Figure 1b), the initial weight loss of the latter up to 200 °C, where both TiO_2 and PAN were quite stable (i.e., weight loss of less than 1%), can be related to the dehydration of $FeSO_4$, while the final residue at 800 °C was due to residual Fe_2O_3 derived from $FeSO_4$ oxidation and TiO_2. On the basis of these considerations, it was possible to assess the relative content of the catalysts deposited on the membranes and the relative mass ratio between the inorganic components of the catalytic systems. Table 2 summarizes the results of these calculations in terms of the catalyst concentration and composition ratios of the different membranes employed in the tests.

Table 2. Residues, relative compositions, and catalyst concentrations derived from the thermogravimetric (TGA) analysis of the different catalysts and membranes employed for the abatement test.

Sample	Residue @200 °C (%)	Residue @800 °C (%)	$FeSO_4$ Content (%)	TiO_2 Content (%)	Fe:TiO_2 Ratio	Catalyst Conc. (mg/cm^2)
TiO_2	98.8	98.0	-	100	-	-
Fe	59.9	32.5	100	-	-	-
PAN_TiO_2	99.8	17.6	-	17.6	-	0.141
PAN_Fe	94.4	4.7	14.5	-	-	0.124
PAN_TiO_2_Fe (2:1)	94.9	16.1	7.7	13.6	0.56	0.135
PAN_TiO_2_Fe (1:1)	94.9	9.8	7.7	7.3	0.94	0.133
PAN_TiO_2_Fe (1:2)	94.6	7.1	8.1	4.5	1.81	0.129

After tailoring the deposition amount for the different catalytic systems, it was possible to control the overall amount deposited on the membranes and the relative composition between the $FeSO_4$ and TiO_2 whenever they were deposited on the same membrane. Since the TGA results in Table 2 show that the overall quantities were quite similar to each other, it was decided to test these systems to compare the abatement of formaldehyde without the necessity of rescaling the amount of pollutant degraded on the catalyst content present on the surface of the membranes.

3.2. Photocatalytic Tests

Following the results of our previous work [45] on the combined use of different AOPs, we tried to improve the HCHO abatement performance of composite nanofibrous membranes by introducing the effect of ultrasound. Such a solution was selected to obtain improvements in terms of the mass transfer on the surface of the catalytic systems employed and to evaluate if ultrasound could also provide a chemical effect to promote the degradation of HCHO. To such an end, the first analysis involved the effect of the neat ultrasound action on HCHO in the absence of any catalyst and membrane support, in comparison to the effect of UV radiation and the combination of these two processes.

The results of these tests are shown in Figure 2, where it is possible to observe that the UV radiation alone did not provide any effect on the residual concentration of HCHO, and therefore, in the absence of any photocatalyst, it was not able to induce photodecomposition of the pollutant. On the other hand, ultrasound showed the ability to induce a limited but constant degradation of HCHO over time, reaching a final conversion of about 20% over the course of 4 h of treatment, quite independently from the presence of UV light. On the basis of these results, it is possible to confirm that there was no relevant synergic effect due to the coupling of the two techniques and that the profile of the concentration of HCHO was only affected by the action of ultrasound. Such an effect, at a relatively low US frequency, was also noted by Navarro et al. [62] for formic acid and was explained as the direct consequence of water sonolysis (Equations (6)–(11), [63]) and the radical scavenging action of the pollutant on the produced $^{\bullet}OH$ species.

$$H_2O +))) \rightarrow H^{\bullet} + {}^{\bullet}OH \tag{6}$$

$${}^{\bullet}OH + {}^{\bullet}OH \rightarrow H_2O_2 \tag{7}$$

$${}^{\bullet}OH + {}^{\bullet}OH \rightarrow H_2 + O_2 \tag{8}$$

$$H^{\bullet} + O_2 \rightarrow HO_2^{\bullet} \tag{9}$$

$$H^{\bullet} + HO_2^{\bullet} \rightarrow H_2O_2 \tag{10}$$

$$HO_2^{\bullet} + HO_2^{\bullet} \rightarrow H_2O_2 + O_2 \tag{11}$$

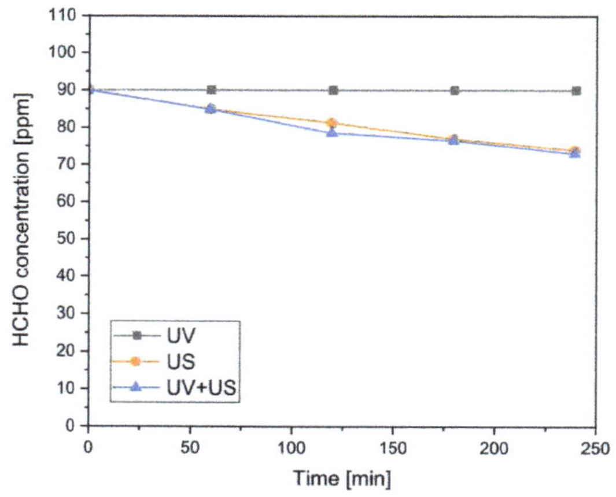

Figure 2. Residual concentration of formaldehyde (HCHO) over time for ultra-violet (UV), ultrasound (US), and UV+US tests (maximum standard deviation, σ_{max} = 1.3 ppm).

A further "blank" test was realized with a PAN-TiO_2 membrane in the absence of US and UV action to probe any effect related to the photocatalyst, but only a slow decrease in the concentration was detected over the course of the test and attributed to surface adsorption of the pollutant.

However, as can be noted in Figure 3, when photocatalysis was coupled with ultrasound (US + UV PAN_TiO_2), the degradation of the pollutant was more efficient since it did not increase the initial concentration of formaldehyde, as was the case with an absence of ultrasound (UV PAN_TiO_2).

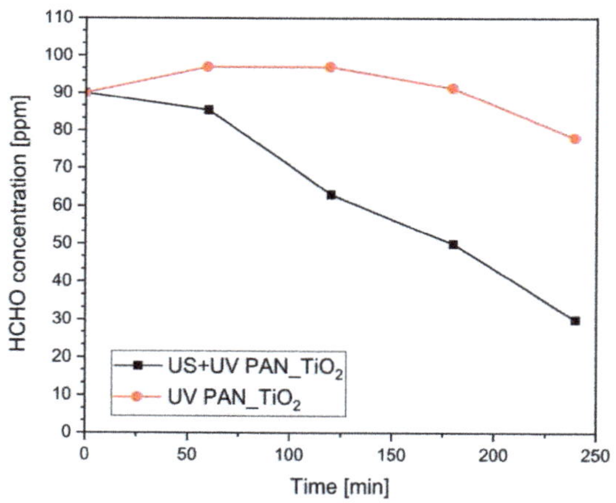

Figure 3. Residual concentration of HCHO for photocatalysis and sono–photocatalysis tests (σ_{max} = 1.9 ppm).

This initial increase in the formaldehyde content during a photocatalytic test, without US, was explained as a consequence of the partial oxidation of methanol, used as a commercial stabilizer from the starting formalin solution, that generates formaldehyde as an intermediate degradation product due to the reaction with dissolved oxygen (Equations (12)–(13), [8]):

$$CH_3OH + {}^{\bullet}OH \rightarrow {}^{\bullet}CH_2OH + H_2O \quad (12)$$

$$^{\bullet}CH_2OH + O_2 \rightarrow CH_2O + HO_2{}^{\bullet} \quad (13)$$

Due to the non-negligible amount of methanol in the formalin solution (about 25 ppm per 90 ppm of formaldehyde), a specific photocatalytic experiment was also carried out to test this hypothesis, using an aqueous solution containing 25 ppm of methanol. The results (not shown) confirm the proposed mechanism with a progressive increase in formaldehyde formation up to a maximum value after 120 min, in accordance with the findings of Araña et al. [8]. This allows for the hypothesis that methanol is preferentially adsorbed on the surface of hydrophilic nanoparticles of P90 and more rapidly photo-oxidized than formaldehyde, increasing its concentration over time. From this point of view, the change in the formaldehyde concentration profile over the course of the reaction promoted by the presence of ultrasounds can be a consequence of a different degradation mechanism. Indeed, as observed by Rasshokin et al. [64], in the sonolysis of diluted aqueous methanol solutions, methanol can be converted into methane during ultrasound treatment, according to the following Equations (14)–(17) in which thermolysis of the C-O bonds is coupled with the radical species produced from water sonolysis [63]:

$$CH_3OH \rightarrow {}^{\bullet}CH_3 + {}^{\bullet}OH \tag{14}$$

$$^{\bullet}CH_3 + H^{\bullet} \rightarrow CH_4 \tag{15}$$

$$^{\bullet}CH_3 + HO_2^{\bullet} \rightarrow CH_4 + O_2 \tag{16}$$

$$^{\bullet}CH_3 + H_2 \rightarrow CH_4 + H^{\bullet} \tag{17}$$

It has to be underlined, however, that those conclusions cannot be strictly applied to the present case as the sonication frequency was significantly different (724 kHz versus 42 kHz) and also because other elementary reactions, which can lead to the formation of formaldehyde, had to be taken into consideration (Equations (18)–(21), [65]):

$$CH_3OH \rightarrow {}^{\bullet}CH_2OH + H^{\bullet} \tag{18}$$

$$CH_3OH + H^{\bullet} \rightarrow {}^{\bullet}CH_2OH + H_2 \tag{19}$$

$$^{\bullet}CH_2OH + H^{\bullet} \rightarrow CH_2O + H_2 \tag{20}$$

$$^{\bullet}CH_2OH + {}^{\bullet}OH \rightarrow CH_2O + H_2O \tag{21}$$

In this sense, it was explained [64] that in diluted solutions of methanol, the possible formation of formaldehyde is only due to the migration of $H^{\bullet}$ and $^{\bullet}OH$ radicals to the bulk of the aqueous solution. If the formaldehyde concentration during the experiments does not increase, as when US and photocatalysis are coupled, it is likely that the extent of migration within the bulk of the liquid phase is limited. As a consequence, the $H^{\bullet}$ and $^{\bullet}OH$ radicals recombine or react with methanol at the bubble interface where the temperature reaction is higher and the $^{\bullet}CH_2OH$ radical produced from methanol can be converted into CO (Equations (22)–(24), [66]):

$$^{\bullet}CH_2OH + M \rightarrow CH_2O + H^{\bullet} + M \tag{22}$$

$$CH_2O + M \rightarrow H^{\bullet} + {}^{\bullet}CHO + M \tag{23}$$

$$^{\bullet}CHO + M \rightarrow CO + H^{\bullet} + M \tag{24}$$

It can therefore be concluded that the high temperatures reached in the cavitation bubbles during the exposition to ultrasound might have provided a different set of reactions involved in methanol degradation, at least in the absence of a concurrent effect of a photocatalytic nature. In the latter case, a further explanation of the result could be related to the TiO_2-promoted production of different active species (i.e., $HO_2^{\bullet}$ and $^{\bullet}OH$) due to the photocatalytic decomposition of H_2O_2 [67] derived from water sonolysis (Equations (6)–(11)), possible even at low ultrasonic frequencies [68]. A higher concentration of such active species can promote the reaction with both methanol and formaldehyde, possibly explaining the lower initial degradation rate observed within the first 60 min of the reaction as a consequence of the balance between formaldehyde degradation and production from methanol oxidation. The relevance of the two proposed mechanisms could be verified with proper analytic techniques, such as electron spin resonance [69,70]; however, this was outside of the main aim of this work. It is, however, clear that such improved reactivity, coupled with the higher mass transfer improvement observed with the use of ultrasound in heterogeneous reacting systems [71,72], can enhance the overall efficiency of the degradation process and possibly explain the constant decrease in the pollutant concentration.

A second set of tests was carried out by combining the ultrasound irradiation with the photo–Fenton process, as shown in Figure 4. It is well documented that UV-C radiation is able to assist the regeneration of the Fenton cycle by the photo-reduction of ferric ions under acidic pH, increasing the efficiency of the process and reducing the consumption of H_2O_2 needed to sustain the process [73–76]. On the other hand, the aforementioned effect of the H_2O_2 concentration increase as a consequence of the introduction of ultrasound might directly promote the Fenton reaction, inducing a higher amount of radical species [77]. For the purpose of the present study, it was decided to take advantage of such an effect as a way to improve the efficiency of the formaldehyde degradation obtained in the absence of any added H_2O_2, as highlighted in our previous study [45].

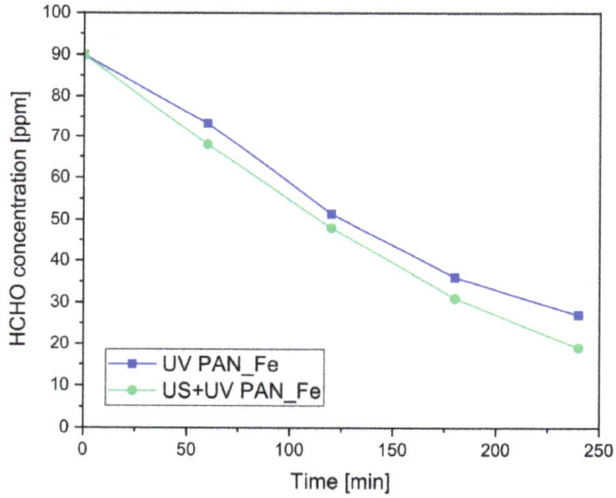

Figure 4. Residual concentration of HCHO over time for photo–Fenton and sono–photo–Fenton tests (σ_{max} = 1.3 ppm).

In this case, the results of the tests in Figure 4 show that the introduction of ultrasound can improve, although to a limited extent, both the rate and the final conversion in comparison to the photo–Fenton process, with an evident additive effect. This can be ascribed to the high initial degradation rate of the pure photo–Fenton reaction, which does not suffer from the limitations shown by the heterogeneous nature of the photocatalysis process. On the basis of these results, it is possible to affirm that the ultrasound had a beneficial effect for both processes, i.e., photocatalysis and photo–Fenton, although to different extents, and therefore, it was plausible to analyze the performance of abatement due to the combination of these different AOPs in a single system, realizing the sono–photocatalytic–Fenton abatement of formaldehyde. The results of these tests are shown in Figure 5, in comparison with other tested systems, where the sono–photocatalytic–Fenton system was employed using a tailored deposition of $FeSO_4$ and TiO_2 in order to respect a Fe/TiO_2 mass composition ratio of 1:1. This may provide a simpler explanation if the combined process shows a synergic or simply additive effect; as already reported before, the total amount of catalysts was about the same in all the tested systems.

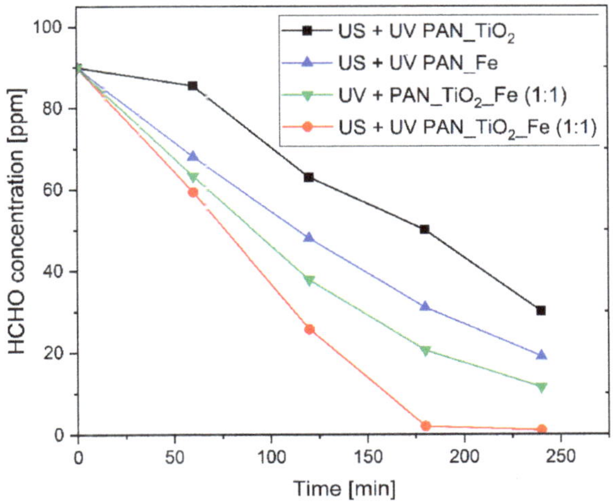

Figure 5. Comparison of the residual concentration trend of HCHO over time for tested advanced oxidation processes (AOPs) (σ_{max} = 1.9 ppm).

As can be seen, there were distinct performances of the different systems. In particular, the trends with the best results were ranked as follows: sono–photocatalytic–Fenton > photocatalytic–Fenton > sono–photo–Fenton > sono–photocatalysis. This ranking reflects the improvements both in the initial degradation rate, calculated as the average value from the data in the 0–120 min range, and the final conversion (Table 3), indicating that the contribution of the ultrasound improved each single AOP.

Table 3. Initial degradation rate and final conversion of HCHO for the different combined AOPs tested.

System	Process	Initial Degradation Rate (ppm/min)	Final Conversion (%)
US + UV PAN_TiO_2	Sono–photocatalysis	0.195	67
US + UV PAN_Fe	Sono–photo–Fenton	0.353	79
UV + PAN_TiO_2_Fe	Photocatalytic–Fenton	0.437	87
US + UV PAN_TiO_2_Fe (1:1)	Sono–photocatalytic–Fenton	0.531	99

A possible explanation of the contribution provided by ultrasounds can be understood not only in terms of higher mass transfer conditions but also as a way for promoting further pollutant degradation by both $^{\bullet}OH$ and H_2O_2 formation [63].

On the one hand, such H_2O_2 can be photo-catalytically decomposed by UV radiation to generate a new active radical species that can oxidize the organic pollutant, and on the other hand, it can be involved in a synergic cascade effect. Indeed, while iron is able to limit the electron–hole recombination of TiO_2, the H_2O_2 derived from water sonolysis can be involved in promoting the iron oxidation from Fe(II) to Fe(III) [78,79], while UV radiation allows for the photo-reduction of Fe(III) to Fe(II) in order to restart the Fenton cycle [80]. This effect could be important for producing H_2O_2 in situ without any addition of the oxidant at the beginning of the process, although the relevance of such an effect has to be carefully evaluated as a function of the pollutant concentration and processing conditions.

A further consideration can be made in terms of the possible products derived from formaldehyde oxidation according to the different findings reported in the literature [40,41,45,70,81]. For the present case, due to the nature of the processes involved and the liquid phase medium, the probable reaction intermediate was formic acid, which can be subsequently oxidized to CO_2. Such an intermediate has commonly been reported for the photocatalysis, Fenton, and photo–Fenton processes in aqueous solutions [40,41] and is consistent with the pH fluctuation observed over the course of the reaction [45]. On the other hand, the effect of ultrasound in combination with photocatalysis on product distribution, although not specifically addressed in the literature, has been proven not to inhibit the degradation of HCHO for low concentrations in aqueous solutions [38] and, therefore, does not represent a main concern for the present study.

Due to the good results obtained with the employment of the sono–photocatalytic–Fenton process, a further analysis regarding the effect of the composition ratio between the TiO_2 and $FeSO_4$ was carried out. According to the previous synergic effect, altering the composition ratio may have led to a different contribution to the catalytic activity of the systems. In this sense, Figure 6 reports the abatement profiles of the sono–photocatalytic–Fenton systems for different Fe/TiO_2 mass ratios in comparison to the 1:1 reference composition.

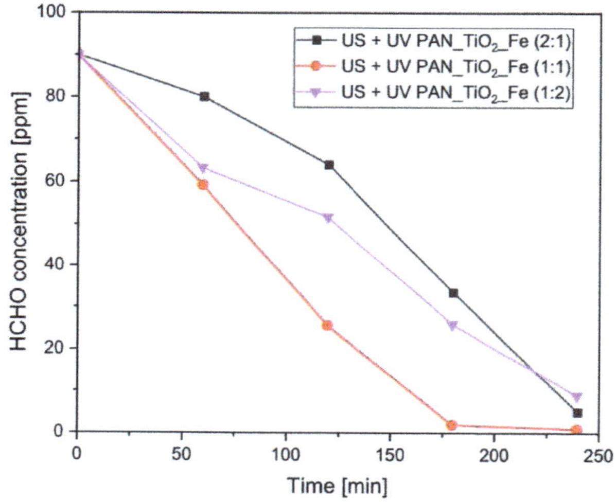

Figure 6. Comparison of the residual concentration trend of HCHO for the sono–photocatalytic–Fenton systems with different $TiO_2/FeSO_4$ mass ratios (σ_{max} = 2.3 ppm).

The results show that the optimal performance, in terms of the abatement rate, occurred when the relative composition ratio was 1:1. The second-highest performing system was the one with a higher $FeSO_4$ content (1:2), which, however, over the course of the experimental time analyzed, provided a similar final conversion as the system with a higher TiO_2 content (2:1). It is evident that altering the composition of the catalytic system influenced the HCHO abatement performance as a consequence of the different kinetics of the involved processes. The lower abatement rate, with a 1:2 mass ratio composition, can be explained by the higher $FeSO_4$ relative content shielding the UV absorption of TiO_2 and reducing its photoactivity, similar to the trend of the sono–photo–Fenton process, but with a more limited final conversion due to the lower iron content. On the other hand, increasing the amount of TiO_2 (a 2:1 mass ratio) worsened the abatement rate because the photocatalytic process, characterized by lower kinetics also in the presence of US, became predominant over the photo–Fenton reaction and slowed down the overall rate of the process.

4. Conclusions

The present work investigated the effects of the combinations of different AOPs on the degradation of aqueous solutions with a low concentration (90 ppm) of formaldehyde. Several strategies based on the employment of UV photocatalysis, the Fenton reaction without added H_2O_2, and ultrasound irradiation were considered with the use of heterogeneous membranes made of electrosprayed TiO_2 and $FeSO_4$ particles supported on electropsun PAN nanofibers. The results show that ultrasound can improve the kinetics and final conversion of formaldehyde, overcoming the limitations of photocatalysis, with a synergic effect possibly related to the higher mass transfer efficiency and combined degradation promoted by sonolysis. This effect was less marked in the case of the sono–photo–Fenton reactions since the photo–Fenton process is characterized by faster kinetics, and the contribution to the trend of conversion over time was rather additive than synergic. The combination of the different AOPs for the sono–photocatalytic–Fenton abatement tests showed that it was possible to combine the previous improvements to obtain a system that was able to provide a 99% decrease in the formaldehyde concentration in 240 min. However, this effect was achieved only when the relative concentrations of the Fe and TiO_2 were balanced to allow for a synergic contribution and avoid a possible photo-shielding effect from the Fe source or predominance of the photocatalysis over the better-performing photo–Fenton reaction. Such findings can provide new detailed insight into the rationale of combining efficiently different AOPs and overcoming the limitations that affect each one of these processes.

Author Contributions: Conceptualization, C.B., M.R., M.M. and A.L.; data curation, M.R.; investigation, C.B. and A.L.; methodology, C.B., M.R., M.M. and A.L.; project administration, A.L.; resources, A.L.; supervision, C.B.; visualization, M.R.; writing—original draft, C.B. and A.L.; writing—review & editing, M.M. All authors have read and agreed to the published version of the manuscript.

Funding: This research received no external funding.

Institutional Review Board Statement: Not applicable.

Informed Consent Statement: Not applicable.

Data Availability Statement: Data are contained within the study.

Conflicts of Interest: The authors declare no conflict of interest.

References

1. Global Industry Analysts Global Formaldehyde Industry. Available online: https://www.reportlinker.com/p05817774/Global-Formaldehyde-Industry.html (accessed on 18 November 2022).
2. United States Department of Labor. Occupational Safety and Health Administration Toxic Industrial Chemicals (TICs) Guide. Available online: https://www.osha.gov/emergency-preparedness/guides/toxic-industrial-chemicals (accessed on 18 November 2022).
3. Bennett, M. *Today's Chemist at Work*; ACS Publications: Washington, DC, USA, 2003; pp. 21–25.
4. Oliveira, S.V.W.B.; Moraes, E.M.; Adorno, M.A.T.; Varesche, M.B.A.; Foresti, E.; Zaiat, M. Formaldehyde Degradation in an Anaerobic Packed-Bed Bioreactor. *Water Res.* **2004**, *38*, 1685–1694. [CrossRef] [PubMed]
5. Lu, Z.; Hegemann, W. Anaerobic Toxicity and Biodegradation of Formaldehyde in Batch Cultures. *Water Res.* **1998**, *32*, 209–215. [CrossRef]
6. Lotfy, H.R.; Rashed, I.G. A Method for Treating Wastewater Containing Formaldehyde. *Water Res.* **2002**, *36*, 633–637. [CrossRef] [PubMed]
7. Silva, A.M.T.; Castelo-Branco, I.M.; Quinta-Ferreira, R.M.; Levec, J. Catalytic Studies in Wet Oxidation of Effluents from Formaldehyde Industry. *Chem. Eng. Sci.* **2003**, *58*, 963–970. [CrossRef]
8. Araña, J.; Martínez Nieto, J.L.; Herrera Melián, J.A.; Doña Rodríguez, J.M.; González Díaz, O.; Pérez Peña, J.; Bergasa, O.; Alvarez, C.; Méndez, J. Photocatalytic Degradation of Formaldehyde Containing Wastewater from Veterinarian Laboratories. *Chemosphere* **2004**, *55*, 893–904. [CrossRef] [PubMed]
9. Yuan, D.; Tian, L.; Gu, D.; Shen, X.; Zhu, L.; Wu, H.; Wang, B. Fast and Efficient Oxidation of Formaldehyde in Wastewater via the Solar Thermal Electrochemical Process Tuned by Thermo-Electrochemistry. *J. Clean. Prod.* **2017**, *156*, 310–316. [CrossRef]
10. Xing, B.; Yang, G.; Chen, H.; Liu, X. Catalytic Wet Oxidation of High Concentration Formaldehyde Wastewater over Pt/Nitrogen-Doped Activated Carbon. *React. Kinet. Mech. Catal* **2019**, *126*, 547–560. [CrossRef]
11. Mei, X.; Guo, Z.; Liu, J.; Bi, S.; Li, P.; Wang, Y.; Shen, W.; Yang, Y.; Wang, Y.; Xiao, Y.; et al. Treatment of Formaldehyde Wastewater by a Membrane-Aerated Biofilm Reactor (MABR): The Degradation of Formaldehyde in the Presence of the Cosubstrate Methanol. *Chem. Eng. J.* **2019**, *372*, 673–683. [CrossRef]
12. Du, L.; Gao, W.; Li, Z.; Jiao, W.; Liu, Y. Oxidative Degradation of Formaldehyde in Wastewater by $MgO/O_3/H_2O_2$ in a Rotating Packed Bed. *Chem. Eng. Process. Process Intensif.* **2020**, *155*, 108053. [CrossRef]
13. Ameta, S.C. Chapter 1—Introduction. In *Advanced Oxidation Processes for Waste Water Treatment*; Ameta, S.C., Ameta, R., Eds.; Academic Press: Cambridge, MA, USA, 2018; pp. 1–12, ISBN 978-0-12-810499-6.

14. Deng, Y.; Zhao, R. Advanced Oxidation Processes (AOPs) in Wastewater Treatment. *Curr. Pollut. Rep.* **2015**, *1*, 167–176. [CrossRef]
15. Ma, D.; Yi, H.; Lai, C.; Liu, X.; Huo, X.; An, Z.; Li, L.; Fu, Y.; Li, B.; Zhang, M.; et al. Critical Review of Advanced Oxidation Processes in Organic Wastewater Treatment. *Chemosphere* **2021**, *275*, 130104. [CrossRef]
16. Iervolino, G.; Zammit, I.; Vaiano, V.; Rizzo, L. Limitations and Prospects for Wastewater Treatment by UV and Visible-Light-Active Heterogeneous Photocatalysis: A Critical Review. *Top Curr. Chem.* **2019**, *378*, 7. [CrossRef] [PubMed]
17. Diya'uddeen, B.H.; Abdul Aziz, A.R.; Daud, W.M.A.W. On the Limitation of Fenton Oxidation Operational Parameters: A Review. *Int. J. Chem. Reactor Eng.* **2012**, *10*, 1. [CrossRef]
18. Bello, M.M.; Abdul Raman, A.A.; Asghar, A. A Review on Approaches for Addressing the Limitations of Fenton Oxidation for Recalcitrant Wastewater Treatment. *Process Saf. Environ. Prot.* **2019**, *126*, 119–140. [CrossRef]
19. Camargo-Perea, A.L.; Rubio-Clemente, A.; Peñuela, G.A. Use of Ultrasound as an Advanced Oxidation Process for the Degradation of Emerging Pollutants in Water. *Water* **2020**, *12*, 1068. [CrossRef]
20. Bagal, M.v.; Gogate, P.R. Wastewater Treatment Using Hybrid Treatment Schemes Based on Cavitation and Fenton Chemistry: A Review. *Ultrason. Sonochem.* **2014**, *21*, 1–14. [CrossRef] [PubMed]
21. Babu, S.G.; Ashokkumar, M.; Neppolian, B. The Role of Ultrasound on Advanced Oxidation Processes. *Top Curr. Chem.* **2016**, *374*, 75. [CrossRef]
22. Gągol, M.; Przyjazny, A.; Boczkaj, G. Wastewater Treatment by Means of Advanced Oxidation Processes Based on Cavitation—A Review. *Chem. Eng. J.* **2018**, *338*, 599–627. [CrossRef]
23. Madhavan, J.; Theerthagiri, J.; Balaji, D.; Sunitha, S.; Choi, M.Y.; Ashokkumar, M. Hybrid Advanced Oxidation Processes Involving Ultrasound: An Overview. *Molecules* **2019**, *24*, 3341.
24. Gogate, P.R. Improvements in Catalyst Synthesis and Photocatalytic Oxidation Processing Based on the Use of Ultrasound. *Top Curr. Chem.* **2020**, *378*, 29. [CrossRef] [PubMed]
25. Anandan, S.; Kumar Ponnusamy, V.; Ashokkumar, M. A Review on Hybrid Techniques for the Degradation of Organic Pollutants in Aqueous Environment. *Ultrason. Sonochem.* **2020**, *67*, 105130. [CrossRef] [PubMed]
26. Titchou, F.E.; Zazou, H.; Afanga, H.; el Gaayda, J.; Ait Akbour, R.; Nidheesh, P.V.; Hamdani, M. Removal of Organic Pollutants from Wastewater by Advanced Oxidation Processes and Its Combination with Membrane Processes. *Chem. Eng. Process. Process Intensif.* **2021**, *169*, 108631. [CrossRef]
27. Subrahmanya, T.M.; Arshad, A.b.; Lin, P.T.; Widakdo, J.; Makari, H.K.; Austria, H.F.M.; Hu, C.-C.; Lai, J.-Y.; Hung, W.-S. A Review of Recent Progress in Polymeric Electrospun Nanofiber Membranes in Addressing Safe Water Global Issues. *RSC Adv.* **2021**, *11*, 9638–9663. [CrossRef]
28. Tijing, L.D.; Yao, M.; Ren, J.; Park, C.-H.; Kim, C.S.; Shon, H.K. Nanofibers for Water and Wastewater Treatment: Recent Advances and Developments. In *Energy, Environment, and Sustainability*; Springer Nature: Singapore, 2019; pp. 431–468.
29. Cui, J.; Li, F.; Wang, Y.; Zhang, Q.; Ma, W.; Huang, C. Electrospun Nanofiber Membranes for Wastewater Treatment Applications. *Sep. Purif. Technol.* **2020**, *250*, 117116. [CrossRef]
30. Zhu, F.; Zheng, Y.-M.; Zhang, B.-G.; Dai, Y.-R. A Critical Review on the Electrospun Nanofibrous Membranes for the Adsorption of Heavy Metals in Water Treatment. *J. Hazard. Mater.* **2021**, *401*, 123608. [CrossRef]
31. Nasir, A.M.; Awang, N.; Jaafar, J.; Ismail, A.F.; Othman, M.H.D.; A. Rahman, M.; Aziz, F.; Mat Yajid, M.A. Recent Progress on Fabrication and Application of Electrospun Nanofibrous Photocatalytic Membranes for Wastewater Treatment: A Review. *J. Water Process Eng.* **2021**, *40*, 101878. [CrossRef]
32. Modesti, M.; Roso, M.; Boaretti, C.; Besco, S.; Hrelja, D.; Sgarbossa, P.; Lorenzetti, A. Preparation of Smart Nano-Engineered Electrospun Membranes for Methanol Gas-Phase Photoxidation. *Appl. Catal. B* **2014**, *144*, 216–222. [CrossRef]
33. Zhang, R.; Ma, Y.; Lan, W.; Sameen, D.E.; Ahmed, S.; Dai, J.; Qin, W.; Li, S.; Liu, Y. Enhanced Photocatalytic Degradation of Organic Dyes by Ultrasonic-Assisted Electrospray TiO2/Graphene Oxide on Polyacrylonitrile/β-Cyclodextrin Nanofibrous Membranes. *Ultrason. Sonochem.* **2021**, *70*, 105343. [CrossRef]
34. Zhao, X.; Dong, Y.; Cheng, B.; Kang, W. Removal of Textile Dyes from Aqueous Solution by Heterogeneous Photo-Fenton Reaction Using Modified PAN Nanofiber Fe Complex as Catalyst. *Int. J. Photoenergy* **2013**, *2013*, 820165. [CrossRef]
35. Zheng, S.; Chen, H.; Tong, X.; Wang, Z.; Crittenden, J.C.; Huang, M. Integration of a Photo-Fenton Reaction and a Membrane Filtration Using CS/PAN@FeOOH/g-C_3N_4 Electrospun Nanofibers: Synthesis, Characterization, Self-Cleaning Performance and Mechanism. *Appl. Catal. B* **2021**, *281*, 119519. [CrossRef]
36. Kajitvichyanukul, P.; Lu, M.-C.; Liao, C.-H.; Wirojanagud, W.; Koottatep, T. Degradation and Detoxification of Formaline Wastewater by Advanced Oxidation Processes. *J. Hazard. Mater.* **2006**, *135*, 337–343. [CrossRef] [PubMed]
37. Kajitvichyanukul, P.; Lu, M.-C.; Jamroensan, A. Formaldehyde Degradation in the Presence of Methanol by Photo-Fenton Process. *J. Environ. Manag.* **2008**, *86*, 545–553. [CrossRef] [PubMed]
38. Sekiguchi, K.; Sasaki, C.; Sakamoto, K. Synergistic Effects of High-Frequency Ultrasound on Photocatalytic Degradation of Aldehydes and Their Intermediates Using TiO_2 Suspension in Water. *Ultrason. Sonochem.* **2011**, *18*, 158–163. [CrossRef]
39. Liang, J.; Liu, X.; Zhang, Z.; Wang, Y. Kinetics and Reaction Mechanism for Formaldehyde Wastewater Using UV-Fenton Oxidation. In Proceedings of the 2010 4th International Conference on Bioinformatics and Biomedical Engineering, Chengdu, China, 18–20 June 2010; pp. 1–5.
40. Liu, X.; Liang, J.; Wang, X. Kinetics and Reaction Pathways of Formaldehyde Degradation Using the UV-Fenton Method. *Water Environ. Res.* **2011**, *83*, 418–426. [CrossRef] [PubMed]

41. Guimarães, J.R.; Turato Farah, C.R.; Maniero, M.G.; Fadini, P.S. Degradation of Formaldehyde by Advanced Oxidation Processes. *J. Environ. Manag.* **2012**, *107*, 96–101. [CrossRef] [PubMed]
42. Tymoyrimoghadam, L.; Momeninejad, H.; Baratpour, P.; Gohari, F.; Ravansalar, B.; Momeninejad, M.; Adibzadeh, A. Advanced Oxidation of Formaldehyde in the Aqueous Solutions Using UVC/$S_2O_8^{2-}$ Process: Degradation and Mineralization TT -. *Environ. Health Eng. Manag. J.* **2019**, *6*, 97–104. [CrossRef]
43. Deniz, F.; Mazmancı, M.A. Advanced Oxidation of High Concentrations of Formaldehyde in Aqueous Solution under Fluorescent and UV Light. *Environ. Health Eng. Manag. J.* **2021**, *8*, 267–276. [CrossRef]
44. Lai, S.; Zhao, H.; Qu, Z.; Tang, Z.; Yang, X.; Jiang, P.; Wang, Z. Promotion of Formaldehyde Degradation by Electro-Fenton: Controlling the Distribution of ·OH and Formaldehyde near Cathode to Increase the Reaction Probability. *Chemosphere* **2022**, *307*, 135776. [CrossRef]
45. Bonora, R.; Boaretti, C.; Campea, L.; Roso, M.; Martucci, A.; Modesti, M.; Lorenzetti, A. Combined AOPs for Formaldehyde Degradation Using Heterogeneous Nanostructured Catalysts. *Nanomaterials* **2020**, *10*, 148. [CrossRef]
46. Keulemans, M.; Verbruggen, S.W.; Hauchecorne, B.; Martens, J.A.; Lenaerts, S. Activity versus Selectivity in Photocatalysis: Morphological or Electronic Properties Tipping the Scale. *J. Catal.* **2016**, *344*, 221–228. [CrossRef]
47. Ameta, R.; Chohadia, A.K.; Jain, A.; Punjabi, P.B. Chapter 3—Fenton and Photo-Fenton Processes. In *Advanced Oxidation Processes for Wastewater Treatment*; Ameta, S.C., Ameta, R., Eds.; Academic Press: Cambridge, MA, USA, 2018; pp. 49–87, ISBN 978-0-12-810499-6.
48. U.S. EPA. *Method 556.1: Determination of Carbonyl Compounds in Drinking Water by Fast Gas Chromatography*; U.S. EPA: Cincinnati, OH, USA, 1999.
49. Grassie, N.; McGuchan, R. Pyrolysis of Polyacrylonitrile and Related Polymers—I. Thermal Analysis of Polyacrylonitrile. *Eur. Polym. J.* **1970**, *6*, 1277–1291. [CrossRef]
50. Fitzer, E.; Müller, D.J. The Influence of Oxygen on the Chemical Reactions during Stabilization of Pan as Carbon Fiber Precursor. *Carbon N. Y.* **1975**, *13*, 63–69. [CrossRef]
51. Biedunkiewicz, A.; Figiel, P.; Sabara, M. Pyrolysis and Oxidation of PAN in Dry Air. Thermoanalytical Methods. *Medziagotyra* **2011**, *17*, 38–42. [CrossRef]
52. Rahaman, M.S.A.; Ismail, A.F.; Mustafa, A. A Review of Heat Treatment on Polyacrylonitrile Fiber. *Polym. Degrad. Stab.* **2007**, *92*, 1421–1432. [CrossRef]
53. Houtz, R.C. "Orlon" Acrylic Fiber: Chemistry and Properties. *Text. Res. J.* **1950**, *20*, 786–801. [CrossRef]
54. Gallagher, P.K.; Johnson, D.W.; Schrey, F. Thermal Decomposition of Iron(II) Sulfates. *J. Am. Ceram. Soc.* **1970**, *53*, 666–670. [CrossRef]
55. Tesfaye, F.; Jung, I.-H.; Paek, M.-K.; Moroz, M.; Lindberg, D.; Hupa, L. *Thermochemical Data of Selected Phases in the FeO_x–$FeSO_4$–$Fe_2(SO_4)_3$ System BT—Materials Processing Fundamentals 2019*; Lambotte, G., Lee, J., Allanore, A., Wagstaff, S., Eds.; Springer International Publishing: Cham, Switzerland, 2019; pp. 227–240.
56. Swamy, M.S.R.; Prasad, T.P. Thermal Decomposition of Iron(II) Sulphate in Air. IV—Heats of Reaction and Heats of Formation. *J. Therm. Anal.* **1981**, *20*, 101–105. [CrossRef]
57. Wang, T.; Debelak, K.A.; Roth, J.A. Dehydration of Iron(II) Sulfate Heptahydrate. *Thermochim. Acta* **2007**, *462*, 89–93. [CrossRef]
58. Masset, P.; Poinso, J.Y.; Poignet, J.C. TG/DTA/MS Study of the Thermal Decomposition of $FeSO_4{\cdot}6H_2O$. *J. Therm. Anal. Calorim.* **2006**, *83*, 457–462. [CrossRef]
59. Mai, C.; Majcherczyk, A.; Schormann, W.; Hüttermann, A. Degradation of Acrylic Copolymers by Fenton's Reagent. *Polym. Degrad. Stab.* **2002**, *75*, 107–112. [CrossRef]
60. Freitas, A.R.; Vidotti, G.J.; Rubira, A.F.; Muniz, E.C. Polychloroprene Degradation by a Photo-Fenton Process. *Polym. Degrad. Stab.* **2005**, *87*, 425–432. [CrossRef]
61. Giroto, J.A.; Guardani, R.; Teixeira, A.C.S.C.; Nascimento, C.A.O. Study on the Photo-Fenton Degradation of Polyvinyl Alcohol in Aqueous Solution. *Chem. Eng. Process. Process Intensif.* **2006**, *45*, 523–532. [CrossRef]
62. Navarro, N.M.; Chave, T.; Pochon, P.; Bisel, I.; Nikitenko, S.I. Effect of Ultrasonic Frequency on the Mechanism of Formic Acid Sonolysis. *J. Phys. Chem. B* **2011**, *115*, 2024–2029. [CrossRef] [PubMed]
63. Mason, T.J. and Lorimer, J.P. Synthesis. In *Applied Sonochemistry: Uses of Power Ultrasound in Chemistry and Processing*; Mason, T.J., Lorimer, J.P., Eds.; Wiley-VCH Verlag: Weinheim, Germany, 2002; pp. 75–130, ISBN 9783527600540.
64. Rassokhin, D.N.; Bugaenko, L.T.; Kovalev, G. v The Sonolysis of Methanol in Diluted Aqueous Solutions: Product Yields. *Radiat. Phys. Chem.* **1995**, *45*, 251–255. [CrossRef]
65. Kerboua, K.; Hamdaoui, O. Oxygen-Argon Acoustic Cavitation Bubble in a Water-Methanol Mixture: Effects of Medium Composition on Sonochemical Activity. *Ultrason. Sonochem.* **2020**, *61*, 104811. [CrossRef]
66. Aronowitz, D.; Naegeli, D.W.; Glassman, I. Kinetics of the Pyrolysis of Methanol. *J. Phys. Chem.* **1977**, *81*, 2555–2559. [CrossRef]
67. Yi, J.; Bahrini, C.; Schoemaecker, C.; Fittschen, C.; Choi, W. Photocatalytic Decomposition of H_2O_2 on Different TiO_2 Surfaces Along with the Concurrent Generation of HO_2 Radicals Monitored Using Cavity Ring Down Spectroscopy. *J. Phys. Chem. C* **2012**, *116*, 10090–10097. [CrossRef]
68. Ziembowicz, S.; Kida, M.; Koszelnik, P. Sonochemical Formation of Hydrogen Peroxide. *Proceedings* **2018**, *2*, 188.
69. Li, R.; Huang, Y.; Zhu, D.; Ho, W.; Cao, J.; Lee, S. Improved Oxygen Activation over a Carbon/Co_3O_4 Nanocomposite for Efficient Catalytic Oxidation of Formaldehyde at Room Temperature. *Environ. Sci. Technol.* **2021**, *55*, 4054–4063. [CrossRef]
70. Li, X.; Li, H.; Huang, Y.; Cao, J.; Huang, T.; Li, R.; Zhang, Q.; Lee, S.; Ho, W. Exploring the Photocatalytic Conversion Mechanism of Gaseous Formaldehyde Degradation on TiO_{2-x}-OV Surface. *J. Hazard. Mater.* **2022**, *424*, 127217. [CrossRef]

71. Gogate, P.R. Cavitational Reactors for Process Intensification of Chemical Processing Applications: A Critical Review. *Chem. Eng. Process. Process Intensif.* **2008**, *47*, 515–527. [CrossRef]
72. Gogate, P.R.; Sutkar, V.S.; Pandit, A.B. Sonochemical Reactors: Important Design and Scale up Considerations with a Special Emphasis on Heterogeneous Systems. *Chem. Eng. J.* **2011**, *166*, 1066–1082. [CrossRef]
73. Ruppert, G.; Bauer, R.; Heisler, G. The Photo-Fenton Reaction—An Effective Photochemical Wastewater Treatment Process. *J. Photochem. Photobiol. A Chem.* **1993**, *73*, 75–78. [CrossRef]
74. Gogate, P.R.; Pandit, A.B. A Review of Imperative Technologies for Wastewater Treatment II: Hybrid Methods. *Adv. Environ. Res.* **2004**, *8*, 553–597. [CrossRef]
75. Pignatello, J.J.; Oliveros, E.; MacKay, A. Advanced Oxidation Processes for Organic Contaminant Destruction Based on the Fenton Reaction and Related Chemistry. *Crit. Rev. Environ. Sci. Technol.* **2006**, *36*, 1–84. [CrossRef]
76. Babuponnusami, A.; Muthukumar, K. A Review on Fenton and Improvements to the Fenton Process for Wastewater Treatment. *J. Environ. Chem. Eng.* **2014**, *2*, 557–572. [CrossRef]
77. Moradi, M.; Elahinia, A.; Vasseghian, Y.; Dragoi, E.-N.; Omidi, F.; Mousavi Khaneghah, A. A Review on Pollutants Removal by Sono-Photo -Fenton Processes. *J. Environ. Chem. Eng.* **2020**, *8*, 104330. [CrossRef]
78. Kim, H.-E.; Lee, J.; Lee, H.; Lee, C. Synergistic Effects of TiO_2 Photocatalysis in Combination with Fenton-like Reactions on Oxidation of Organic Compounds at Circumneutral PH. *Appl. Catal. B* **2012**, *115–116*, 219–224. [CrossRef]
79. Seo, J.; Park, S.Y.; Cho, J.; Lee, J.; Kim, H.-H.; Lee, K.-M.; Pham, A.L.-T.; Lee, C. Synergistic Effects between the S-TiO_2 Photocatalyst and the Fenton-like Reagent: Enhanced Contaminant Oxidation under Visible Light Illumination. *J. Environ. Chem. Eng.* **2021**, *9*, 104598. [CrossRef]
80. Kim, S.-M.; Vogelpohl, A. Degradation of Organic Pollutants by the Photo-Fenton-Process. *Chem. Eng. Technol.* **1998**, *21*, 187–191. [CrossRef]
81. Shin, E.; Senthurchelvan, R.; Munoz, J.; Basak, S.; Rajeshwar, K.; Benglas-Smith, G.; Howell, B.C. Photolytic and Photocatalytic Destruction of Formaldehyde in Aqueous Media. *J. Electrochem. Soc.* **1996**, *143*, 1562. [CrossRef]

nanomaterials

MDPI

Article

Bidimensional $SnSe_2$—Mesoporous Ordered Titania Heterostructures for Photocatalytically Activated Anti-Fingerprint Optically Transparent Layers

Jessica De Santis [1,2], Valentina Paolucci [1,2], Luigi Stagi [3], Davide Carboni [2,3], Luca Malfatti [2,3], Carlo Cantalini [1,2,*] and Plinio Innocenzi [2,3,4,*]

1 Department of Industrial and Information Engineering and Economics, 67100 L'Aquila, Italy; jessica.desantis@graduate.univaq.it (J.D.S.); valentina.paolucci2@univaq.it (V.P.)
2 National Interuniversity Consortium of Materials Science and Technology (INSTM), 50121 Florence, Italy; dcarboni@uniss.it (D.C.); lucamalfatti@uniss.it (L.M.)
3 Laboratory of Materials Science and Nanotechnology (LMNT), Department of Biomedical Sciences, CR-INSTM, University of Sassari, 07100 Sassari, Italy; lstagi@uniss.it
4 Department of Chemistry, College of Science, United Arab Emirates University, Al Ain 15551, United Arab Emirates
* Correspondence: carlo.cantalini@univaq.it (C.C.); plinio@uniss.it (P.I.)

Abstract: The design of functional coatings for touchscreens and haptic interfaces is of paramount importance for smartphones, tablets, and computers. Among the functional properties, the ability to suppress or eliminate fingerprints from specific surfaces is one of the most critical. We produced photoactivated anti-fingerprint coatings by embedding 2D-$SnSe_2$ nanoflakes in ordered mesoporous titania thin films. The $SnSe_2$ nanostructures were produced by solvent-assisted sonication employing 1-Methyl-2-pyrrolidinone. The combination of $SnSe_2$ and nanocrystalline anatase titania enables the formation of photoactivated heterostructures with an enhanced ability to remove fingerprints from their surface. These results were achieved through careful design of the heterostructure and controlled processing of the films by liquid phase deposition. The self-assembly process is unaffected by the addition of $SnSe_2$, and the titania mesoporous films keep their three-dimensional pore organization. The coating layers show high optical transparency and a homogeneous distribution of $SnSe_2$ within the matrix. An evaluation of photocatalytic activity was performed by observing the degradation of stearic acid and Rhodamine B layers deposited on the photoactive films as a function of radiation exposure time. FTIR and UV-Vis spectroscopies were used for the photodegradation tests. Additionally, infrared imaging was employed to assess the anti-fingerprinting property. The photodegradation process, following pseudo-first-order kinetics, shows a tremendous improvement over bare mesoporous titania films. Furthermore, exposure of the films to sunlight and UV light completely removes the fingerprints, opening the route to several self-cleaning applications.

Keywords: $SnSe_2$; 2D materials; titania; photocatalysis; anti-fingerprint

Citation: De Santis, J.; Paolucci, V.; Stagi, L.; Carboni, D.; Malfatti, L.; Cantalini, C.; Innocenzi, P. Bidimensional $SnSe_2$—Mesoporous Ordered Titania Heterostructures for Photocatalytically Activated Anti-Fingerprint Optically Transparent Layers. *Nanomaterials* **2023**, *13*, 1406. https://doi.org/10.3390/nano13081406

Academic Editor: Zhixing Gan

Received: 30 March 2023
Revised: 14 April 2023
Accepted: 17 April 2023
Published: 19 April 2023

1. Introduction

The photocatalytic activity of anatase nanocrystalline titania is a well-known phenomenon with manifold applications, such as energy conversion, photonics, and photoinduced antimicrobial activity [1]. In particular, the technologies based on thin films are successful examples of applications of the photocatalytic properties of titania [2]. A thin layer of transparent TiO_2 deposited on window glasses or lenses enables the so-called self-cleaning effect [3], which causes the decomposition of the organic compounds on the glass surface thanks to the sunlight absorption [4]. Several products based on this phenomenon are already on the market; however, their ability to completely decompose solid pollutants on lenses and glass facades has not yet fully developed, and the main benefit is the need of less-frequent cleanings of the surfaces. On the other hand, efficient self-cleaning materials can also be applied as anti-fingerprint and anti-smudge surfaces [5,6] which are highly requested in many applications, such as in touchscreens for smart phones [7], photovoltaic devices, and anti-smudge eyeglasses [8]. At the moment, the technological challenge is still the fabrication of thin photoactive layers that do not modify some key properties of the surface, such as optical transparency and reflectance. Nonetheless, the most common coatings developed for this purpose are fluorinated [9] or polymeric films [10], which have adhesion and/or durability problems on glass surfaces.

The use of an ordered mesoporous titania layer instead of a dense thin film provides a strong improvement of the photocatalytic self-cleaning effect, which is up to four times faster than that measured on the nonporous coating [11,12]. Further developments are now envisaged thanks to the possibility of fabricating titania-based heterostructures in mesoporous layers. The potential of 2D layered materials, especially metal chalcogenides, has been investigated across many fields, such as photocatalysis [13–15], optoelectronics [16], and sensing [17]. The formation of heterostructures using 2D materials has been confirmed as a promising strategy to enhance the photocatalytic response in thin transparent mesoporous ordered titania films [18,19]. Several 2D layered materials, depending on their bandgap, can be combined with nanocrystalline titania to yield photoactive heterostructures with a tailored light-harvesting capacity. The fabrication of optically transparent thin films integrating 2D materials is a challenging goal. For several applications, as we previously underlined, the photoactive layer must preserve the transparency. This is a mandatory property for photovoltaic devices, self-cleaning coatings on window glass, or touch screens. Titania-based heterostructures that employ metal nanoparticles (i.e., gold or silver) also work well to improve the photocatalytic performances, but, regrettably, they reduce the overall transmittance by coloring the films. In previous works, we successfully formed photoactive heterostructures by fabricating mesoporous titania nanocomposites embedding 2D materials [20–22]. Graphene, boron nitride, and WS_2 materials have been incorporated into mesoporous titania thin films. By adopting a one-pot synthesis, the method we devised enables the fabrication of transparent films that include 2D components as the dispersed phase.

Here we report the synthesis of $SnSe_2$-loaded mesoporous ordered TiO_2 films through liquid phase deposition. The choice of $SnSe_2$ is based on its intrinsic ability to trigger surface chemical reactions, as in the case of chemical sensors [23], and its narrow bandgap that ensures a broad absorption in the whole solar spectrum. Moreover, theoretical studies have shown that $SnSe_2$ can efficiently act as co-catalyst in water-splitting reactions, as it promotes charge separation and provides trapping and reduction sites for protons. $SnSe_2$, however, also exhibits a fast electron recombination and has a low conduction band (CB) level (-0.16 eV vs. the normal nitrogen electrode (NHE)); therefore, it has to be coupled with other semiconductors, such as TiO_2, to reduce the recombination of charge carriers and increase the rate of radical formation [24]. The formation of $SnSe_2/TiO_2$ heterostructures in thin mesoporous films allows us to improve the self-cleaning capability while keeping the level of transparency the same. Achieving such a condition, however, requires a careful design of the coating to avoid $SnSe_2$ aggregation. Furthermore, the as-deposited titania films, after dip- or spin-coating, must be thermally annealed to form the anatase crystallites. The annealing process, however, has to be finely tuned to avoid oxidation of the $SnSe_2$. The addition of few-layer $SnSe_2$ into the titania precursor sol allowed us to achieve great control over the final chemo-physical properties of the nanocomposites with a remarkable improvement of the photocatalytic performance. The resulting heterostructure was used to fabricate highly efficient self-cleaning and anti-fingerprint optically transparent thin films.

2. Materials and Methods

2.1. Materials

Powder $SnSe_2$ crystals (Ossila Ltd—Sheffield, UK, CAS 20770-09-6), 1-Methyl-2-pyrrolidinone (NMP), (Sigma-Aldrich, Milano, Italy, CAS 872-50-4), titanium(IV) chloride ($TiCl_4$) (Sigma-Aldrich, Milano, Italy, CAS 7550-45-0), ethanol (EtOH), (Sigma-Aldrich,

Milano, Italy, CAS 64-17-5), Pluronic F-127 (~12,600 g·mol^{-1}, Sigma-Aldrich, Milano, Italy CAS 9003-11-6), stearic acid (Sigma-Aldrich, Milano, Italy, CAS 57-11-4), Rhodamine B (RhB), (Sigma-Aldrich, Milano, Italy, CAS 81-88-9), and deionized water were used as received, without further purification. (100) oriented, P-type/Boron-doped silicon wafers and synthetic fused silica slides (Suprasil®, Hanau, Germany) were employed as the substrates.

2.2. $SnSe_2$-Nanomaterial Synthesis

A total of 0.020 g of $SnSe_2$ powder was dispersed into 40 mL of NMP used as solvent and sonicated by a bath sonicator for 6 h (Elmasonic P, working at 37 kHz) in a thermostat bath to prevent temperature rise (T 25 °C). After sonication, the suspension was centrifuged at 500 rpm for 20 min to precipitate the non-exfoliated material. The supernatant was then filtered (nitrocellulose filter with 0.45 µm pore size) and subsequently washed several times with ethanol to remove the residual NMP solvent. Finally, the filtered $SnSe_2$ powder was dried at 60 °C overnight. For microstructural characterization of the exfoliated flakes, $SnSe_2$ nanosheets were redispersed in ethanol and spin-coated on silicon substrates.

2.3. Synthesis of Mesoporous Ordered $SnSe_2$-TiO_2 Heterostructures

An evaporation-induced self-assembly method was used to synthesize the mesoporous films. Initially, a titania–stock solution (Ti-stk) was prepared by adding 4.4 mL of $TiCl_4$ in 40 mL of EtOH. Then 248 mg of Pluronic F-127 was dissolved in 5.07 mL of ethanol, and after 15 min of stirring, 4.23 mL of Ti-stk and 0.687 mL of deionized water were dropped into the mixture, following this sequence. The final solution (10 mL) was stirred for 2 h to obtain the precursor sol. Then 0.02 g of powdered $SnSe_2$ nanoflakes was added to the precursor sol.

Silicon wafers and silica glass were used as substrates for the deposition of thin films via dip-coating. Before the film deposition, the silicon wafers were rinsed with ethanol and dried with compressed air. The fused silica glass slides were cleaned with an aqueous solution containing an ionic detergent, rinsed with acetone, washed with ethanol, and then dried with compressed air.

The substrates were immersed in the $SnSe_2$-titania sol for 30 s and withdrawn at a 10 cm min^{-1} rate. During the process, the relative humidity was kept below 25% by a dried air stream within the deposition chamber. Lastly, the obtained films were thermally annealed in air at 400 °C for 3 h in an oven; the samples were inserted directly into the oven at 400 °C. The overall process is illustrated in Figure 1.

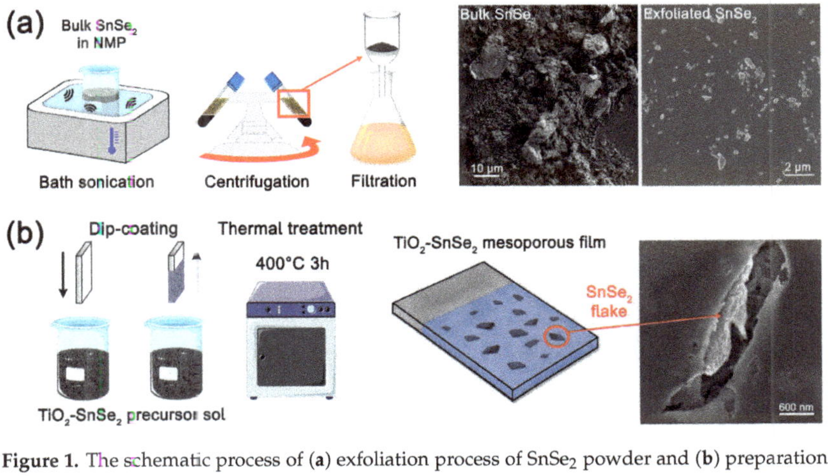

Figure 1. The schematic process of (**a**) exfoliation process of $SnSe_2$ powder and (**b**) preparation of mesoporous ordered titania films containing dispersed nanoflakes of $SnSe_2$.

In the article, the mesoporous ordered films are indicated as m-TiO_2, and the mesoporous ordered films containing the $SnSe_2$ nano flakes as m-TiO_2-$SnSe_2$.

2.4. Characterizations of the Materials

The microstructure of bidimensional $SnSe_2$ flakes and mesoporous films was investigated by scanning electron microscopy (SEM) (Gemini FESEM 500), working at an accelerating voltage of 2 kV. An element analysis was performed by using energy-dispersive X-ray spectroscopy (EDS).

Transmission electron microscopy (TEM) with an accelerating voltage of 100 kV and 0.34 nm resolution (TEM Philips CM100) was used to investigate the morphology and mesoporosity of TiO_2 and TiO_2-$SnSe_2$ films.

Flake thickness was estimated by Atomic Force Microscopy (AFM); the measures were performed in air by using a NT-MDT Ntegra AFM operating in tapping mode.

The N_2 sorption isotherms of the dried and 400 °C annealed sol were obtained using a Quantachrome Nova 1200e, and the specific surface area was determined with the Brunauer-Emmett-Teller (BET) method.

X-Ray Diffraction was performed by XRD-PAN Analytical X'PERT Pro, using Cu $K_{\alpha 1}$ radiation (λ = 1.5406 Å) with an angle step of 0.013°.

Raman spectra were acquired using a Senterra confocal Raman microscope in the 65–1555 cm^{-1} range, with a laser excitation wavelength of 532 nm, 50X objective, and nominal power of 12.5 mW.

An α-SE-Wollam spectroscopic ellipsometer with fixed-angle geometry (Complete EASE v. 4.2 software) was used to measure the refractive index, thickness, porosity, and $SnSe_2$ loading of the films. Two or three component models with void and Cauchy film were used to fit the data.

Dataphysics OCA 20 was used to evaluate the contact angle. For the measure, 2 µL water droplets deposited on the film surface were employed. The final value was obtained by averaging three measurements.

Bruker infrared Vertex 70 V interferometer was used to obtain Fourier-transform infrared (FTIR) absorption spectra. The spectra were acquired in the range 400–4000 cm^{-1} (scans average of 128 and 4 cm^{-1} of resolution).

UV-Vis spectra in the 200–800 nm range were recorded by using a Nicolet Evolution 300 UV-Vis spectrometer, with an integration time of 1.5 s per step, using a fused-silica glass substrate.

Micro-FTIR imaging of fingerprints deposited on films was performed using a Thermo Fisher Nicolet iN10M spectrometer in transmission mode, equipped with an MCT-A detector (7800–650 cm^{-1} spectral range). The fingerprints were obtained using an aqueous solution of stearic acid (5 g L^{-1}) to wet the fingers. The area used of the collected image was 150 × 150 µm, with a spectral resolution of 8 cm^{-1} for 128 scans; the spectra were normalized using the stearic acid band peaking at 2919 cm^{-1} versus the Si-O-Si substrate band peaking at 1111 cm^{-1}.

2.5. Evaluation of Photocatalytic Activity of the Films

Stearic acid and Rhodamine B were chosen as target molecules to evaluate the room-temperature photocatalytic activity and anti-fingerprint property of TiO_2 and TiO_2-$SnSe_2$ mesoporous films deposited on Si substrates.

Stearic acid was dissolved in EtOH (5.0 g L^{-1}), and 100 µL of solution was deposited on the titania films by spin-coating at 1500 rpm for 30 s. The backside of the slide was cleaned with ethanol to remove residual solution. After deposition, the films were irradiated using a UV lamp (365 nm) (nominal power density of 470 µW cm^{-2} at 15 cm) at a distance of 1 cm for different illumination times, 0-15 min. FTIR spectra were recorded after each 2.5 min illumination step.

The photodegradation of stearic acid on the films was evaluated by FTIR analysis by considering the change of the vibrational modes at 2945 and 2845 cm^{-1} (-CH_2 and -CH_3

stretching, respectively) by averaging 128 scans with 4 cm^{-1} of resolution. The change in the corresponding area of the infrared bands as a function of irradiation time was estimated using the FTIR spectra. The integral of the areas, ranging from 2800 to 3010 cm^{-1}, were used for the evaluation.

Rhodamine B (RhB) was also used as a target molecule to evaluate the photocatalytic activity of films deposited in fused silica slides. The integral of the absorption bands, ranging from 450 to 625 nm, was used to monitor the changes induced by the UV exposure.

The Rhodamine B solution for the photocatalytic test was prepared by dissolving RhB powder in EtOH ($5{\cdot}10^{-3}$ mol L^{-1}); 100 µL of solution was spin-coated at 1500 rpm for 30 s on the selected substrate. The backside part of the films was cleaned with ethanol to remove the eventual presence of the residual solution. After the deposition, the films were irradiated under 365 nm, using a UV lamp for 0–50 min, and the UV-Vis spectra were re-corded immediately after illumination. To obtain a precise readout on the photodegradation kinetics of RhB, a baseline subtraction was carried out with the use of a spline fit. This procedure was employed to suppress the interference pattern produced by the thin films. Calculation of the baseline is critical because it affects the evaluation of the photocatalytic activity.

2.6. Evaluation of Rhodamine B Degradation in Liquid Phase

The photodegradation test of Rhodamine B in water ($2.5{\cdot}10^{-6}$ mol L^{-1}) was carried out in a UV-Vis cuvette. The degradation of Rhodamine B in aqueous solution caused by UV light exposure was compared to that resulting from the addition of bidimensional $SnSe_2$ flakes to the solution. For the test, a 3 mL aqueous solution and a concentration of 0.07 g L^{-1} $SnSe_2$ were employed. The UV lamp (365 nm, Led Zolix instruments, M365L 420 mW, 90% power) was positioned at a distance of 3 cm. To avoid sedimentation of $SnSe_2$ flakes, the cuvettes were stirred during the UV exposure. The integral of Rhodamine B absorption spectra in the range of 400–650 nm was used to evaluate the effect of UV irradiation; the data were smoothed using the Savitzky–Golay algorithm.

3. Results and Discussion

The synthesis of m-TiO_2-$SnSe_2$ heterostructures requires well-exfoliated $SnSe_2$ nanostructures, which can be uniformly dispersed in the titania precursor sol. Among the several exfoliation processes used to obtain 2D materials, liquid-phase exfoliation (LPE) stands out due to its scalability and reproducibility. Small fragments of $SnSe_2$, in the shape of flakes, were obtained by LPE. These nanostructures were characterized by combining different techniques.

3.1. Microstructural Characterization

Figure 2a shows the high-resolution SEM image of representative 2D-$SnSe_2$ stacked flakes exfoliated in NMP solvent. The statistical analysis (Figure 2b) displays that the flakes have an average lateral dimension of 0.738 ± 0.022 µm, similar to what reported in previous works [25]. The thickness of the flakes, as evaluated by AFM (Figure 2c), follows a log-normal distribution peaking at $\sim$ 30 nm (Figure 2d). Considering a theoretical thickness of a $SnSe_2$ monolayer at room temperature of 1.2 nm [26], the nanoflakes consist of roughly 25 layers. The width-to-thickness ratio is 24.6, which is consistent with previous reports for bidimensional $SnSe_2$ flakes synthesized in a similar manner [27,28].

The Raman analysis was used to study the structure of bulk and exfoliated $SnSe_2$ samples (Figure 3a). The $SnSe_2$ bulk shows a strong band peaking at 181.5 cm^{-1} due to out-plane stretching of the A_{1g} mode and a second band of smaller intensity at 108.8 cm^{-1} assigned to the E_g mode of the in-plane stretching [29]. The position of the $SnSe_2$ E_g Raman shift is indicative of the polytypes [30]: 2H-$SnSe_2$ crystals have an E_g mode located at ~108 cm^{-1}, while 1T-$SnSe_2$ crystals exhibit an E_g mode located at ~118 cm^{-1} [31]. The observed Raman shift of the Eg peak at 108.8 cm^{-1} indicates that, in the present case, the bulk is a 2H-$SnSe_2$ polytype.

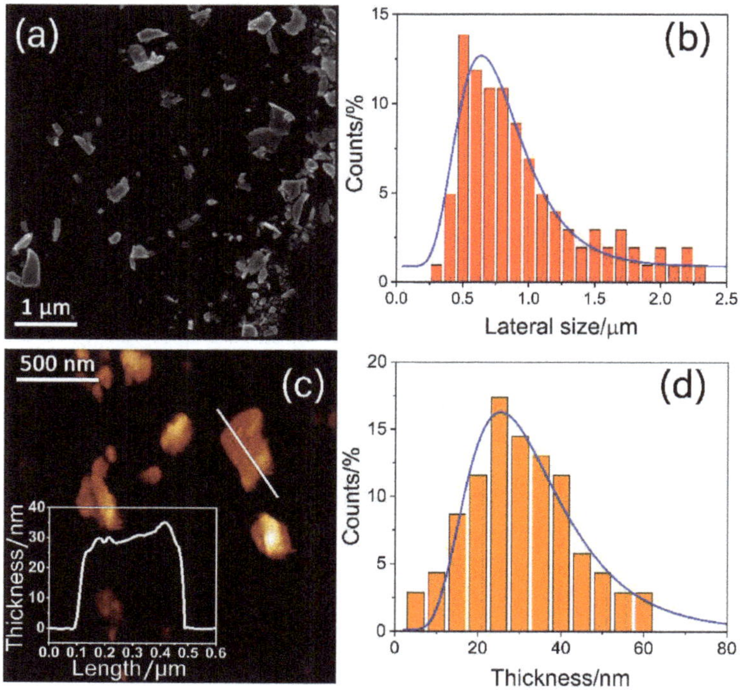

Figure 2. (**a**) High-resolution SEM image of representative $SnSe_2$ nanoflakes. (**b**) Statistical analysis of lateral dimension calculated from SEM images. The blue curve represents the Gaussian fit. (**c**) Representative AFM image of $SnSe_2$ nanoflakes. The inset shows the height profile of the flake indicated by the segment in white. (**d**) Statistical analysis of the $SnSe_2$ flake thickness distribution calculated from AFM images. The blue curve is the log-normal distribution fit.

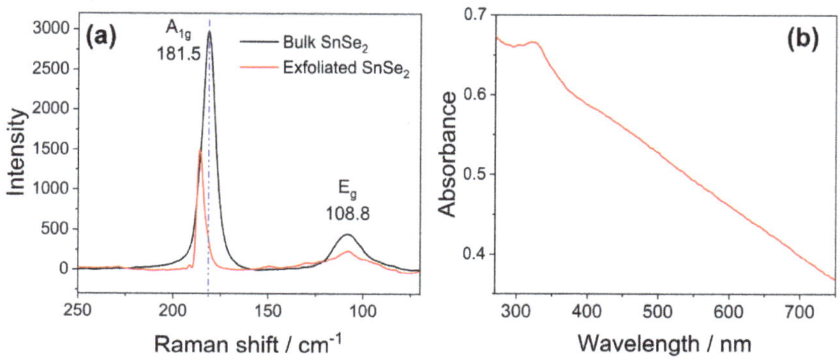

Figure 3. (**a**) Raman spectra of bulk $SnSe_2$ (black line) and exfoliated $SnSe_2$ nanoflakes (red line) (λ_{ex} = 532 nm). The blue dot line is a guide for eyes. (**b**) UV-Vis absorption spectrum $SnSe_2$ flakes dispersed in ethanol (0.83 g L^{-1}).

Previous research has shown a correlation between the Eg band strength and the number of $SnSe_2$ layers. Nevertheless, this relationship does not follow a linear trend; in fact, for flakes thicker than ten layers, the intensity does not grow any longer or even decreases due to the interference effect. [32] The ratio of A_{1g} and E_g band intensities is consistent with a flake thickness greater than ten layers in agreement with SEM observations.

A comparison between the Raman spectra after and before exfoliation process also reveals a ~3.5 cm^{-1} redshift of the A_{1g} band. The shift is compatible with a reduced number of layers with respect to the bulk, in accordance with previous findings [24].

Figure 3b displays the UV-Vis absorption spectra of exfoliated $SnSe_2$ nanoflakes in ethanol between 270 and 800 nm. The spectrum shows an absorption band at ~350 nm [33] emerging from an intense broadband absorption. The extended interval of absorption suggests that $SnSe_2$-based materials can be efficiently used as a light absorber in a wide range from UV-Vis up to the red region [34,35]. The band at 350 nm, in accordance with previous findings, is attributed to the transition from the crystal field split selenium p_{xy}-like levels into tin p-like levels [36].

The film thickness was evaluated by spectroscopic ellipsometry, which allows us to also calculate the porosity and the loading percentage of $SnSe_2$. The data were collected from a series of three samples prepared in the same conditions to evaluate the reproducibility. The data were fit while keeping the mean squared error (MSE) below 25. The average thickness of bare m-TiO_2 films is ~120 nm, which is ~20 nm lower than that of the $SnSe_2$-loaded samples (Table 1). The change in the sol viscosity induced by the addition of 2D nanoflakes explains the increase in thickness. The residual porosity in the m-TiO_2 -$SnSe_2$ nanocomposites is around 10% lower than that of the m-TiO_2 films; the difference is attributed to the incorporation of ~ 7% of exfoliated $SnSe_2$ nanoflakes.

Table 1. Thickness, porosity and $SnSe_2$ loading measured by ellipsometry for three m-TiO_2 and m-TiO_2-$SnSe_2$ films (Samples 1, 2, and 3).

	m-TiO_2		**m-TiO_2-$SnSe_2$**		
	Thickness (nm)	**Porosity (%)**	**Thickness (nm)**	**Porosity (%)**	**$SnSe_2$ (%)**
Sample 1	121.3 ± 0.3	37.1 ± 0.1	138.7 ± 0.3	27.5 ± 0.1	6.5 ± 0.1
Sample 2	117.9 ± 0.6	38.8 ± 0.2	142.0 ± 0.3	26.4 ± 0.1	8.1 ± 0.1
Sample 3	124.1 ± 0.9	39.5 ± 0.2	139.9 ± 0.4	27.3 ± 0.2	8.9 ± 0.1

Figure 4a shows the transmittance spectra of m-TiO_2 and m-TiO_2-$SnSe_2$ films. The spectra perfectly overlap with each other, with a cutoff around 350. The transmittance is higher than 85%, indicating that loading the mesoporous films with ~7% of bidimensional $SnSe_2$ flakes (see Table 1) does not cause any sensible change in the transparency of the film, as shown in the insets in Figure 4a [19,37]. This result is not surprising considering the low percentage of loading and the low film thickness. The dispersion curves of the refractive index, n, also confirm that the loading process does not affect the optical properties (Figure 4b). The value of n is 2.25 for both samples at 550 nm, and this is in line with previous works that reported for anatase TiO_2 films a refractive index in the range of 2.10–2.50 [31].

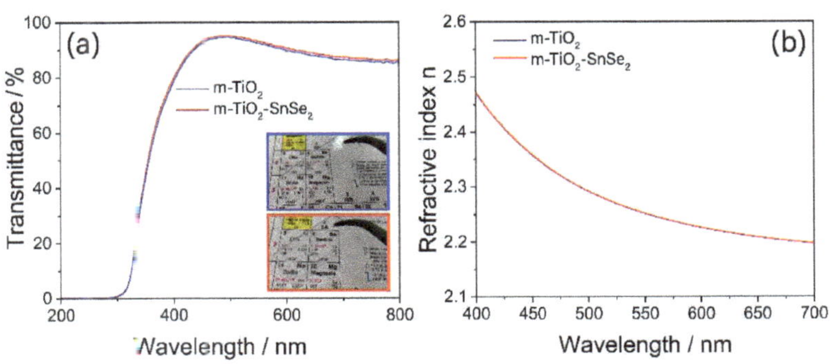

Figure 4. (**a**) UV-Vis spectra and refractive index (**b**) of m-TiO_2 film (blue line) and m-TiO_2-$SnSe_2$ film (red line). The two curves overlap. The inset in (**a**) shows the snapshots of the films deposited on silica glass slides.

The high optical transparency is a critical requirement for most applications, such self-cleaning and anti-fingerprint layers that also need significant photocatalytic activity.

The surface morphology of the nanocomposite films was investigated by SEM analysis (Figure 5a). The picture displays that the titania self-assembled films achieve a well-organized mesoporous structure (surface area of 112.60 m^2/g, obtained by the BET equation fit for N_2 gas adsorption isotherm reported in Supplementary Figure S4), and the organization is not disrupted by the presence of the $SnSe_2$ flakes. These results have already been observed for mesoporous titania films embedding exfoliated graphene [19] and can be explained by considering that the mesopores size (5–6 nm) is much smaller than the lateral size of 2D-$SnSe_2$ flakes ($\approx$700 nm). The self-assembly process of the block copolymer occurs on a smaller scale with respect to the flake size and, therefore, is not affected by the addition of 2D-$SnSe_2$ to the sol. On the other hand, the presence of a bidimensional material in the precursor sol favors the micelle nucleation lowering the interfacial energy through heterogeneous nucleation [38]. The mesopore organization is clearly visible from the SEM surface images and shows the typical grid-like structure, with partial merging of some pore walls on the topmost layer, in agreement with previous results [39]. The images obtained by the TEM analysis (Figure 5b) confirm the formation of spherical mesopores with a body-centered cubic ordered structure (Im3m) within the titania films [40]. The appearance of the channel-like pore arrangement is an effect generated by the transmission microscope that projects the Im3m spherical pore array in the [110] direction.

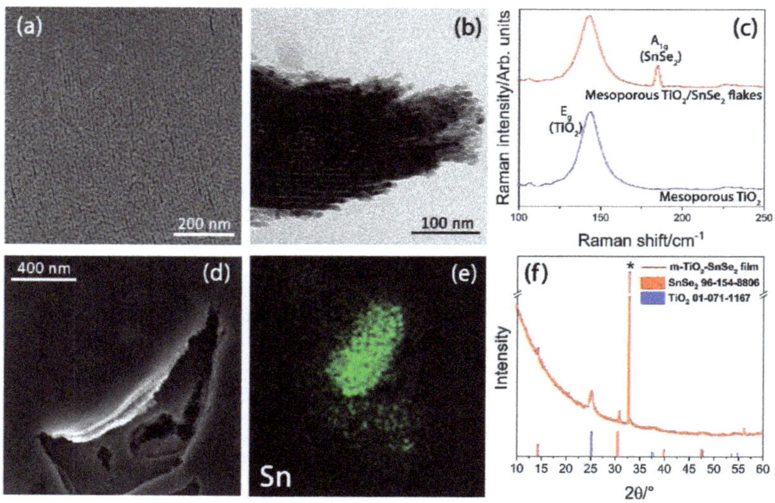

Figure 5. (**a**) FESEM and (**b**) TEM images of m-TiO_2-$SnSe_2$ film. (**c**) Raman comparison of m-TiO_2 film (blue line) and m-TiO_2-$SnSe_2$ film (red line). (**d**,**e**) FESEM image and EDS analysis of $SnSe_2$ flakes into m-TiO_2 film and (**f**) GI-XRD pattern of m-TiO_2-$SnSe_2$ film. The asterisk indicates the (200) diffraction of the Si substrate.

The existence of an organized mesoporosity is crucial, as the large surface increases the photocatalytic activity, while the pore organization promotes the diffusion processes [41]. The highest diffusivity within the mesoporous matrix is achieved when the pores are ordered in a body-centered cubic fashion, whereas 2D hexagonal arrays of cylindrical channels provide limited diffusion due to pore packing defects [42,43]. The presence of $SnSe_2$ within the titania matrix was confirmed by the FESEM analysis of a scratched area of the film (Figure 5d). The SEM picture confirms the presence of exfoliated $SnSe_2$ flakes, as also confirmed by the EDS analysis (Figure 5e), which detects Sn and Se atoms (see Supplementary Figure S2). The successful incorporation of the $SnSe_2$ nanoflakes into m-TiO_2 film is also supported by the Raman analysis (Figure 5c). The pure mesoporous titania films show a Raman mode located at 144 cm^{-1} that is the signature of the typical E_g vibration mode of TiO_2 anatase phase (blue line) [44]. This mode shows the same intensity and position in the nanocomposite m-TiO_2-$SnSe_2$ films (red line). Additionally, the Raman spectra show a new mode located at ~183 cm^{-1} that is attributed to the A_{1g} mode of $SnSe_2$ nanoflakes (Figure 3a) [26].

The microstructure of the m-TiO_2-$SnSe_2$ films was further assessed by GI-XRD analysis (Figure 5f). The X-ray pattern indicates the crystallization of titania in anatase, as shown by the main diffraction peak at 25.3 attributed to the (110) plane of TiO_2 anatase phase (# JCPDS No. 01-071-1167) [45]. The average crystallite size of the TiO_2 nanoparticles is ~15.5 nm, calculated using the Scherrer equation. Additionally, the diffraction peaks detected at 14.3 and 31.1° in 2θ h were assigned to the (001) and (101) planes of the $SnSe_2$ nanostructures (# JCPDS No. 96-154-8806) [46].

The XRD data, in combination with SEM, FESEM, TEM, and Raman, allow us to obtain an overall characterization of the nanocomposite mesoporous films. After the thermal treatment at 400 °C, titania crystallizes into anatase, which is the active photocatalytic phase. At the same time, the phase transformation keeps the organization of the mesopores, while the $SnSe_2$ flakes are homogeneously distributed within the matrix. It is important to underline that the introduction of the bidimensional $SnSe_2$ flakes via the one-pot route does not interfere with the self-assembly process, particularly with the organization of the supramolecular template that triggers the formation of the mesophase (see Supplementary Figure S3).

Contact angle measurements were used to investigate the surface wettability of the films (see Supplementary Figure S5), since this is, in fact, an important property to be assessed for anti-fingerprint coatings. Despite the presence of loaded $SnSe_2$ nanoflakes, the contact angle remains the same as that of the unloaded film (~23°) [17], indicating that the surface keeps the hydrophilic properties.

3.2. Photocatalytic Degradation of Rhodamine B by $SnSe_2$ Nanoflakes

The self-cleaning properties of the nanocomposites films were tested by following the photocatalytic degradation under UV irradiation of two different molecules, Rhodamine B (RhB) and stearic acid (Figure 6). The RhB degradation kinetic was tracked by UV-Vis spectroscopy, adopting the main absorption band with a maximum around 550 nm as a reference. Furthermore, the decomposition of stearic acid was monitored using the CH_2 stretching mode of FTIR.

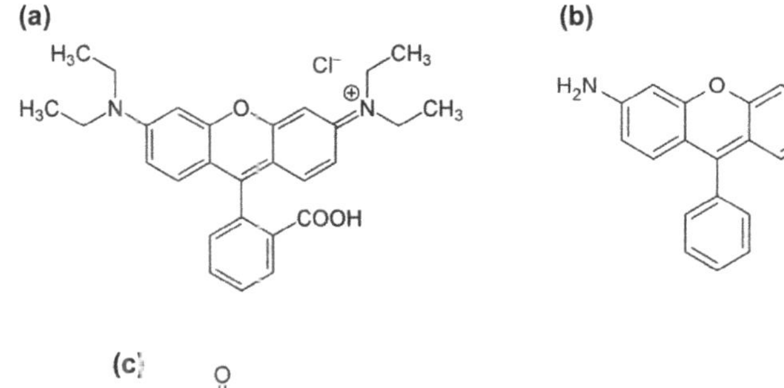

Figure 6. Chemical structure of Rhodamine B (**a**), Rhodamine 110 (**b**), and stearic acid (**c**).

The photocatalytic property of bare $SnSe_2$ was verified before testing that of the m-TiO_2-$SnSe_2$ nanocomposite films. As already mentioned, nanostructured $SnSe_2$ was rarely studied as a photocatalyst because of the fast electrons–holes recombination [47,48]. As a control, an aqueous solution of RhB was exposed to UV radiation for increasing periods of time. The results were compared to those obtained using another aqueous RhB solution, which also contained the $SnSe_2$ flakes (Figure 7a,b).

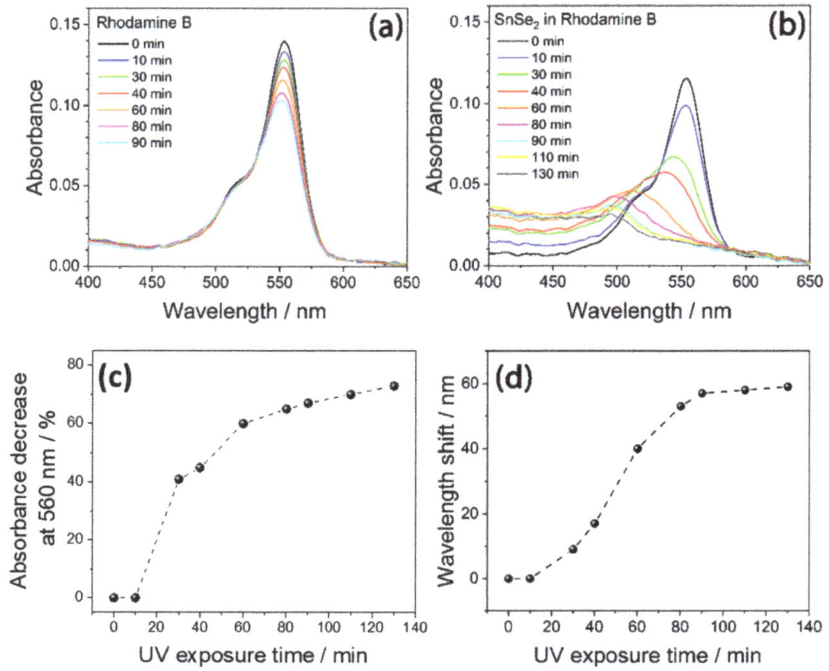

Figure 7. (**a**) UV–Vis absorption spectra of an aqueous solution of RhB (2.5×10^{-6} mol L^{-1}) as a function of UV (365 nm) exposure time. (**b**) UV-Vis absorption spectra of the aqueous RhB solution containing dispersed $SnSe_2$ flakes (0.07 g L^{-1}) as a function of UV exposure time. (**c**,**d**) Percentage absorbance decrease and wavelength shift as a function of the UV exposure time. The data were taken from (**b**), using the maxima at 560 nm as reference for the calculation of the percentage decrease and wavelength shift. The line is a guide for the eyes.

Rhodamine B is a xanthene dye that is characterized by a strong absorption in the visible range, with a band around 550 nm [49] attributed to $\pi \rightarrow \pi^*$ transitions. We used a 2.5×10^{-6} M concentration to avoid dimerization of the dye, which has a strong tendency to aggregate with the increase of the concentration, whereas monomers and dimers degrade at a different rate [50].

The absorption spectra show that only the monomeric form of Rhodamine B is present in solution (Figure 7a); RhB dimers have a strong absorption peaking around 530 nm that is not detected. The spectra in the 450–650 nm range are characterized by a broad and intense absorption band with a maximum around 550 nm, due to the delocalization, which involves the amines groups of the dye, and a vibronic shoulder peaking around 510 nm. The RhB control solution shows about 15% degradation after exposure to UV light for 90 min, in agreement with the photostability reported in the literature. The 550 nm absorbance measured in the RhB aqueous solution containing dispersed $SnSe_2$ flakes decreased by 73% after 90 min of exposure, demonstrating that $SnSe_2$ has a significant role in the photocatalytic activity (Figure 7c). At the same time, the maximum of the spectrum gradually shifts to around 500 nm (60 nm of blueshift) when the exposure time is longer than 90 min. (Figure 7d). The solution gradually turns from intense orange to green (Figure 8). The direct observation of the color change in the RhB aqueous solutions (Figure 8a,b) shows no differences after 90 min of exposure, confirming that RhB is not significantly degraded by UV in this exposure time. On the other hand, the color changes in the presence of bidimensional $SnSe_2$ flakes indicate that the dye was chemically modified by the irradiation process (Figure 8c–e).

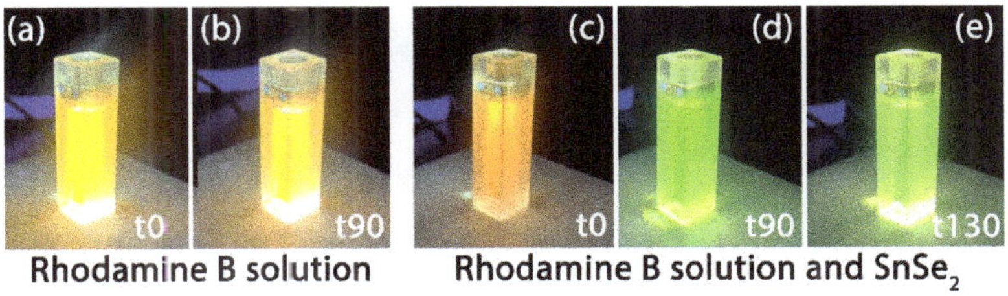

Figure 8. Optical images of RhB aqueous solutions (**a**,**b**) and RhB solution containing dispersed $SnSe_2$ flakes (**c**–**e**) at different UV illumination times, t0 (before exposition), t90 (90 min), and t130 (130 min).

The decrease in intensity and the blueshift of the absorption maximum with the irradiation time indicate that RhB degradation occurs via de-ethylation [51,52]. As previously reported, the gradual blueshift (Figure 7b) suggests that the de-ethylation process activated by $SnSe_2$ occurs stepwise, with the ethyl groups of RhB being removed in different stages [53]. When all the four ethyl groups have been removed from the dye, the molecule has the physical properties and chemical structure of Rhodamine 110 (Figure 6b). The final green color appearance of the dye suggests that an almost complete conversion of RhB into Rhodamine 110 has been achieved [54]. It has been observed that this peculiar photoactivated reaction is mostly related to a surface process which occurs when RhB is absorbed on the photocatalyst and allows radical water species (especially $^{\bullet}OH$) to directly attack the ethyl groups of the dye structure. On the contrary, when the RhB is not absorbed on the photocatalyst, oxidative degradation of the dye chromophore prevails in solution, causing a general decrease of the absorbance, with no significant blueshift of the main absorption band [48]. Hence, the effective UV-photodegradation of RhB can be explained by the strong affinity between the $SnSe_2$ flakes and the dye molecule.

Rhodamine B [55] has also been used as a test molecule to demonstrate the photocatalytic efficiency of the loaded film in comparison to the unloaded titania film. In both m-TiO_2 and m-TiO_2-$SnSe_2$ systems, we observe that the RhB absorbance decreases and blueshifts as a function of the UV exposure time (Figure 9). On the nanocomposite film, the dye absorption band completely disappears after 30 min of irradiation; however, on the m-TiO_2 film, a faint absorbance is detectable even after 70 min of UV exposure.

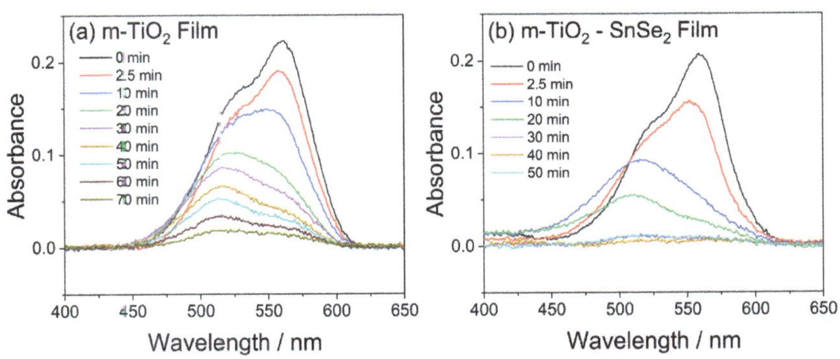

Figure 9. UV-Vis spectra of Rhodamine B deposited on m-TiO_2 (**a**) and m-TiO_2-$SnSe_2$ (**b**) films as a function of UV exposure time.

To achieve a more detailed analysis of the photochemical changes occurring to the RhB deposited on the samples and UV-irradiated, we fitted the absorption in the 400–650 nm region, with two Gaussian functions taking into consideration two main contributions: one centered on 520 nm and the other with a maximum around 565 nm. (Figure 10a).

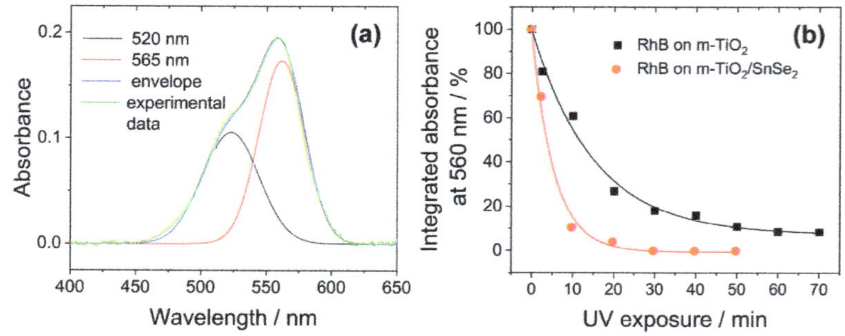

Figure 10. (**a**) Optical absorption spectrum of RhB deposited on m-TiO_2 film and curve fit with 2 Gaussian functions. (**b**) Decrease of the integrated absorbance calculated from the Gaussian function peaked at 565 from the spectra of RhB deposited on m-TiO_2 and m-TiO_2-$SnSe_2$ films (black squares and red dots, respectively). The lines depict the exponential decay fit of the experimental data.

The contribution at shorter wavelengths is attributed both to the vibronic shoulder of the Rhodamine B and the de-ethylated rhodamine forms that show a blue-shifted main absorption, as seen before. The contribution at 565 nm, on the other hand, is unambiguously correlated with the electronic transition of not-photodegraded RhB and can thus be monitored to determine a more accurate photodegradation kinetic. Figure 10b is a plot of the integrated absorbance of the Gaussian function centered at 565 nm vs. the irradiation time. To estimate the photocatalytic efficiency, we used a decay law which follows pseudo-first-order kinetics [56–58]:

$$I(t) = I_0 \cdot e^{-kt} \quad (1)$$

where k is the degradation rate of pollutant molecules. The k-value of loaded films is 3 times higher than that of the unloaded sample (~0.182 min^{-1} vs. ~0.066 min^{-1} for m-TiO_2-$SnSe_2$ and m-TiO_2, respectively). These results suggest that the heterostructure between $SnSe_2$ and the nanocrystalline anatase titania increases the dye photodegradation.

Notably, the UV-Vis spectroscopy monitoring of RhB photodegradation does not effectively offer information on the total removal of the organic molecule but rather on the chemical changes occurring on the dye molecule as a result of photodegradation. To acquire a better understanding of the self-cleaning properties of the nanocomposite films, stearic acid solution (5.0 g L^{-1}) was spin-coated onto the films as a model pollutant to be decomposed by UV irradiation. FTIR spectroscopy was used to evaluate the photocatalytic degradation of stearic acid by monitoring the intensity of the -CH_2 stretching mode, as shown in Figure 11a,b for m-TiO_2 and m-TiO_2-$SnSe_2$, respectively.

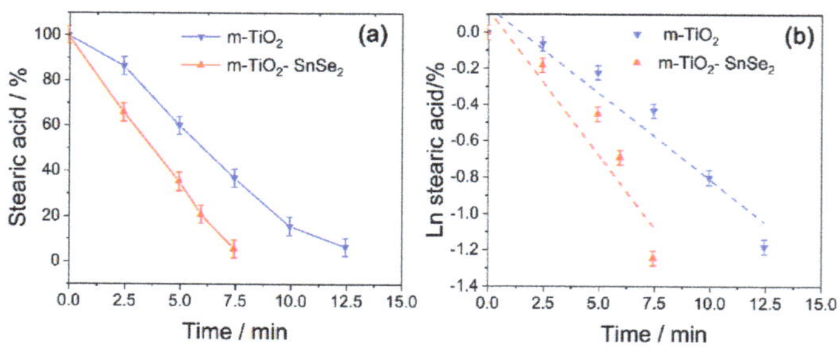

Figure 11. (**a**) Degradation rate (lines are a guide for eyes) obtained by measuring the decrease in intensity of the 2945 cm^{-1} C-H_2 stretching mode and (**b**) decay law (dashed lines represent the decay fit) of stearic acid deposited on the different samples.

Figure 11 displays the degradation rate of stearic acid (stearic acid/%), which is evaluated by the following:

$$\text{Stearic acid}/\% = I_t/I_0 \cdot 100 \quad (2)$$

where I_t/I_0 is the ratio between the infrared absorption intensity at t irradiation time, I_t, and the absorption intensity before exposure, I_0.

m-TiO_2 (blue line in Figure 11a) is able to remove more than 80% of stearic acid from the surface after 10 min of exposure. Nevertheless, the m-TiO_2-$SnSe_2$ film degrades the pollutant at a quicker rate, achieving a 90% degradation after only 7.5 min. The dashed straight lines in Figure 11b represent the decay law that follows a pseudo-first-order kinetics (1).

The calculated k-value for m-TiO_2 film is ~0.092 min^{-1}, while for m-TiO_2-$SnSe_2$ film, it is ~0.160 min^{-1}. This confirms that the $SnSe_2$ flakes enhance the photoactivity of titania mesoporous film by increasing the degradation rate more than 70% with only 7% of loading (see Table 1).

The experiments with the stearic acid are complementary to those with the organic dye. The measurement obtained by FTIR on stearic acid provides a quantitative indication of the organic removal from the surface, simulating a real self-cleaning process. Moreover, the absorbance variations of Rhodamine B provide evidence for the hypothesis that interactions between the dye and the nanocomposite films' surfaces are critical for boosting photocatalysis.

The photocatalytic activity shown by the m-TiO_2-$SnSe_2$ film can be applied to produce anti-fingerprint coatings. This function has been evaluated by immersing the finger in a stearic acid solution, pressing the finger to the substrate, and observing its removal under exposure to sunshine or ultraviolet radiation [59,60]. We selected this testing method to obtain a direct comparison and more reproducible findings in contrast to direct human fingerprints, whose chemical composition is more challenging to duplicate.

Figure 12a,b,f,g display some representative images of m-TiO_2 and m-TiO_2-$SnSe_2$ films before and after fingerprint degradation under ambient conditions (sunlight). The observation of a latent fingerprint on the surface of the m-TiO_2-$SnSe_2$ film (Figure 12g), as opposed to the clear residual pattern detected in the m-TiO_2 film (Figure 12b), indicated that the fingerprint was fully degraded after 24 h. The test validates the nanocomposite's improved photocatalytic degradation performance compared to that of the bare m-TiO_2 film.

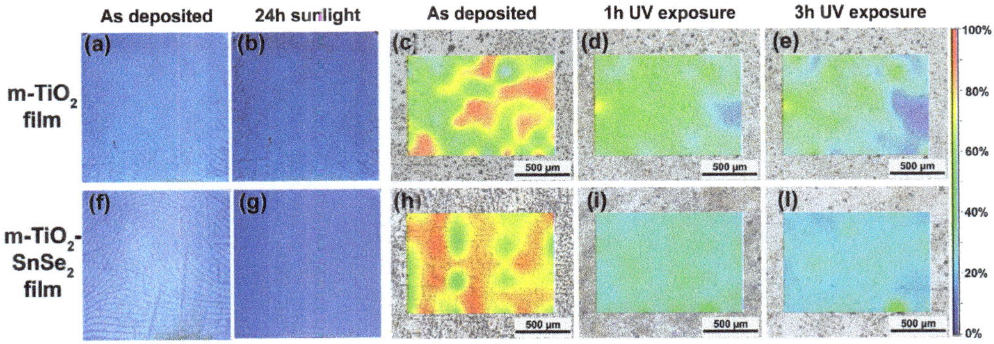

Figure 12. Photographs of human fingerprints on m-TiO_2 (**a**,**b**) and m-TiO_2-$SnSe_2$ (**f**,**g**) films before and after exposure to sunlight in air for 24 h. Infrared images of a human fingerprint deposited on m-TiO_2 (**c**–**e**) and m-TiO_2-$SnSe_2$ (**h**–**l**) films. The images were taken by integrating the area of the -CH_2 stretching bands of stearic acid. The fingerprint was produced by immersing the finger in a stearic acid solution. The infrared images were detected at different times: as deposited, after 1 h and after 3 h of UV exposure, respectively. The color scale bar shows the intensity scale in false colors; the red and blue colors represent the highest and the lowest absorbance of the infrared signal, respectively.

Figure 12c–e,h–l show the infrared images performed to evaluate the degradation of a fingerprint deposited on m-TiO_2 and m-TiO_2-$SnSe_2$ films at different UV exposure times (Figure 12). The infrared images were taken by integrating the area of the -CH_2 stretching bands of stearic acid. After 1 h of UV exposure, the infrared images showed that the m-TiO_2-$SnSe_2$ film almost completely cancelled the fingerprint that is still detected on the m-TiO_2 film (Figure 12d).

Figure 13 outlines the possible mechanism of photocatalysis occurring in the nanocomposite films [61]. When the UV light is absorbed by the titania matrix, the valence band (VB) electrons are excited to the conduction band (CB) and then transferred to that of $SnSe_2$. The UV light also interacts with $SnSe_2$ nanoflakes, which have a broad absorption up to the infrared range. However, we observed that a 7% $SnSe_2$ loading has no effect on the films' absorbance. This implies that the primary impact of $SnSe_2$ is not to boost the light-harvesting efficiency but rather to decrease the total hole and electron recombination rate in the heterojunction. As a consequence of the light absorption, photogenerated electrons (e^-) and holes (h^+) on the surface of the photo-catalyst generate free radicals such as hydroxyl radicals (•OH) and superoxide radical anions ($\bullet O_2^-$). The •OH radicals are produced by the reaction between e^-/h^+ pairs and OH^-/H_2O in the VB. At the same time, the excited electrons in the CB react with the dissolved oxygen in the solution to form $\bullet O_2^-$, which reacts with H^+ to form other hydroxyl radicals (•OH and $O_2\bullet^-$). These highly active hydroxyl radical species can oxidize the organic molecules, hence promoting the self-cleaning effect [62].

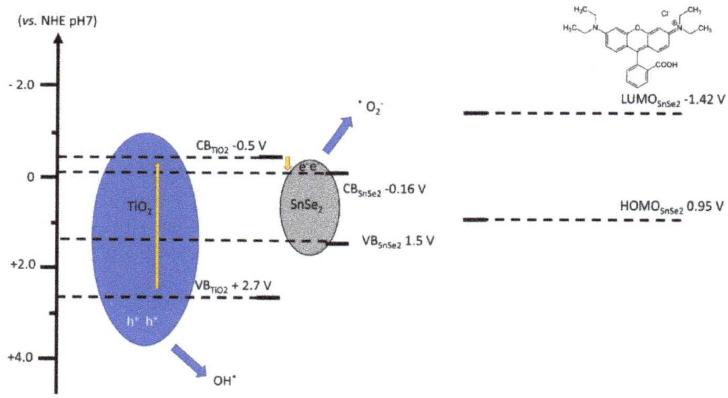

Figure 13. Hypothesis of the Rhodamine B degradation mechanism in presence of $SnSe_2$ nanoflakes under UV-Visible light. NHE, normal nitrogen electrode; CB, conduction band; VB, valence band; LUMO, lowest unoccupied molecular orbital; HOMO, highest unoccupied molecular orbital.

4. Conclusions

The photophysical properties of bidimensional $SnSe_2$ make it an excellent candidate for the development of photoactive heterostructures in combination with titania. Bidimensional $SnSe_2$ flakes in solution have a high photocatalytic activity towards xanthene dyes such as Rhodamine B. The flakes can be directly added to a titania precursor sol to fabricate nanocomposite mesoporous ordered films via one-pot self-assembly without disruption of the mesophase organization. Remarkably, the optical quality of the films is not affected by the $SnSe_2$ addition, and the difference in the film transmittance is negligible, ensuring the transparency of the coatings in the whole visible range. The nanocomposite films show a higher photocatalytic activity in comparison to mesoporous titania films. The cooperative effect of the two photocatalysts reduces the recombination rate of the photogenerated electron/hole couple. The effective degradation of human fingerprints on the film surface confirms that the specifications for anti-fingerprint coatings are met excellently by titania mesoporous nanocomposite films.

Supplementary Materials: The following supporting information can be downloaded at https://www.mdpi.com/article/10.3390/nano13081406/s1, Figure S1. Calculation of the band gap from UV-Vis spectra; Figure S2. EDS analysis of $SnSe_2$ flakes into m-TiO_2 film; Figure S3. SEM characterization of the m-TiO_2-$SnSe_2$ samples; Figure S4: N2 gas adsorption isotherm of the m-TiO_2-$SnSe_2$ nanocomposite; Figure S5. Contact angle images. References [63,64] is cited in the supplementary materials.

Author Contributions: Conceptualization, L.S., D.C., L.M., C.C. and P.I.; Validation, L.S., D.C., L.M. and P.I.; Formal analysis, L.M. and P.I.; Investigation, J.D.S., V.P. and D.C.; Data curation, J.D.S. and V.P.; Writing—original draft, J.D.S., V.P., L.M. and P.I.; Writing—review & editing, L.M., C.C. and P.I.; Supervision, L.M., C.C. and P.I. All authors have read and agreed to the published version of the manuscript.

Funding: This research was funded by the Ministry of Education, University, and Researcher (MIUR), through the PRIN 2017 grant n.2017W75RAE; and by the Italian Ministry of Foreign Affairs and International Cooperation, through the grant PGR07324. The University of Sassari is also acknowledged for funding through "fondo di Ateneo per la ricerca 2020".

Data Availability Statement: Original data are available upon request to the corresponding author.

Conflicts of Interest: The authors declare no conflict of interest.

Abbreviations

m-TiO_2, mesoporous ordered titania films; m-TiO_2-$SnSe_2$, mesoporous ordered titania films containing dispersed $SnSe_2$ nanomaterials; RhB, Rhodamine B; LPE, liquid phase exfoliation; NHE, normal nitrogen electrode; CB, conduction band; VB, valence band; LUMO, lowest unoccupied molecular orbital; HOMO, highest unoccupied molecular orbital.

References

1. Rong, W.; Kazuhito, H.; Akira, F.; Makoto, C.; Eiichi, K.; Astushi, K.; Mitsuhide, S.; Toshiya, W. Light-Induced Amphiphilic Surfaces. *Nature* **1997**, *338*, 431–432. [CrossRef]
2. Taga, Y. Titanium Oxide Based Visible Light Photocatalysts: Materials Design and Applications. *Thin Solid Films* **2009**, *517*, 3167–3172. [CrossRef]
3. Zhao, X.; Zhao, Q.; Yu, J.; Liu, B. Development of Multifunctional Photoactive Self-Cleaning Glasses. *J. Non-Cryst. Solids* **2008**, *354*, 1424–1430. [CrossRef]
4. Yates, J.T. Photochemistry on TiO_2: Mechanisms behind the Surface Chemistry. *Surf. Sci.* **2009**, *603*, 1605–1612. [CrossRef]
5. Luo, H.; Wei, H.; Wang, L.; Gao, Q.; Chen, Y.; Xiang, J.; Fan, H. Anti-smudge and self-cleaning characteristics of waterborne polyurethane coating and its construction. *J. Colloid Interface Sci.* **2022**, *628*, 1070–1081. [CrossRef] [PubMed]
6. Belhadjamor, M.; El Mansori, M.; Belghith, S.; Mezlini, S. Anti-Fingerprint Properties of Engineering Surfaces: A Review. *Surf. Eng.* **2018**, *34*, 85–120. [CrossRef]
7. Galvez, M. New Innovations in Coating Technologies for Display Glass. *Inf. Disp.* **2020**, *36*, 23–26. [CrossRef]
8. Gopal, L.; Sudarshan, T. Surface Engineering for Anti-Fingerprint Applications. *Surf. Eng.* **2022**, *38*, 571–575. [CrossRef]
9. Fu, K.; Tian, Y.; Zhu, Z.; Zhang, G.; Xie, J.; Zhang, Q. Robust and Highly Transparent Photocurable Fluorinated Polyurethane Coating Prepared via Thiol-Click Reactions and What Essentially Influences Omniphobic Coating's Anti-Graffiti Properties. *ACS Appl. Polym. Mater.* **2022**, *4*, 8386–8395. [CrossRef]
10. Zhong, X.; Hu, H.; Yang, L.; Sheng, J.; Fu, H. Robust Hyperbranched Polyester-Based Anti-Smudge Coatings for Self-Cleaning, Anti-Graffiti, and Chemical Shielding. *ACS Appl. Mater. Interfaces* **2019**, *11*, 14305–14312. [CrossRef]
11. Mellott, N.P.; Durucan, C.; Pantano, C.G.; Guglielmi, M. Commercial and Laboratory Prepared Titanium Dioxide Thin Films for Self-Cleaning Glasses: Photocatalytic Performance and Chemical Durability. *Thin Solid Films* **2006**, *502*, 112–120. [CrossRef]
12. Carboni, D.; Marongiu, D.; Rassu, P.; Pinna, A.; Amenitsch, H.; Casula, M.; Marcelli, A.; Cibin, G.; Falcaro, P.; Malfatti, L.; et al. Enhanced Photocatalytic Activity in Low-Temperature Processed Titania Mesoporous Films. *J. Phys. Chem. C* **2014**, *118*, 12000–12009. [CrossRef]
13. Li, Y.; Gao, C.; Long, R.; Xiong, Y. Photocatalyst design based on two-dimensional materials. *Mater. Today Chem.* **2019**, *11*, 197–216. [CrossRef]
14. Xie, L.; Hao, J.-G.; Chen, H.-Q.; Li, Z.-X.; Ge, S.-Y.; Mi, Y.; Yang, K.; Lu, K.-Q. Recent advances of nickel hydroxide-based cocatalysts in heterogeneous photocatalysis. *Catal. Commun.* **2022**, *162*, 106371. [CrossRef]
15. Chen, H.-Q.; Hao, J.-G.; Wei, Y.; Huang, W.-Y.; Zhang, J.-L.; Deng, T.; Yang, K.; Lu, K.-Q. Recent Developments and Perspectives of Cobalt Sulfide-Based Composite Materials in Photocatalysis. *Catalysts* **2023**, *13*, 544. [CrossRef]
16. Cheng, J.; Wang, C.; Zou, X.; Liao, L. Recent Advances in Optoelectronic Devices Based on 2D Materials and Their Heterostructures. *Adv. Opt. Mater.* **2019**, *7*, 1800441. [CrossRef]

17. Zhang, L.; Guo, C.; Kuo, C.-N.; Xu, H.; Zhang, K.; Ghosh, B.; De Santis, J.; Boukhvalov, D.W.; Vobornik, I.; Paolucci, V.; et al. Terahertz Photodetection with Type-II Dirac Fermions in Transition-Metal Ditellurides and Their Heterostructures. *Phys. Status Solidi Rapid Res. Lett.* **2021**, *15*, 2100212. [CrossRef]
18. Ho, W.; Yu, J.C.; Lin, J.; Yu, J.; Li, P. Preparation and Photocatalytic Behavior of MoS2 and WS2 Nanocluster Sensitized TiO_2. *Langmuir* **2004**, *20*, 5865–5869. [CrossRef]
19. Lettieri, S.; Pavone, M.; Fioravanti, A.; Amato, L.S.; Maddalena, P. Charge Carrier Processes and Optical Properties in TiO_2 and TiO_2-Based Heterojunction Photocatalysts: A Review. *Materials* **2021**, *14*, 1645. [CrossRef]
20. Ren, J.; Stagi, L.; Malfatti, L.; Paolucci, V.; Cantalini, C.; Garroni, S.; Mureddu, M.; Innocenzi, P. Improving the Photocatalytic Activity of Mesoporous Titania Films through the Formation of WS2/TiO_2 Nano-Heterostructures. *Nanomaterials* **2022**, *12*, 1074. [CrossRef]
21. Malfatti, L.; Falcaro, P.; Pinna, A.; Lasio, B.; Casula, M.F.; Loche, D.; Falqui, A.; Marmiroli, B.; Amenitsch, H.; Sanna, R.; et al. Exfoliated Graphene into Highly Ordered Mesoporous Titania Films: Highly Performing Nanocomposites from Integrated Processing. *ACS Appl. Mater. Interf.* **2014**, *6*, 795–802. [CrossRef] [PubMed]
22. Ren, J.; Stagi, L.; Malfatti, L.; Garroni, S.; Enzo, S.; Innocenzi, P. Boron Nitride-Titania Mesoporous Film Heterostructures. *Langmuir* **2021**, *37*, 5348–5355. [CrossRef] [PubMed]
23. Boukhvalov, D.W.; Paolucci, V.; D'Olimpio, G.; Cantalini, C.; Politano, A. Chemical Reactions on Surfaces for Applications in Catalysis, Gas Sensing, Adsorption-Assisted Desalination and Li-Ion Batteries: Opportunities and Challenges for Surface Science. *Phys. Chem. Chem. Phys.* **2021**, *23*, 7541–7552. [CrossRef] [PubMed]
24. Nasir, M.S.; Yang, G.; Ayub, I.; Wang, S.; Yan, W. Tin Diselinide a Stable Co-Catalyst Coupled with Branched TiO2 Fiber and g-C3N4 Quantum Dots for Photocatalytic Hydrogen Evolution. *Appl. Catal. B Environ.* **2020**, *270*, 118900. [CrossRef]
25. Paolucci, V.; De Santis, J.; Lozzi, L.; Giorgi, G.; Cantalini, C. Layered Amorphous A-SnO_2 Gas Sensors by Controlled Oxidation of 2D-$SnSe_2$. *Sens. Actuators B Chem.* **2022**, *350*, 130890. [CrossRef]
26. Zhou, W.; Yu, Z.; Song, H.; Fang, R.; Wu, Z.; Li, L.; Ni, Z.; Ren, W.; Wang, L.; Ruan, S. Lattice Dynamics in Monolayer and Few-Layer $SnSe_2$. *Phys. Rev. B* **2017**, *96*, 1–6. [CrossRef]
27. Nguyen, E.P.; Carey, B.J.; Daeneke, T.; Ou, J.Z.; Latham, K.; Zhuiykov, S.; Kalantar-Zadeh, K. Investigation of Two-Solvent Grinding-Assisted Liquid Phase Exfoliation of Layered MoS_2. *Chem. Mater.* **2015**, *27*, 53–59. [CrossRef]
28. Paolucci, V.; Emamjomeh, S.M.; Nardone, M.; Ottaviano, L.; Cantalini, C. Two-Step Exfoliation of WS_2 for NO_2, H_2 and Humidity Sensing Applications. *Nanomaterials* **2019**, *9*, 1363. [CrossRef]
29. Smith, A.J.; Meek, P.E.; Liang, W.Y. Raman Scattering Studies of SnS_2 and $SnSe_2$. *J. Phys. C Solid State Phys.* **1977**, *10*, 1321. [CrossRef]
30. Acharva, S.; Srivastava, O. Occurrence of polytypism in $SnSe_2$. *J. Cryst. Growth* **1981**, *55*, 395–397. [CrossRef]
31. Gonzalez, J.M.; Oleynik, I.I. Layer-Dependent Properties of SnS_2 and $SnSe_2$ Two-Dimensional Materials. *Phys. Rev. B* **2016**, *94*, 1–10. [CrossRef]
32. Li, S.L.; Miyazaki, H.; Song, H.; Kuramochi, H.; Nakaharai, S.; Tsukagoshi, K. Quantitative Raman Spectrum and Reliable Thickness Identification for Atomic Layers on Insulating Substrates. *ACS Nano* **2012**, *6*, 7381–7388. [CrossRef] [PubMed]
33. Wu, J.J.; Tao, Y.R.; Wu, X.C.; Chun, Y. Nonlinear optical absorption of SnX_2 (X = S, Se) semiconductor nanosheets. *J. Alloys Compd.* **2017**, *713*, 38–45. [CrossRef]
34. Guan, H.; Li, H.; Ming, J.; Hong, J.; Dong, J.; Qiu, W.; Zhu, W.; Yu, J.; Chen, Z.; Peng, G.; et al. Broadband Light Amplitude Tuning Characteristics of SnSe2 coated Microfiber. *J. Light. Technol.* **2020**, *38*, 6089–6096. [CrossRef]
35. Chen, Z.; Xiong, L.; Li, G.; Wei, L.; Yang, C.; Chen, M. Wafer-Scale Growth of Vertical-Structured $SnSe_2$ Nanoflakes for Highly Sensitive, Fast-Response UV-Vis–NIR Broadband Photodetectors. *Adv. Opt. Mater.* **2022**, *10*, 2102250. [CrossRef]
36. Camassel, J.; Schlüter, M.; Kohn, S.; Voitchovsky, J.P.; Shen, Y.R.; Cohen, M.L. Wavelength Modulation Spectra and Electronic Band Structure of SnS_2, and $SnSe_2$. *Phys. Stat. Sol. B* **1976**, *75*, 303–314. [CrossRef]
37. Hasan, M.M.; Haseeb, A.S.M.A.; Saidur, R.; Masjuki, H.H. Effects of annealing treatment on optical properties of anatase TiO_2 thin films. *Int. J. Chem. Biol. Eng.* **2009**, *40*, 221–225. [CrossRef]
38. Wang, Z.M.; Wang, W.; Coombs, N.; Soheilnia, N.; Ozin, G.A. Graphene Oxide Periodic Mesoporous Silica Sandwich Nanocomposites with Vertically Oriented Channels. *Acs Nano* **2012**, *12*, 7437–7450. [CrossRef]
39. Tan, P.; Chen, X.; Wu, L.; Shang, Y.Y.; Liu, W.; Pan, J.; Xiong, X. Hierarchical Flower-like $SnSe_2$ Supported Ag_3PO_4 Nanoparticles: Towards Visible Light Driven Photocatalyst with Enhanced Performance. *Appl. Catal. B Environ.* **2017**, *202*, 326–334. [CrossRef]
40. Malfatti, L.; Falcaro, P.; Amenitsch, H.; Caramori, S.; Argazzi, R.; Bignozzi, C.A.; Enzo, S.; Maggini, M.; Innocenzi, P. Mesostructured Self-Assembled Titania Films for Photovoltaic Applications. *Microp. Mesop. Mater.* **2006**, *88*, 304–311. [CrossRef]
41. Yu, G.; Chen, Z.; Zhang, Z.; Zhang, P.; Jiang, Z. The Photocatalytic Activity and Stability of a Nanosized TiO_2 Film Prepared by Carbon Black Modified Method. *Catal. Today* **2004**, *90*, 305–312. [CrossRef]
42. Innocenzi, P. Mesoporous ordered films via self-assembly: Trends and perspectives. *Chem. Sci.* **2022**, *13*, 13264–13279. [CrossRef] [PubMed]
43. Soler-Illia, G.J.A.A.; Angelomé, P.C.; Fuertes, M.C.; Grosso, D.; Boissiere, C. Critical Aspects in the Production of Periodically Ordered Mesoporous Titania Thin Films. *Nanoscale* **2012**, *4*, 2549–2566. [CrossRef] [PubMed]
44. Stagi, L.; Carbonaro, C.M.; Corpino, R.; Chiriu, D.; Ricci, P.C. Light Induced TiO_2 Phase Transformation: Correlation with Luminescent Surface Defects. *Phys. Status Solidi Basic Res.* **2015**, *252*, 124–129. [CrossRef]

45. Wojcieszak, D.; Mazur, M.; Kaczmarek, D.; Poniedziałek, A.; Domanowski, P.; Szponar, B.; Czajkowska, A.; Gamian, A. Effect of the Structure on Biological and Photocatalytic Activity of Transparent Titania Thin-Film Coatings. *Mater. Sci. Pol.* **2016**, *34*, 856–862. [CrossRef]
46. Arokiya Mary, T.; Fernandez, A.C.; Sakthivel, P.; Jesudurai, J.G.M. A Study on the Role of Surfactant on the Layered Growth of SnSe2 for Electrical Applications. *J. Mater. Sci. Mater. Electron.* **2016**, *27*, 11041–11048. [CrossRef]
47. Wu, G.; Xiao, L.; Gu, W.; Shi, W.; Jiang, D.; Liu, C. Fabrication and Excellent Visible-Light-Driven Photodegradation Activity for Antibiotics of $SrTiO_3$ Nanocube Coated CdS Microsphere Heterojunctions. *RSC Adv.* **2016**, *6*, 19878–19886. [CrossRef]
48. Ashiq, M.N.; Irshad, S.; Ehsan, M.F.; Rehman, S.; Farooq, S.; Najam-Ul-Haq, M.; Zia, A. Visible-Light Active Tin Selenide Nanostructures: Synthesis, Characterization and Photocatalytic Activity. *New J. Chem.* **2017**, *41*, 14689–14695. [CrossRef]
49. Lin, L.; Yang, Y.; Men, L.; Wang, X.; He, D.; Chai, Y.; Zhao, B.; Ghoshroy, S.; Tang, Q. A Highly Efficient TiO_2@ZnO N-p-n Heterojunction Nanorod Photocatalyst. *Nanoscale* **2013**, *5*, 588–593. [CrossRef]
50. Anh Tran, V.; Vu, K.B.; Thi Vo, T.T.; Thuan Le, V.; Do, H.H.; Bach, L.G.; Lee, S.W. Experimental and Computational Investigation on Interaction Mechanism of Rhodamine B Adsorption and Photodegradation by Zeolite Imidazole Frameworks-8. *Appl. Surf. Sci.* **2021**, *538*, 148065. [CrossRef]
51. Lasio, B.; Malfatti, L.; Innocenzi, P. Photodegradation of Rhodamine 6G Dimers in Silica Sol-Gel Films. *J. Photochem. Photobiol. A Chem.* **2013**, *271*, 93–98. [CrossRef]
52. Wu, T.; Liu, G.; Zhao, J.; Hidaka, H.; Serpone, N. Photoassisted Degradation of Dye Pollutants. V. Self-Photosensitized Oxidative Transformation of Rhodamine B under Visible Light Irradiation in Aqueous TiO_2 Dispersions. *J. Phys. Chem. B* **1998**, *102*, 5845–5851. [CrossRef]
53. Cui, Y.; Goldup, S.M.; Dunn, S. Photodegradation of Rhodamine B over Ag Modified Ferroelectric BaTiO3 under Simulated Solar Light: Pathways and Mechanism. *RSC Adv.* **2015**, *5*, 30372–30379. [CrossRef]
54. Jakimińska, A.; Pawlicki, M.; Macyk, W. Photocatalytic Transformation of Rhodamine B to Rhodamine-110—The Mechanism Revisited. *J. Photochem. Photobiol. A Chem.* **2022**, *433*, 114176. [CrossRef]
55. Snare, M.J.; Treloar, F.E.; Ghiggino, K.P.; Thistlethwaite, P.J. The Photophysics of Rhodamine B. *J. Photochem.* **1982**, *18*, 335–346. [CrossRef]
56. Purcar, V.; Caprarescu, S.; Donescu, D.; Petcu, C.; Stamatin, I.; Ianchis, R.; Stroescu, H. Degradation of TiO_2 and/or SiO_2 hybrid films doped with different cationic dyes. *Thin Solid Films* **2013**, *534*, 301–307. [CrossRef]
57. Song, X.M.; Wu, J.M.; Yan, M. Photocatalytic degradation of selected dyes by titania thin films with various nanostructures. *Thin Solid Films* **2009**, *517*, 4341–4347. [CrossRef]
58. Sawunyama, P.; Jiang, L.; Fujishima, A.; Hashimoto, K. Photodecomposition of a Langmuir-Blodgett film of stearic acid on TiO_2 film observed by in situ atomic force microscopy and FT-IR. *J. Phys. Chem. B* **1997**, *101*, 11000–11003. [CrossRef]
59. Wu, L.Y.L.; Ngian, S.K.; Chen, Z.; Xuan, D.T.T. Quantitative test method for evaluation of anti-fingerprint property of coated surfaces. *Appl. Surf. Sci.* **2011**, *257*, 2965–2969. [CrossRef]
60. Luda, M.P.; Li Pira, N.; Trevisan, D.; Pau, V. Evaluation of Antifingerprint Properties of Plastic Surfaces Used in Automotive Components. *Int. J. Polym. Sci.* **2018**, *2018*, 1895683. [CrossRef]
61. Revathi, B.; Balakrishnan, L.; Pichaimuthu, S.; Nirmala Grace, A.; Krishna Chandar, N. Photocatalytic Degradation of Rhodamine B Using BiMnO3 Nanoparticles under UV and Visible Light Irradiation. *J. Mater. Sci. Mater. Electron.* **2020**, *31*, 22487–22497. [CrossRef]
62. Gligorovski, S.; Strekowski, R.; Barbati, S.; Vione, D. Environmental Implications of Hydroxyl Radicals (•OH). *Chem. Rev.* **2015**, *115*, 13051–13092. [CrossRef] [PubMed]
63. Aadim, K.A.; Haneen, K.; Hadi, Q.M. Effect of Annealing Temperature on the Optical Properties of TiO_2 Thin Films Prepared by Pulse Laser Deposition. *Int. Lett. Chem. Phys. Astron.* **2015**, *56*, 63–70. [CrossRef]
64. Rahman, A.; Kim, H.J.; Noor-A-Alam, M.; Shin, Y.H. A theoretical study on tuning band gaps of monolayer and bilayer SnS_2 and $SnSe_2$ under external stimuli. *Curr. Appl. Phys.* **2019**, *19*, 709. [CrossRef]

nanomaterials

MDPI

Article

Hardness, Modulus, and Refractive Index of Plasma-Assisted Atomic-Layer-Deposited Hafnium Oxide Thin Films Doped with Aluminum Oxide

Mikk Kull, Helle-Mai Piirsoo, Aivar Tarre, Hugo Mändar, Aile Tamm and Taivo Jõgiaas *

Department of Materials Science, Institute of Physics, University of Tartu, W. Ostwald Str. 1, EE50411 Tartu, Estonia; hugo.mandar@ut.ee (H.M.)
* Correspondence: taivo.jogiaas@ut.ee

Abstract: Coatings with tunable refractive index and high mechanical resilience are useful in optical systems. In this work, thin films of HfO_2 doped with Al_2O_3 were deposited on silicon at 300 °C by using plasma-enhanced atomic layer deposition (PE-ALD). The mainly amorphous 60–80 nm thick films consisted Al in the range of 2 to 26 at.%. The refractive indexes varied from 1.69 to 2.08 at the wavelength of 632 nm, and they consistently depended on the composition. The differences were higher in the UV spectral region. At the same time, the hardness of the films was from 12–15 GPa; the modulus was in the range of 160–180 GPa; and the mechanical properties did not have a good correlation with the deposited compositions. The deposition conditions, element contents, and refractive indexes at respective wavelengths were correlated. The results indicated that it is possible to tune optical properties and retain mechanical properties of atomic layer-deposited thin films of HfO_2 with Al_2O_3 as doping oxide. Such films could be used as mechanically resilient and optically tunable coatings in, for instance, micro- or nano-electromechanical systems or transparent displays.

Keywords: atomic layer deposition; thin film; mechanical properties; nanoindentation; hardness; modulus; refractive index

Citation: Kull, M.; Piirsoo, H.-M.; Tarre, A.; Mändar, H.; Tamm, A.; Jõgiaas, T. Hardness, Modulus, and Refractive Index of Plasma-Assisted Atomic-Layer-Deposited Hafnium Oxide Thin Films Doped with Aluminum Oxide. *Nanomaterials* **2023**, *13*, 1607. https://doi.org/10.3390/nano13101607

Academic Editor: Jordi Sort

Received: 14 April 2023
Revised: 8 May 2023
Accepted: 9 May 2023
Published: 10 May 2023

1. Introduction

Hafnium oxide is a scientifically and technologically important material. It has been researched or applied as an optical coating, a dielectric material in capacitors, transistors, and processors; a material in bio-sensors, a resistive switching medium, ferroelectric or pyroelectric material, etc. [1–15]. Among other applications, such coatings could be used as optical interference filter material in lab-on-chip for targeted biological sensors [16]. HfO_2 has been proposed as an environmentally protective and reflectivity-enhancing coating for aluminum micromirrors [17]. A thin film applied to an optical system is expected to possess certain mechanical stability, or, generally, resilience [18].

Several different methods have been used to produce HfO_2. For instance, hydrothermal or wet-chemical routes [8,9,19], magnetron sputtering [3,4,10,15,20] and atomic layer deposition (ALD) [1,5,7,8,21,22]. It has appeared as fibers [9], nanoparticles [19], and thin films [1,3–5,8,10,12]. HfO_2 can possess amorphous [4,23] or nanocrystalline forms [3,8,10,19,21]. HfO_2 has also been produced in various doped variants with Si, Zr, Y, or Al as the dopant [8,12,23–27].

Different precursor combinations have been used for ALD of hafnia. Typical are halides of Hf [21,22,27–29] and metal-organic precursors, such as amino compounds [1,23,26,30]. Water vapor, ozone, O_2, or O_2 plasma have been used as oxygen sources [7,22,23,26,27,30]. The variety in crystallographic phases of ALD hafnia includes (X-ray) amorphous, monoclinic, tetragonal, and cubic phases [8,21,27,29–31]. The phase composition tends to depend on the used precursors and deposition temperature. Other phases, such as orthorhombic, have been reported for Si-doped and post-deposition annealed HfO_2 thin films [23].

HfO_2 has good transparency in the infrared to ultraviolet spectral range [32]. The transmittance of ALD hafnia has been shown to be near 80%, but it depends on the film growth temperature [1,29]. The refractive index for ALD hafnium oxide is usually above 2. The thin-film growth temperature has a minute influence on the refractive index [21].

Hardness values of HfO_2 on silicon have been reported to be 8.3–9.7 GPa and 9.5 ± 2 GPa [30,33]. It has also been shown that rapid annealing can increase the hardness up to 14.4 GPa [30]. As for the elastic modulus, the results varied from 163–165 GPa to 220 ± 40 GPa [30,33]. In these examples, the film thicknesses ranged from 60 to 100 nm. HfO_2 was deposited on silicon substrates using tetrakisdimethylamidohafnium/water and tetrakisethylmethylaminohafnium/ozone processes at 300 °C or less. In a previous work, HfO_2 on glass substrates (deposited using a $HfCl_4$/water process) had a modulus of 111(11) GPa and a hardness of 9.1(0.7) GPa. The thin-film thickness was approximately 160 nm [34].

Despite the fact that for years ALD has been used to deposit HfO_2 and its doped versions, including Al as the dopant, there are fewer reports on mechanical properties of such films. In a previous work about Al_2O_3-doped ZrO_2, it was shown that Al_2O_3 can induce and stabilize a metastable tetragonal phase of zirconia [35]. The same has been claimed about Al_2O_3-doped HfO_2 [8]. Metastable phases might have higher hardness and modulus and, therefore, could improve the mechanical resilience of thin films.

In the current work, plasma-enhanced atomic layer deposition was used to produce HfO_2 thin films doped with Al_2O_3. The aim was to determine the mechanical and optical properties of the films in relation to their elemental and phase compositions. Such thin films could be used as functional coatings in micro- or nano-electromechanical devices [36,37].

2. Materials and Methods

The HfO_2 and Al_2O_3 thin films were deposited using an R200 Advanced (Picosun, Masala, Finland) ALD reactor equipped with a load-lock and oxygen plasma generator. Hf was deposited from tetrakisethylmethylaminohafnium (TEMAH, 99% purity, Strem, Bischheim, France), and aluminum was deposited from trimethylaluminum (TMA, 99% purity, Volatec, Porvoo, Finland). The pulse sequence for HfO_2 deposition was 0.3/4/15/4 s (TEMAH/N_2/O_2 plasma/N_2), and for Al_2O_3 it was 0.1/4/15/4 s (TMA/N_2/O_2 plasma/N_2). The substrate Si (100) temperatures during depositions were 300 °C; two samples were deposited at 200 °C and 400 °C. The total cycle count was 700 cycles, or the closest possible to that. The deposition rate of HfO_2 was 0.113 ± 0.002 nm/cycle, and that for Al_2O_3 was 0.091 ± 0.002 nm/cycle.

In the following text, HfO_2 is abbreviated as H and Al_2O_3 is abbreviated as A. The respective number in front of the designation is the deposition cycle count of a single oxide. The first number in front of the parenthesis indicates the repetition of hafnia and alumina cycles (sometimes called super-cycles). Also, samples designated as 233 × (1H + 2A) and 233 × (2H + 1A) were deposited and measured after the first batch of samples had been characterized, aiming to verify the developed model for refractive index calculation with regard to elemental compositions.

The crystallographic phase composition and the thickness of the films were determined using X-ray diffraction (XRD) or reflectivity (XRR) methods with a diffractometer (Rigaku SmartLab™, Rigaku, Tokyo, Japan) working at 8.1 kW (Cu Kα radiation, λ = 0.154178 nm). An asymmetric 2θ-scan at a fixed grazing incidence (GIXRD) angle of ω = 0.42° was used for phase composition analysis. GlobalFit 2.0 was used for XRR model fitting. The element content was measured using X-ray fluorescence (Rigaku ZSX-400).

A spectroscopic ellipsometer (Semilab Sopra GES-5E, Semilab, Budapest, Hungary) with arms at 70 degrees, was used to measure thin film refractive indexes and thicknesses. Since the measured material systems are simple enough (extinction coefficients were $k = 0$), a Sellmeier double-parameter model was used for theoretical fitting (WinElli II v2.2.0.7 software, Semilab, Budapest, Hungary) of the measurements [38]. The worst fit had a coefficient of determination of $R^2 = 0.989$.

The hardness and elastic modulus of the Al_2O_3-doped HfO_2 films were investigated by using nanoindentation (Bruker Hysitron Triboindenter TI 980, Billerica, MA, USA; Minneapolis, MN, USA). The samples were measured using a Berkovich diamond tip. The device was calibrated using a fused silica standard (Bruker) with a defined hardness of 9.25 GPa and a reduced modulus of 69.6 GPa. The 20 calibration measurements in dynamic mode (continuous stiffness measurements) and subsequent testing on the silica indicated that the valid displacement range was approximately 8 nm to 75 nm. A single measurement in dynamic mode includes 60 separate data points. The standard deviations for a single measurement remained near 3 GPa for modulus and 0.3 GPa for hardness at displacements of approximately 10 nm. The deviations were somewhat reduced towards higher displacements. Thirty separate continuous stiffness measurements were performed on each prepared sample. Any low-quality measurements were removed, reducing the count of acceptable measurements by 3 to 8 measurements. Scanning probe microscopy (SPM) was performed with the same device using the same Berkovich-type diamond tip for probing. In addition to SPM images, a scanning electron microscope (SEM, FEI Helios™ NanoLab 600 DualBeam, FEI, Hillsboro, OR, USA) was used to investigate the surface morphology.

3. Results

3.1. The Element Content

The X-ray fluorescence analysis (Table 1) showed that there was no residual nitrogen detected at trustable levels in the thin films. The carbon content was, in most cases, near the respective detection limit, but only the films deposited at 200 °C and 400 °C had a carbon level 4–5 times higher than the limit; for the latter, the detection limit was 0.2 and the measure was 0.83 $\mu g/cm^2$. Therefore, the carbon content of those films could be considered trustworthy. The noticeable carbon levels in 200 °C and 400 °C deposited films could come, respectively, from incomplete reactions at 200 °C or some thermal decomposition of TEMAH or TMA at 400 °C. However, as the thin films deposited at these temperatures actually did not create any remarkable variety in any of the discussed properties, most of the attention was later turned to samples deposited at 300 °C.

Table 1. The elemental contents of Al_2O_3-doped HfO_2 thin films (deposited at 300 °C if not noted otherwise) measured by XRF.

Deposition Scheme	Amount of an Element, $\mu g/cm^2$			Al at.%
	Hf	Al	O	
700 × A	0	8.66 ± 0.02	7.7 ± 0.2	40.1
233 × (1H + 2A)	18.3 ± 0.2	7.00 ± 0.02	10.0 ± 0.2	25.9
350 × (1H + 1A)	29.7 ± 0.2	6.86 ± 0.02	12.1 ± 0.2	21.6
233 × (2H + 1A)	46.9 ± 0.2	5.09 ± 0.02	14.7 ± 0.3	13.8
140 × (4H + 1A)	55.8 ± 0.4	2.22 ± 0.03	13.9 ± 0.5	6.5
70 × (9H + 1A), 200 °C	60.2 ± 0.4	1.34 ± 0.02	14.8 ± 0.5	3.8
70 × (9H + 1A)	65.3 ± 0.3	1.16 ± 0.02	14.8 ± 0.5	3.2
70 × (9H + 1A), 400 °C	53.8 ± 0.4	1.24 ± 0.02	12.7 ± 0.5	4.1
47 × (14H + 1A)	64.8 ± 0.3	0.74 ± 0.02	13.8 ± 0.5	2.2
35 × (19H + 1A)	66.6 ± 0.4	0.59 ± 0.02	14.3 ± 0.6	1.7
28 × (24H + 1A)	64.6 ± 0.3	0.70 ± 0.02	13.8 ± 0.5	2.1
700 × H	78.15 ± 0.3	0	16.21 ± 0.6	

The amount of Al atoms in the first deposited batch varied from 2 at.% up to 26 at.% out of total amount of elements (Table 1). The error for an element's content is given as plus–minus the triple of the respective detection limit. Checking the stoichiometry indicated that, in pure Al_2O_3, the ratio of O/Al was 1.5, which can be considered a fully stoichiometric oxide. Pure HfO_2 had some excess of oxygen with an O/Hf ratio of 2.3, which is approximately 15% more oxygen than stoichiometrically expected. The samples with alumina addition had O/Hf ratios, mostly in the 2.1–2.2 range. The actual reason for such an inconsistency was not investigated but taken as is. The excess oxygen presence in the same precursors has been reported in a previous study [31], where the O/Hf ratio varied from 2.0 to 2.2 and depended on the deposition temperature. The ratio of 2.0 was obtained for samples deposited at 350 °C. At the same time, it was shown that the deposition rate increased at higher substrate temperatures due to some pyrolysis of the Hf precursor. In another study, it was shown that HfO_2, using the same ALD process but deposited at 160 °C, had over 10 at.% of impurities [39]. That amount did not change during post-deposition annealing (in nitrogen or air) at 300 °C. X-ray photospectroscopy showed that the films had a certain amount of oxygen bound to hydrogen and carbon. During annealing in air, the O/Hf ratio had an increasing tendency, over the stochiometric value (approx. 2.05). Sub-stochiometric HfO_2 from tetrakis(dimethylamino)hafnium and O_2 plasma, deposited at 250 °C, has also been reported [40]. Therefore, with this precursor combination, some excess elements and imbalanced ratios of those can be expected.

The anion/cation ratio in metal oxides is important. Copper oxides are a typical example. CuO is an n-type conductor, whereas Cu_2O is a p-type conductor [41]. As the O/Cu ratios change from 1 (CuO) to 0.75 (Cu_4O_3) to 0.5 (Cu_2O), the optical band gaps change by 2.11 eV, 1.34 eV, or 2.47 eV (depending on the analysis method) and 2.45 eV, respectively. A similar trend occurs in electrical resistivity, which reduces in the same row as 10×10^8 µΩcm, 6.2×10^8 µΩcm, and 3×10^8 µΩcm, respectively [42]. In the case of HfO_2 thin films, Hildebrandt et al. have shown that the band gap for a stochiometric compound is 5.7 eV, but it reduces if oxygen vacancies or interstitials are present [43]. They achieved p-type conductivity in oxygen-vacant HfO_2 thin films with a band gap down to 4.5 eV and resistivity of 400 µΩcm. Unfortunately, they did not report the elemental composition of their thin films or point out any other influence on properties caused by oxygen interstitials.

3.2. *Film Thickness and Phase Composition*

3.2.1. Film Thickness, Density, and Roughness

The results of film thickness and density are summarized in Table 2. Examples of XRR measurement and fitting results can be seen in Figure 1. A set of XRR measurements and respective fits can be found in Supplementary part S1 (Figure S1). The density value for pure hafnia corresponds to the density of monoclinic HfO_2, the phase confirmed by the diffraction pattern. The alumina density corresponds approximately to a previously published value [35]. The deviations for thickness values were 1 nm or less and 0.2 g/cm^3 or less for density. Roughness values for alumina and doped films were approximately 1 nm and approximately 3.5 nm for pure HfO_2.

Table 2. Al_2O_3-doped HfO_2 thin-film thickness and density, derived from XRR measurements.

Formula	Thickness, nm	Density, g/cm^3
700 × A	63.8	3.08
233 × (1H + 2A)	60.8	5.24
350 × (1H + 1A)	62.4	6.30
233 × (2H + 1A)	69.1	7.75
140 × (4H + 1A)	70.2	8.52
70 × (9H + 1A), 200 °C	75.5	8.66
70 × (9H + 1A)	73.5	9.10
70 × (9H + 1A), 400 °C	62.2	9.14
47 × (14H + 1A)	70.9	9.12
35 × (19H + 1A)	69.2	9.13
28 × (24H + 1A)	67.3	9.32
700 × H	78.8	9.68

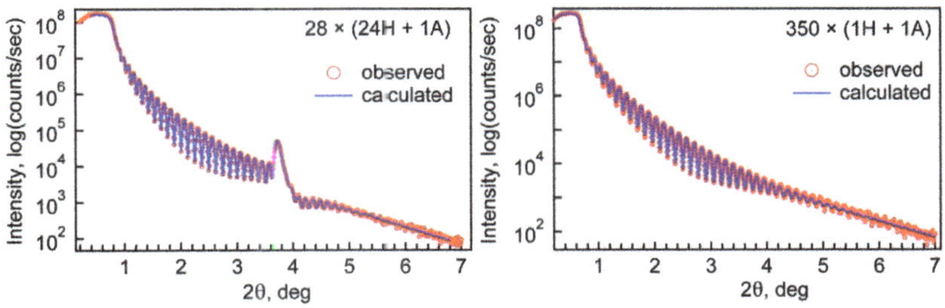

Figure 1. Examples of XRR measurements and fitting results.

Table 2 indicates that the deposition rate of HfO_2 was 0.113 ± 0.002 nm/cycle and that for Al_2O_3 was 0.091 ± 0.002 nm/cycle.

3.2.2. Phase Composition

The HfO_2 diffraction pattern shows that it deposited mainly in the monoclinic phase (ICDD PDF-2 card #01-075-6426, Figure 2).

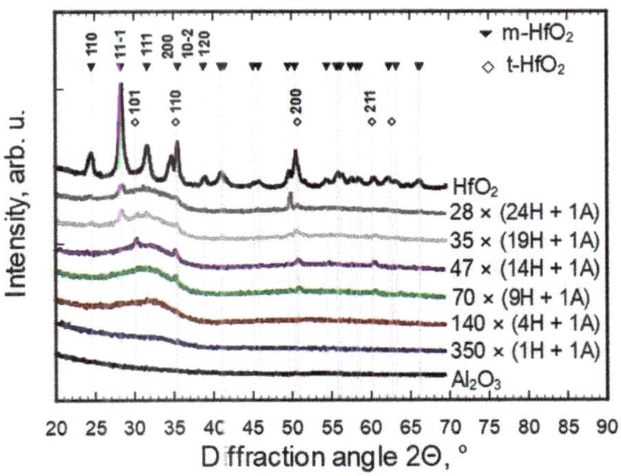

Figure 2. Diffraction patterns of thin films deposited at 300 °C.

Only a few reflections could be observed for thin films with a hafnia to alumina cycle ratio of 24 ÷ 1 to 9 ÷ 1. All four reflections on the patterns of samples 47 × (14H + 1A) and 70 × (9H + 1A) match with the strongest reflections of cubic or tetragonal metastable phases of hafnia (ICDD PDF-2 cards #53-0560 and #01-078-5756, correspondingly). The same could be claimed for 70 × (9H + 1A) deposited at T_g = 400 °C (Figure 3). The strongest reflections of monoclinic hafnia were present in the case of samples 35 × (19H + 1A) and 28 × (24H + 1A). The metastable phases of samples 47 × (14H + 1A) and 70 × (9H + 1A), detectable by a reflection at $2\Theta \approx 30.2°$ and indexed as 111 of the cubic or 101 of the tetragonal HfO_2 phase, disappeared when the concentration of hafnium increased or decreased. The samples with an Al content of ≈6 at.% and more were X-ray amorphous and did not show any diffraction reflections. It was undetermined how much of the amorphous nature was caused by interstitial oxygen. The pure HfO_2 pattern indicates a reasonably well-crystallized thin film, even though it had the highest O/Hf ratio, but adding Al in a few percent levels changed that remarkably.

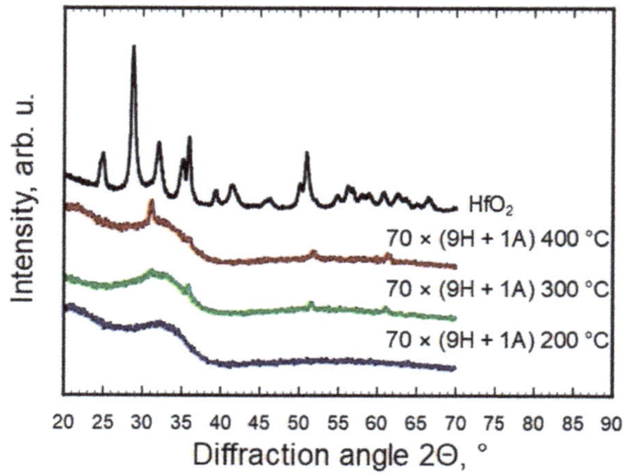

Figure 3. Phase composition evolution for the same deposition scheme but different growth temperatures.

3.3. *Refractive Index*

Spectroscopic ellipsometry (SE) allows the determination of the refractive indexes n of the thin films for a range of wavelengths. The measurements show that most of the change in n between wavelengths of 225 and 770 nm (photon energies of 5.5–1.6 eV) appears for thin films with HfO_2 to Al_2O_3 cycle ratios of 1 ÷ 1 and 4 ÷ 1 (Figure 4). The difference is marginal between pure HfO_2 and films with HfO_2 cycle ratios of 9 and up. Throughout the compositions, the refractive indexes change consistently, with alumina possessing the lowest and hafnia the highest values. (See also Supplementary part S2, Figure S2)

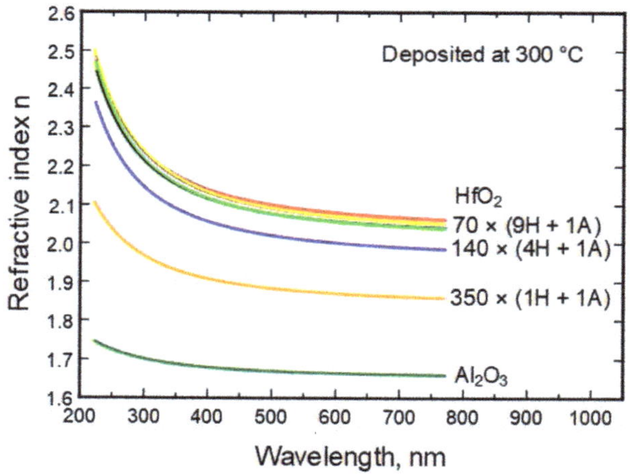

Figure 4. Refractive index calculated from ellipsometry data in relation to the cycles of HfO_2 and Al_2O_3.

Sun et al. have shown that the refractive index does not depend much on the O/Hf ratio [44]. Their samples had O/Hf ratios ranging from 1.98 to 2.09, and the highest difference in refractive indexes was approximately 0.02 at 775 nm. They also showed that the extinction coefficient k is 0 for the samples, which is a presumption in the discussion below.

Deposition temperature had a low effect on the refractive index compared to composition change (Figure 5).

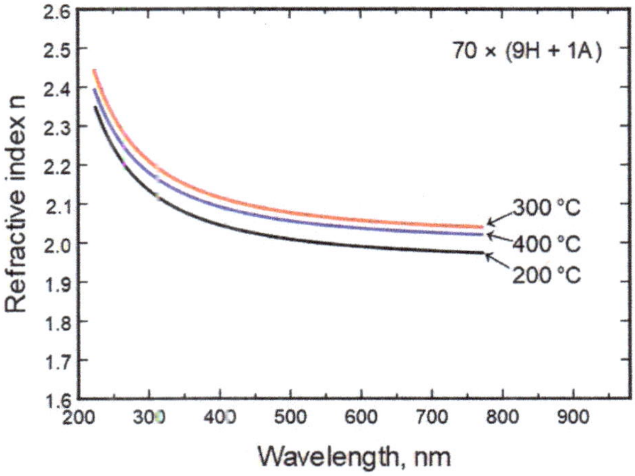

Figure 5. Refractive index relation to thin film deposition temperature.

3.4. Hardness and Modulus of Thin Films

3.4.1. The Substrate and Single Oxides

The nanoindentation of Si (100) indicated that the substrate hardness was approximately 13 ± 1 GPa and the (reduced) modulus was 145 ± 10 GPa.

The alumina indentation results show the non-uniform behavior of the material system (Figure 6). Initially, the hardness was approximately 10.5 ± 0.8 GPa, and the (reduced) modulus was 137 ± 7 GPa. (Hereafter, the values are given at ≈10 nm of displacement, if not stated otherwise.) These values are similar to previously published results [35].

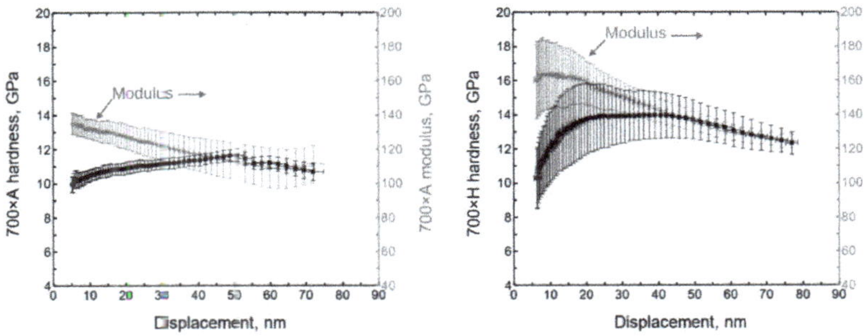

Figure 6. Hardness and modulus of Al_2O_3 and HfO_2 on silicon deposited at 300 °C, as measured by nanoindentation.

The initial hardness of hafnia was approximately 10.5 ± 1.5 GPa, and the (reduced) modulus was 160 ± 20 GPa (Figure 6). The initial lower hardness value might be influenced by surface roughness since the X-ray reflectivity showed hafnia to have the highest roughness.

3.4.2. The Doped Films

The averaged moduli and hardness values are presented in Figure 7. (The full set is presented in Supplementary part S3, Figure S3) It can be seen that the moduli were in the approximate range of 150 to 190 GPa and the hardness values ranged from 11 to 14 GPa. Somewhat exceptional is sample 140 × (4H + 1A) with a hardness value of approximately 15 GPa.

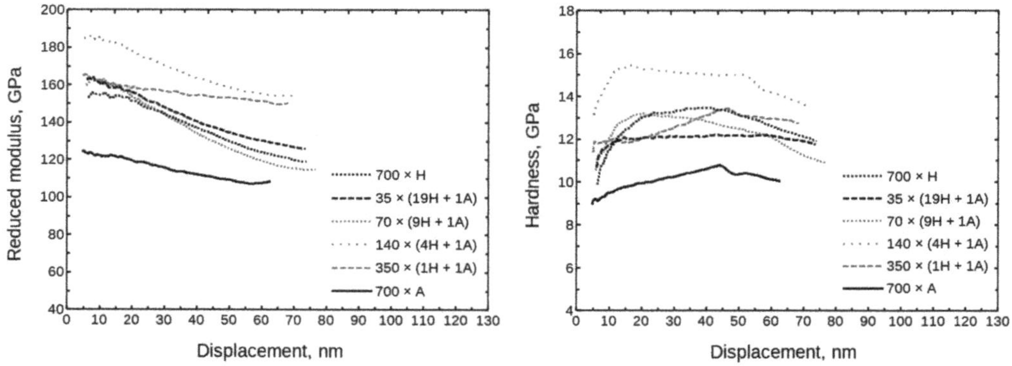

Figure 7. Averaged modulus and hardness values of Al_2O_3 doped HfO_2 thin films as measured by nanoindentation.

The growth temperature did not have a large effect on the modulus for samples deposited at 300 °C and 400 °C (formula 70 × (9H + 1A)). At 10 nm of displacement, the moduli remained between 160 and 170 GPa, and the hardness values were 12 and 13 GPa, respectively. The thin film deposited at 200 °C had a modulus of ≈150 GPa, and the hardness was approximately 10 GPa.

3.5. Surface Morphology

SPM images (created by using the Gwyddion 2.62 freeware program) sized 1 × 1 μm and 5 × 5 μm were taken of some samples to confirm or deny any pile-up or surface roughness effects. In Figure 8, it can be seen that the amount of surface features increased with the relative increase in HfO_2 content. (A closer look at 28 × (24H + 1A) can be seen in Supplementary part S4, Figure S4) The profiles, presented at the bottom, show that pure, pristine Al_2O_3 had a smooth surface (black line), and the surface roughness increased with the increase in HfO_2 content. Profiles were taken across a sample. The residual depth of the indent on Al_2O_3/Si was approximately 30 nm (respective indentation force 1.5 mN) and was similar to depths for other objects. In the case of alumina thin film, a rise on the indent (red line) edges can be seen, which is an indication of material pile-up around the indent mark. The height of the piled-up material on the left side of the indent was approximately 9 nm. This is a considerable amount compared to the total indentation depth of 70 nm. Therefore, in the case of alumina, only the first few indentation datapoints could be representative of the material system. (Note that the dimensions of the x and y axes in the profile images differ; y is in tens of nanometers, x in hundreds.)

As the tip of the triboindenter is rather blunt (radius approximately 100–150 nm), the SPM images provide a surface morphology description that is rather like a pseudo-surface. Scanning electron microscope images of the same samples were taken to supplement SPM (Figure 9). It can be seen that the surface gets rougher as the amount of deposited HfO_2 increases. This confirms the previously deduced trend. Micrographs indicate that, in the case of hafnia, for instance, there are grains of crystallites rather than three-sided pyramids on the surface. It is reasonable to consider the triangular features on SPM images as having a convoluted indenter tip shape.

The doped thin films have different morphology compared to pure hafnia surface. Note that the alumina-containing samples have grains alternated with smoother regions.

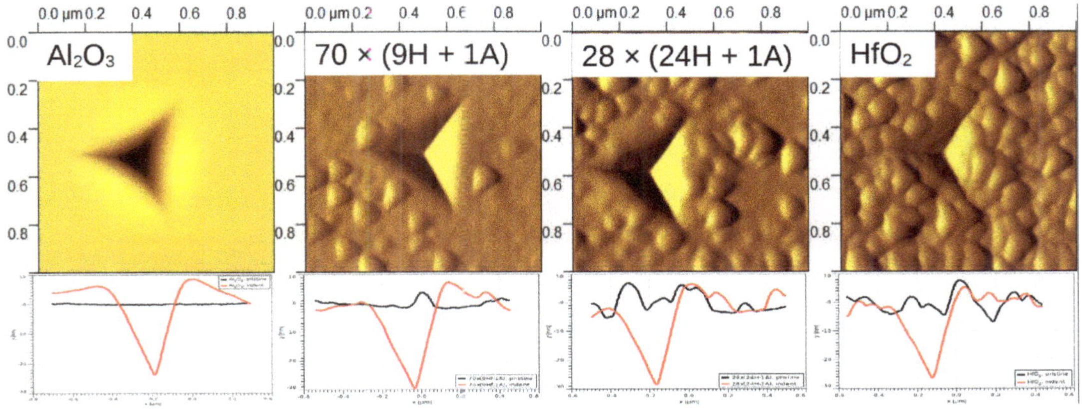

Figure 8. SPM images and respective horizontal profiles of films deposited at 300 °C. The black line indicates plane surface, while red is horizontal profile across the indent. The residual depth of the indent on Al_2O_3 was approximately 30 nm, and was similar to depths for other objects.

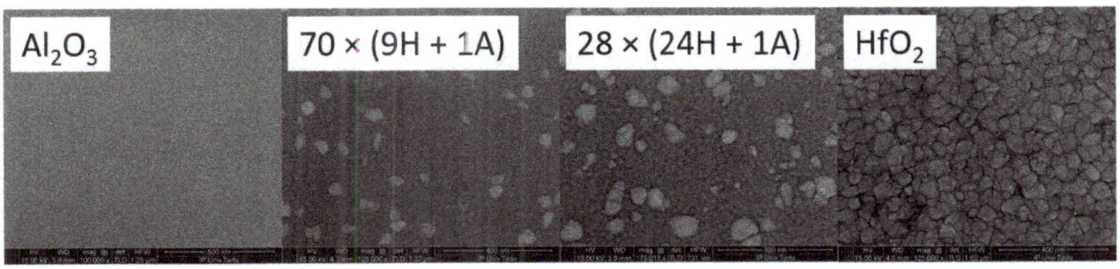

Figure 9. SEM images of selected thin films deposited at 300 °C. The presented area is approx. 1 × 1 micrometer for all samples

4. Discussion

The densities of the thin films correlate linearly to the compositions (Figure 10). The best fit had a coefficient of determination over 0.99. This is visualized by the narrow band of deviations in fit parameter errors, also plotted in the image. A model that has hafnia and alumina densities of 9.68 g/cm^3 and 3.08 g/cm^3, respectively, as inputs is also shown. The model itself is a derivation of a simple rule of mixture known from composite material theory. The Voigt rule of mixture could be used to predict properties of material systems (mixtures), including density.

During nanoindentation, some influence from the substrate could be expected at 10 nm of tip displacement. At the same time, the properties of the substrate are similar to those of the thin films. Therefore, it could be proposed that the substrate influence was reduced compared to, for instance, that of a substrate possessing a hardness of 20 GPa or a modulus of 340 GPa. This suggests that the values given at approximately 10 nm displacement actually closely resemble those of a thin film.

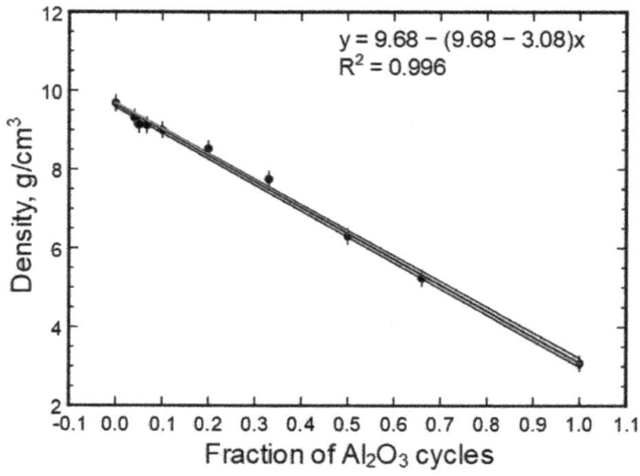

Figure 10. Thin film XRR densities in relation to the Al_2O_3 deposition cycle fractions.

In several cases, a sudden reduction in hardness can be seen. For pure alumina, this happened at approximately 50 nm of displacement. It has been shown in previous publications that this could be a consequence of coating delamination or cracking [45–48]. Considering that the pop-in happened during indentation and the SPM images show some material pile-up near the indents, it is likely that the pile-up was caused by thin film delamination and buckling.

Nanocrystals could reinforce a material and increase its hardness, modulus, and fracture toughness. These effects have long been known, for instance, in glass-ceramics research [49–55]. The amount of crystalline phase and actual phase composition also influence mechanical properties [49–52]. It has been shown that the appearance of softer phases can reduce hardness but might not ultimately affect the modulus [49]. Also, the modulus can reduce with an increase in crystallinity without having a significant influence on hardness [51].

Overall, the deposition recipe or respective elemental composition did not uniformly correlate with hardness or modulus (Figure 11), especially if measurement errors were considered. The general trend seems to be that the doped films had a higher modulus/hardness than pure single oxides.

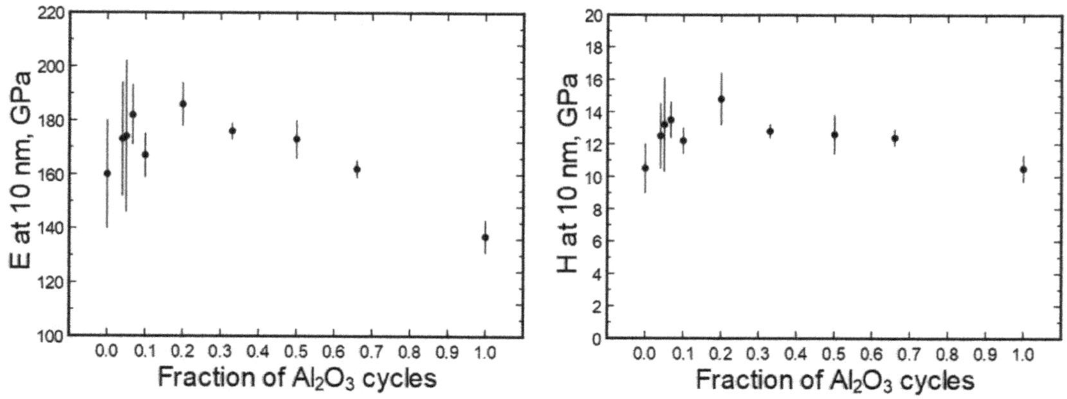

Figure 11. Thin-film moduli in relation to the alumina cycle fractions.

On average, 140 × (4H + 1A) has a hardness of approximately 15 GPa and a modulus of 184 GPa, making it the hardest and stiffest thin film. Currently, the cause is not clear. Based on effects seen in glass-ceramics, it could be speculated that 140 × (4H + 1A) has a higher hardness and modulus due to the appearance of some short-ordered crystallinity. In the case of 47 × (14H + 1A), it could be that there was some increase in the crystalline phase volume part, compared, for instance, with 70 × (9H + 1A), which was softer. The next mixtures already show different phase compositions based on diffraction patterns. The appearance of different phases, such as monoclinic and cubic/tetragonal in sample 35 × (19H + 1A), could create additional defective interphases around crystallite grains, which could have a detrimental effect on the mechanical properties. This would explain the reduction in hardness/modulus in samples 35 × (19H + 1A) and 28 × (24H + 1A) compared to 47 × (14H + 1A).

A better correlation could be drawn between the specific modulus (modulus divided by density) or specific hardness and the alumina cycle fraction (Figure 12). Here again, a linear correlation was suitable to model the data. In principle, the best-fit line in both cases goes through all data points, but the scatter is wider. Therefore, the equations shown in Figure 12 have approx. 5% error.

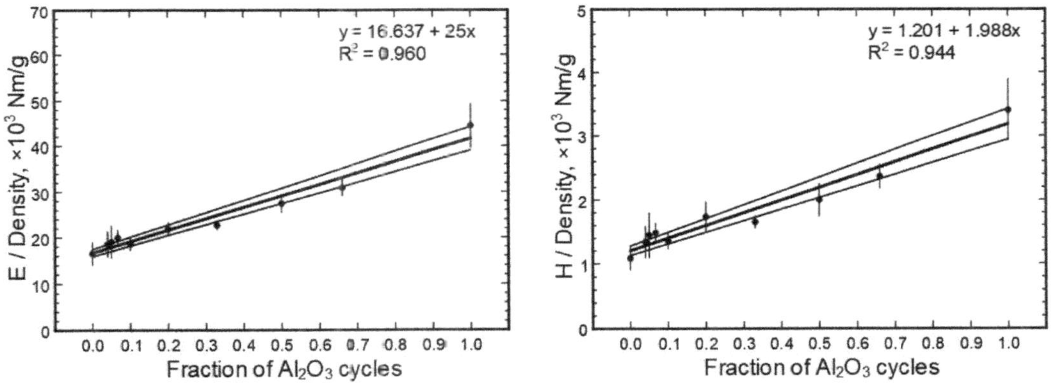

Figure 12. Specific modulus and hardness in relation to the deposited Al_2O_3 cycle fractions.

4.1. Surface Morphology Influence on Nanoindentation

Several areas with different morphologies were selected by using the indenter SPM capability. Indentations were performed on grainy and other areas of sample 70 × (9H + 1A), grown at 300 °C (Figure 13). The paired images show the surface before and after indentation, respectively. It can be seen that in the case in which the indent is made on a plane surface, the moduli are rather high (near 250 GPa). If the indent is basically on top of the grains, the moduli drop to the 120 GPa level. If the indent is close to some grains but not directly on top, the results tend to be intermediate. The defect density is probably higher near grains. Therefore, a reduction in mechanical resilience should be expected. Any enhancing effects that could be attributed to, for instance, a dislocation movement impediment, were not noticed in any of the samples.

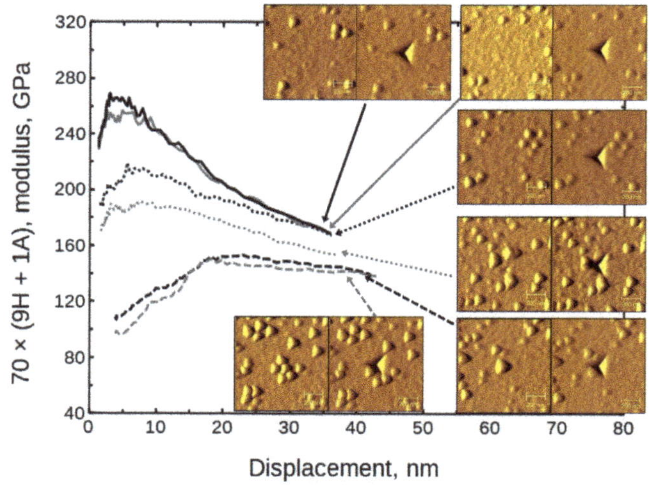

Figure 13. The influence of surface morphology to modulus; sample 70 × (9H + 1A). The paired SPM images show surface before and after indentation.

The corresponding hardness values have a similar tendency (Figure 14), but not as consistent as in the case of the moduli. Note that the presented results are from the same locations.

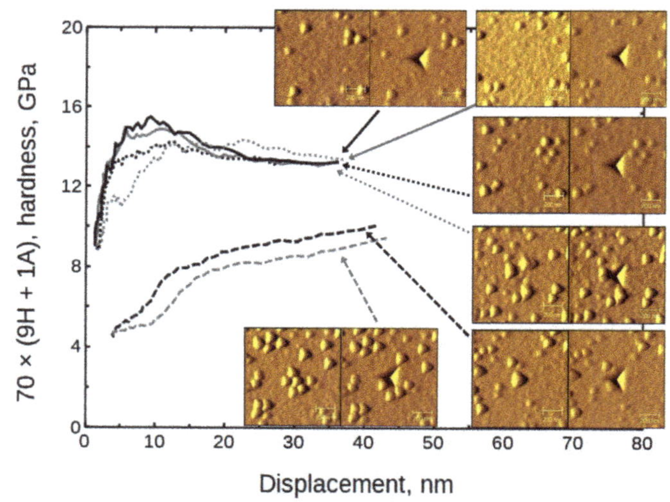

Figure 14. The influence of surface morphology to hardness; sample 70 × (9H + 1A). The paired SPM images show surface before and after indentation.

4.2. Correlation of Refractive Index to Composition

The refractive indexes n and compositions correlate well within the initial batch of thin films deposited at 300 °C. That batch was used to create a relationship between the two-term Sellmeier formula (for refractive index) and the compositions. In WinElli II v2.2.0.7 software, the Sellmeier parameters A and B are given as:

$$n(\lambda) = \sqrt{1 + \frac{(A^2 - 1)\lambda^2}{\lambda^2 - B}}$$

where λ is wavelength in micrometers; the formula considers the imaginary part to be zero.

The relations of A and B to their respective compositions can be seen in Figure 15. The best-fit line is given as the respective formula on a graph. The error range of a fit was calculated from the formula parameter errors and is depicted as minimum and maximum lines (thinner lines on the graphs).

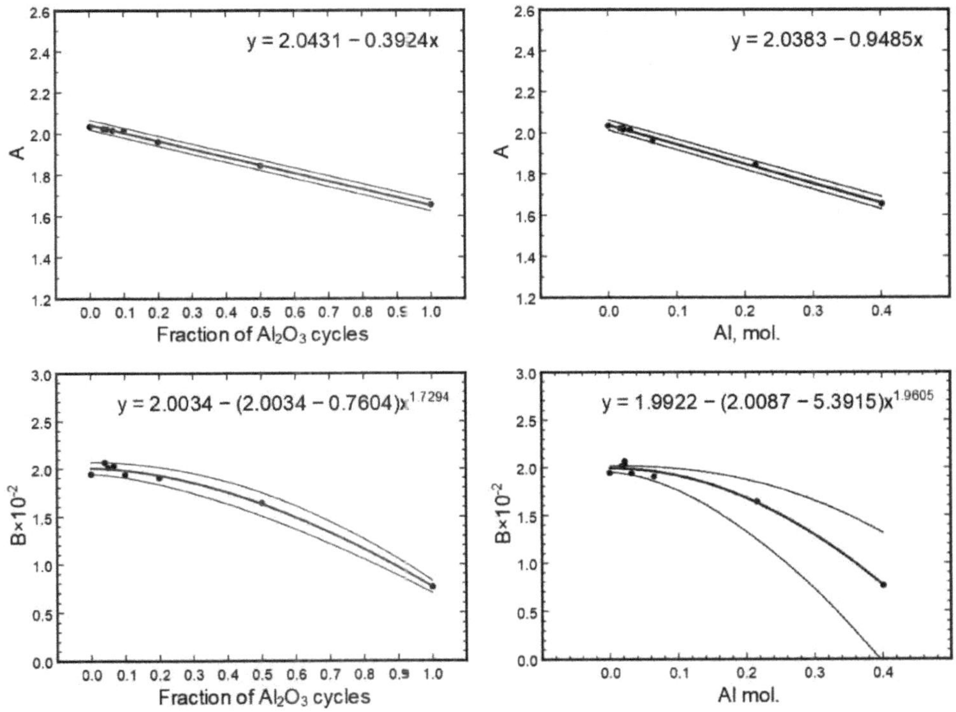

Figure 15. Sellmeier parameters in relation to the thin film compositions.

Based on the equations shown in Figure 15, parameters for samples 233 × (1H + 2A) and 233 × (2H + 1A) were calculated (using the cycle part formulas), and respective indexes n versus wavelength were constructed and compared to direct measurements to check the predictive power of the equations. The results are summarized in Table 3.

Table 3. Sellmeier parameters and respective limits from the A and B correlations to the compositions.

Deposition Scheme	Best Fit A	Min/Max A	Best Fit B ($\times 10^{-2}$)	Min/Max B ($\times 10^{-2}$)
233 × (1H + 2A)	1.782	1.774/1.789	1.387	1.293/1.474
233 × (2H + 1A)	1.912	1.907/1.918	1.830	1.739/1.884

The initial derivation of A and B values from measurements had an accuracy of approximately 1%, which has to be added/subtracted from the Table 3 minimum/maximum values. Figure 16 shows the results of the measured data (black and grey crosses) and the predictions from models (black and grey lines) as minimum and maximum values. It can be seen that the measurements fit the model within error.

as a porogen, directing the structure forming a micelle-forming structure and interacting with the precursor. After the bulk structure was assembled, the surfactant was removed, leaving pores or channels embedded in the material framework. The formation of hybrid OSG films is based on the hydrolysis and condensation of MTMS and BTMSE precursors:

$$\equiv\text{Si-OCH}_3 + \text{H}_2\text{O} \rightarrow \equiv\text{Si-OH} + \text{CH}_3\text{-OH (hydrolysis)} \quad (1)$$

$$\equiv\text{Si-OH} + \text{HO-Si}\equiv \rightarrow \equiv\text{Si-O-Si}\equiv +\text{H}_2\text{O (condensation)} \quad (2)$$

As-deposited (AD) films contain silanol groups formed as a result of precursor hydrolysis (Reaction (1)) and some unreacted methoxy (–OCH_3) groups from precursors. The concentration of these groups can still be significant after the soft bake at 150 °C (SB) but almost not detectable after the final curing at 400 °C (HB). SB films mainly reflect the non-relaxed structure determined by reactions (1) and (2). HB increases the degree of cross-linking of the organosilicate matrix with a significant improvement in the bonding structure of the Si–O–Si network [8,41]. At this stage, the matrix compaction occurs, accompanied by the segregation of unassembled porogen residues from the low-k matrix. Photoluminescence was measured from AD samples, SB and HB samples. Samples termed "as-deposited (AD)" were dried at room temperature and 60–80 °C for 10 min. Destruction of the ethylene bridge during AD and SB is not expected, and these samples cannot contain ODC centers.

ii. Pure 1,4-benzene-bridged films (2-2 and 2-3 in Figure 1) were deposited using 1,4-bis(triethoxysilyl)benzene (BTESB). The "dense" version of the films had an EP-measured porosity (free volume) of about 10% (1,4-BB), while the sample deposited with 30 wt% porogen had a porosity of about 29% (1,4-BB-p). In contrast to the ethylene-bridged and methyl-terminated samples, the benzene-bridged films had an extremely small pore size of about 0.5 nm against 3 nm in MTMS-p (3-2) samples. It also had much better mechanical properties, as demonstrated in the paper [42]. Furthermore, all the details related to the film deposition, including curing conditions and measurement procedures, can be found in Ref. [42].

iii. Recently, hyperconnected structures developed by groups at IBM and Stanford University have attracted particular interest [33]. These films were deposited using a mixture of 1,3,5-tris(triethoxysilyl)benzene and 1,3-bis(triethoxysilyl)benzene as described in Refs. [43,44]. The architecture of the hyperconnected network has been achieved through the use of 1,3,5-silyl benzene precursors, where each silicon atom can be connected to the five other nearest silicon neighbors. Thus, the 1,3,5-benzene bridging group structure connects each silicon atom to two others via carbon bridges that share one common Si–C bond while maintaining the ability of a silicon atom to connect with three others via Si–O–Si bonds (Figure 1b in Ref. [33]). The films with different ratio of 1,3,5- and 1,3-benzene bridges were deposited and analyzed.

iv. The films containing only terminal methyl groups (3-1 and 3-2 in Figure 1) were deposited using pure MTMS precursor. Film 3-1 was deposited as "dense" (without sacrificial porogen) and film 3-2 by the addition of 30 wt% Brij®30 surfactant $C_{12}H_{25}(OCH_2CH_2)_4OH$. The samples deposited with porogen had an open porosity of about 33% as measured by ellipsometric porosimetry (EP) [45], while the "dense" sample had micropores with an open free volume of about 8% [42].

The physical properties of all deposited materials used are summarized in Table S1. All samples were thermally cured; UV light and plasma were not used for curing.

2.2. Analysis

The chemical compositions of the films (1-1, 1-2, 2-1) were analyzed by Nicolet 6700 (Thermo Electron Corporation, Waltham, MA, USA) Fourier transform infrared spectroscopy (FTIR) in the range 4000–400 cm^{-1} with a resolution of 4 cm^{-1} (64 scans) in transmission mode. The optical characteristics of the films, including thickness and refrac-

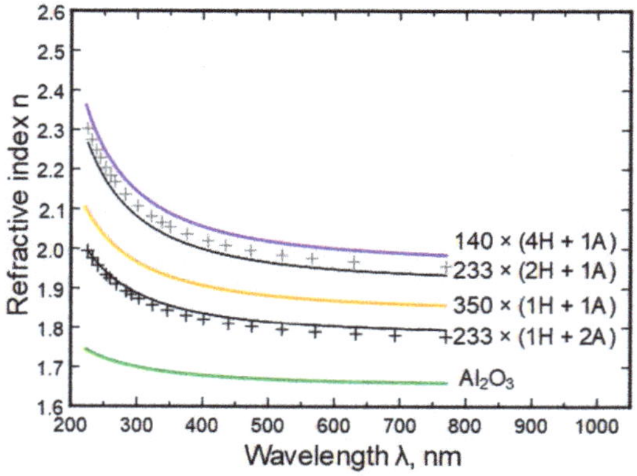

Figure 16. Measured data (crosses) and modeled results (black and grey lines) based on derived Sellmeier parameters.

5. Conclusions

Aluminum and hafnium oxides were atomic-layer-deposited using metal–organic precursors with oxygen plasma. Plasma ALD of aluminum oxide resulted in stochiometric oxide. ALD of hafnium oxide had excess oxygen in the thin film.

The linear correlation between density and composition means that a simple rule of mixture, known from composite material theory, could also be used to model the results. The agreement between experimental and theoretical approaches was very high. This is an indication that a simple rule of mixture can be used at the nanoscale for density predictions/calculations. In principle, the same applies to specific modulus and hardness, but with a somewhat higher error.

The alumina addition changes crystallography and surface morphology even at 2 mol.% levels. If metastable phases (cubic, tetragonal) are of interest, the hafnium to aluminum cycle ratio should be from approximately 8 ÷ 1 to 15 ÷ 1. At lower ratios, the films tend to be X-ray amorphous, and at higher ratios, hafnia starts to form a monoclinic phase.

The doped thin films had ≈30% higher hardness and modulus values compared to those for alumina. The highest hardness and modulus values were measured for thin films over with 6 mol.% of Al in the thin film. Higher surface roughness hindered analysis and generated larger deviations in the modulus and hardness results. The areas in-between grains had approximately 30% higher modulus (≈240 GPa) and approximately 25% higher hardness (≈15 GPa) compared to the averaged results (≈170 GPa and ≈12 GPa, respectively) of the same sample. The change in material phase composition, increase in interfaces between them, have detrimental effects on the mechanical properties.

Aluminum oxide has a refractive index of approximately 1.70 ± 0.04 in the spectral range of 770–225 nm. This steady value could be employed in the practical design of optical coatings. HfO_2 had a refractive index in the same wavelength range of 2.07 to 2.33, respectively. A change in the hafnia to alumina deposition cycle ratio changed the refractive index but had a minute effect if the alumina amount was below 4 at.%. The indexes can be predictively computed from composition data by using Sellmeier's formula, which helps select a suitable coating for an optical system. The models have errors of up to ±3%.

Overall, HfO_2 with Al_2O_3 dopant provides the possibility to develop relatively hard and stiff thin films with tunable optical properties, which could be used in microelectromechanical systems, for instance.

Supplementary Materials: The following supporting information can be downloaded at: https://www.mdpi.com/article/10.3390/nano13101607/s1, Figure S1: X-ray reflectance (XRR) of all samples; Figure S2: Refractive indexes of all samples; Figure S3: Nanoindentation modulus and hardness results of all samples; Figure S4: Close-up view of sample 28 × (24H + 1A) SPM image and respective cross-profiles.

Author Contributions: Conceptualization, T.J.; methodology, T.J.; validation, T.J. and H.M.; formal analysis, T.J., M.K., H.-M.P., A.T. (Aivar Tarre) and H.M.; investigation, T.J., M.K., H.-M.P., A.T. (Aivar Tarre) and H.M.; data curation, T.J., M.K., H.-M.P., A.T. (Aivar Tarre) and H.M.; writing—original draft preparation, T.J.; writing—review and editing, M.K., H.-M.P., A.T. (Aivar Tarre), H.M. and A.T. (Aile Tamm); visualization, T.J.; supervision, T.J. and A.T. (Aile Tamm); project administration, T.J.; funding acquisition, A.T. (Aile Tamm). All authors have read and agreed to the published version of the manuscript.

Funding: This work was funded by the Estonian Research Agency project PRG4 "Emerging novel phases in strongly frustrated quantum magnets" and by the European Regional Development Fund project TK134 "Emerging orders in quantum and nanomaterials".

Institutional Review Board Statement: Not applicable.

Informed Consent Statement: Not applicable.

Data Availability Statement: The data that support the findings of this study are available from the corresponding author upon reasonable request.

Acknowledgments: This work was supported by ERDF project Centre of Technologies and Investigations of Nanomaterials (NAMUR+, project number 2014-2020.4.01.16-0123).

Conflicts of Interest: The authors declare no conflict of interest.

References

1. Gao, J.; He, G.; Zhanf, J.W.; Liu, Y.M.; Sun, Z.Q. Deposition temperatuure dependent optical and electrical properties of ALD HfO_2 gate dielectrics prepared with tetrakisethylmethyl amino hafnium. *Mater. Res. Bull.* **2015**, *70*, 840–846. [CrossRef]
2. Choi, J.H.; Mao, Y.; Chang, J.P. Development of hafnium based high-k materials–A review. *Mat. Sci. Eng. R* **2011**, *72*, 97–136. [CrossRef]
3. Wiatrowski, A.; Obstarczyk, A.; Mazur, M.; Kaczmarek, D.; Wojcieszak, D. Characterization of HfO_2 Optical Coatings Deposited by MF Magnetron Sputtering. *Coatings* **2019**, *9*, 106. [CrossRef]
4. Khoshman, J.M.; Kordesch, M.E. Optical properties of a-HfO_2 thin films. *Surf. Coat. Tech.* **2006**, *201*, 3530–3535. [CrossRef]
5. Kumar, R.; Chauhan, V.; Koratkar, N.; Kumar, S.; Sharma, A.; Chae, K.-H.; Won, S.O. Influence of high energy ion irradiation on structural, morphological and optical properties of high-k dielectric hafnium oxide (HfO_2) thin films grown by atomic layer deposition. *J. All. Comp.* **2020**, *831*, 154698. [CrossRef]
6. Fahrenkopf, N.M.; Rice, P.Z.; Bergkvist, M.; Deskins, N.A.; Cady, N.C. Immobilization Mechanisms of Deoxyribonucleic Acid (DNA) to Hafnium Dioxide (HfO_2) Surfaces for Biosensing Applications. *ACS Appl. Mater. Interfaces* **2012**, *4*, 5360–5368. [CrossRef] [PubMed]
7. Kahro, T.; Tarre, A.; Käämbre, T.; Piirsoo, H.-M.; Kozlova, J.; Ritslaid, P.; Kasikov, A.; Jõgiaas, T.; Vinuesa, G.; Dueñas, S.; et al. Hafnium Oxide/Graphene/Hafnium Oxide-Stacked Nanostructures as Resistive Switching Media. *ACS Appl. Nano Mater.* **2021**, *4*, 5152–5163. [CrossRef]
8. Ahn, J.-H.; Lee, M.-J. Enhanced electrical properties of Hf-aluminate thin films by crystal structure modulation. *Mater. Lett.* **2015**, *157*, 215–218. [CrossRef]
9. Jung, J.-W.; Kim, G.-Y.; Lee, N.-W.; Ryu, W.-H. Low-temperature synthesis of tetragonal phase of hafnium oxide using polymer-blended nanofiber precursor. *Appl. Surf. Sci.* **2020**, *533*, 147496. [CrossRef]
10. Tang, Z.-X.; Tang, W.-W.; Tang, X.-G.; Liu, Q.-X.; Jiang, Y.-P.; Li, W.-H.; Tang, Z.-H.; Guo, X.-B.; Tang, Z.-F. Analog-type resistive switching behavior of Au/HfO_2/ZnO memristor fabricated on flexible Mica substrate. *Phys. E* **2020**, *120*, 114047. [CrossRef]
11. Lee, M.Y.P.Y.H.; Mikolajick, T.; Schroeder, U.; Hwang, C.S. Review and perspective on ferroelectric HfO2-based thin films for memory applications. *MRS Comm.* **2018**, *8*, 795–808. [CrossRef]
12. Ryu, T.-H.; Yoon, S.-J.; Na, S.-Y.; Yoon, S.-M. Crystallization annealing effects on ferroelectric properties of Al-doped HfO_2 thin film capacitors using indium-tin-oxide electrodes. *Curr. Appl. Phys.* **2019**, *19*, 1383–1390. [CrossRef]
13. Uthra, B.; Bang, V.; Anantha SS, V.; Agarwal, P.B. Effect of lanthanide dopants on stability of orthorhombic hafnium oxide for pyroelectric applications. *Mater. Lett.* **2022**, *317*, 132097. [CrossRef]
14. Sharma, B.; Thapa, A.; Sarkar, A. Ab-initio study of LD-HfO_2, Al_2O_3, La_2O_3 and h-BN for applications as dielectric in MTJ memory device. *Superlattices Microstruct.* **2021**, *150*, 106753. [CrossRef]
15. Liu, C.-F.; Tang, X.-G.; Guo, X.-B.; Liu, Q.-X.; Jiang, Y.-P.; Tang, Z.-H.; Li, W.-H. Photodiode characteristics of HfO_2 thin films prepared by magnetron sputtering. *Mater. Des.* **2020**, *188*, 108465. [CrossRef]

16. Nikolaidou, K.; Condelipes, P.G.M.; Caneira, C.R.F.; Krack, M.; Fontes, P.M.; Oliveira, H.M.; Kovačič, M.; Topič, M.; Chu, V.; Conde, J.P. Monolithically integrated optical interference and absorption filters on thin film amorphous silicon photosensors for biological detection. *Sensors Actuators B Chem.* **2022**, *356*, 131330. [CrossRef]
17. Cianci, E.; Lamperti, A.; Tallarida, G.; Zanuccoli, M.; Fiegna, C.; Lamagna, L.; Losa, S.; Rossini, S.; Vercesi, F.; Gatti, D.; et al. Advanced protective coatings for reflectivity enhancement by low temperature atomic layer deposition of HfO_2 on Al surfaces for micromirror applications. *Sensors Actuators A Phys.* **2018**, *282*, 124–131. [CrossRef]
18. Macleod, H.A. *Thin-Film Optical Filters*, 3rd ed.; Institute of Physics Publishing, Dirac House, Temple Back: Bristol, UK, 2017; ISBN 0-7503-0688-2.
19. Wan, Y.; Zhou, X. Formation mechanism of hafnium oxide nanoparticles by a hydrothermal route. *RSC Adv.* **2017**, *7*, 7763–7773. [CrossRef]
20. Park, J.-C.; Yoon, Y.-S.; Kang, S.-J. Structural and Optical Properties of HfO_2 Films on Sapphire Annealed in O_2 Ambient. *J. Korean Ceram. Soc.* **2016**, *53*, 563–567. [CrossRef]
21. Rammula, R.; Aarik, J.; Mändar, H.; Ritslaid, P.; Sammelselg, V. Atomic layer deposition of HfO_2: Effect of structure development on growth rate, morphology and optical properties of thin films. *Appl. Surf. Sci.* **2010**, *257*, 1043–1052. [CrossRef]
22. Aarik, J.; Sundqvist, J.; Aidla, A.; Lu, J.; Sajavaara, T.; Kukli, K.; Hårsta, A. Hafnium tetraiodide and oxygen as precursors for atomic layer deposition of hafnium oxide thin films. *Thin Solid Films* **2002**, *418*, 69–72. [CrossRef]
23. Böscke, T.S.; Teichert, S.; Bräuhaus, D.; Muller, J.; Schröder, U.; Bottger, U.; Mikolajick, T. Phase transitions in ferroelectric silicon doped hafnium oxide. *Appl. Phys. Lett.* **2011**, *99*, 112904. [CrossRef]
24. Liang, H.; Zhang, B.; Zhou, D.; Guo, X.; Li, Y.; Lu, Y.; Guo, Y. Effect of Y concentration and film thickness on microstructure and electrical properties of HfO_2 based thin films. *Ceram. Int.* **2021**, *47*, 12137–12143. [CrossRef]
25. Noh, Y.; Jung, M.; Yoon, J.; Hong, S.; Park, S.; Kang, B.S.; Ahn, S.-E. Switching Dynamics and modeling of multi-domain Zr-doped HfO_2 ferroelectric thin films. *Curr. Appl. Phys.* **2019**, *19*, 486–490. [CrossRef]
26. Tamm, A.; Kukli, K.; Niinistö, J.; Lu, J.; Ritala, M.; Leskelä, M. Properties of HfO_2 and HfO_2:Y films grown by atomic layer deposition in an advanced monocyclopentadienyl-based process. In *IOP Conference Series: Materials Science and Engineering*; IOP Publishing: Bristol, UK, 2010; Volume 8. [CrossRef]
27. Piirsoo, H.-M.; Jõgiaas, T.; Mändar, H.; Ritslaid, P.; Kukli, K.; Tamm, A. Microstructure and mechanical properties of atomic layer deposited alumina doped zirconia. *AIP Adv.* **2021**, *11*, 055316. [CrossRef]
28. Jõgiaas, T.; Kull, M.; Seemen, H.; Ritslaid, P.; Kukli, K.; Tamm, A. Optical and mechanical properties of nanolaminates of zirconium and hafnium oxides grown by atomic layer deposition. *J. Vac. Sci. Technol. A* **2020**, *38*, 022406. [CrossRef]
29. Aarik, J.; Mändar, H.; Kirm, M.; Pung, L. Optical characterization of HfO_2 thin films grown by atomic layer deposition. *Thin Solid Films* **2004**, *466*, 41–47. [CrossRef]
30. Berdova, M.; Liu, X.; Wiemer, C.; Lamperti, A.; Tallarida, G.; Cianci, E.; Fanciulli, M.; Franssila, S. Hardness, elastic modulus, and wear resistance of hafnium oxide-based films grown by atomic layer deposition. *J. Vac. Sci. Technol. A Vac. Surf. Films* **2016**, *34*, 051510. [CrossRef]
31. Heil, S.B.S.; Hemmen, J.L.; Hodson, C.J.; Singh, N.; Klootwijk, J.H.; Roozeboom, F.; van de Sanden, M.C.M.; Kessels, W.M. Deposition of TiN and HfO_2 in a commercial 200 mm remote plasma atomic layer deposition reactor. *J. Vac. Sci. Technol. A* **2007**, *25*, 1357. [CrossRef]
32. Lauria, A.; Villa, I.; Fasoli, M.; Niederberger, M.; Vedda, A. Multifunctional Role of Rare Earth Doping in Optical Materials: Nonaqueous Sol-Gel Synthesis of Stabilized Cubic HfO_2 Luminescent Nanoparticles. *ACS Nano* **2013**, *7*, 7041–7052. [CrossRef]
33. Tapily, K.; Jakes, J.E.; Stone, D.S.; Shrestha, P.; Gu, D.; Baumgart, H.; Elmustafa, A.A. Nanoindentation Investigation of HfO_2 and Al_2O_3 Films Grown by Atomic Layer Deposition. *J. Electrochem. Soc.* **2008**, *155*, H545–H551. [CrossRef]
34. Jõgiaas, T.; Zabels, R.; Tamm, A.; Merisalu, M.; Hussainova, I.; Heikkilä, M.; Mändar, H.; Kukli, K.; Ritala, M.; Leskelä, M. Mechanical properties of aluminum, zirconium, hafnium and tantalum oxides and their nanolaminates grown by atomic layer deposition. *Surf. Coat. Technol.* **2015**, *282*, 36–42. [CrossRef]
35. Jõgiaas, T.; Zabels, R.; Tarre, A.; Tamm, A. Hardness and modulus of elasticity of atomic layer deposited Al_2O_3-ZrO_2 nanolaminates and mixtures. *Mater. Chem. Phys.* **2020**, *240*, 122270. [CrossRef]
36. Meng, Q.; Huang, G.; Lai, J.; Fan, D.; Zhang, K. Fabrication of 128 $\times$ 128 MEMS tunable F-P cavity optical filter array with surface micromachining. *Infrared Phys. Technol.* **2020**, *105*, 103199. [CrossRef]
37. Liu, Y.; Xu, J.; Zhong, S.; Wu, Y. Large size MEMS scanning mirror with vertical comb drive for tunable optical filter. *Opt. Lasers Eng.* **2013**, *51*, 54–60. [CrossRef]
38. Fujiwara, H. *Spectroscopic Ellipsometry: Principles and Applications*; John Wiley & Sons Ltd.: Hoboken, NJ, USA, 2007; ISBN 978-0-470-01608-4/0-470-01608-6.
39. Lin, Z.; Zhu, M.; Song, C.; Liu, T.; Yin, C.; Zeng, T.; Shao, J. Effect of annealing on the properties of plasma-enhanced atomic layer deposition grown HfO_2 coatings for ultraviolet laser applications. *J. Alloy. Compd.* **2023**, *946*, 169443. [CrossRef]
40. Martínez-Puente, M.; Horley, P.; Aguirre-Tostado, F.; López-Medina, J.; Borbón-Nuñez, H.; Tiznado, H.; Susarrey-Arce, A.; Martínez-Guerra, E. ALD and PEALD deposition of HfO_2 and its effects on the nature of oxygen vacancies. *J. Mater. Sci. Eng. B* **2022**, *285*, 115964. [CrossRef]
41. Shukor, A.H.; Alhattab, H.A.; Takano, I. Electrical and optical properties of copper oxide thin films prepared by DC magnetron sputtering. *J. Vac. Sci. Technol. B* **2020**, *38*, 012803. [CrossRef]

42. Pierson, J.F.; Thobor-Keck, A.; Billard, A. Cuprite, paramelaconite and tenorite films deposited by reactive magnetron sputtering. *Appl. Surf. Sci.* **2003**, *210*, 359–367. [CrossRef]
43. Hildebrandt, E.; Kurian, J.; Müller, M.M.; Schroeder, T.; Kleebe, H.-J.; Alff, L. Controlled oxygen vacancy induced p-type conductivity in HfO_{2-x} thin films. *Appl. Phys. Lett.* **2011**, *99*, 112902. [CrossRef]
44. Sun, L.; Jones, J.G.; Grant, J.T.; Murphy, N R.; Ramana, C.V.; Eyink, K.G.; Vernon, J.P.; Stevenson, P.R. Nanoscale Thick Thin Films of High-Density HfO_2 for Bulk-like Optical Responses. *ACS Appl. Nano Mater.* **2021**, *4*, 10836–10844. [CrossRef]
45. Chen, L.; Yeap, K.B.; She, C.M.; Liu, G.R. A computational and experimental investigation of three-dimensional micro-wedge indentation-induced interfacial delamination in a soft-film-on-hard-substrate system. *Eng. Struct.* **2011**, *33*, 3269–3278. [CrossRef]
46. Xiao, Y.; Shi, W.; Luo, J. Indentation for evaluating cracking and delamination of thin coatings using finite element analysis. *Vacuum* **2015**, *122*, 17–30. [CrossRef]
47. She, C.; Zhang, Y.-W.; Zeng, K.-Y. A three-dimensional finite element analysis of interface delamination in ductile film/hard substrate system induced by wedge indentation. *Eng. Fract. Mech.* **2009**, *76*, 2272–2280. [CrossRef]
48. Li, W.; Siegmund, T. An analysis of a size effect in indentation delamination of a ductile film on an elastic substrate. *Scr. Mater.* **2003**, *49*, 497–502. [CrossRef]
49. Hunger, A.; Carl, G.; Gebhardt, A.; Rüssel, C. Ultra-high thermal expansion glass-ceramics in the system $MgO/Al_2O_3/TiO_2/ZrO_2/SiO_2$ by volume crystallization of cristobalite. *J. Non-Cryst. Solids* **2008**, *354*, 5402–5407. [CrossRef]
50. Hunger, A.; Carl, G.; Rüssel, C. Young's moduli and microhardness of glass-ceramics in the system $MgO/Al_2O_3/TiO_2/ZrO_2/SiO_2$ containing quartz nanocrystals. *Mater. Chem. Phys.* **2010**, *122*, 502–506. [CrossRef]
51. Hunger, A.; Carl, G.; Rüssel, C. Formation of nano-crystalline quartz crystals from $ZnO/Al_2O_3/TiO_2/ZrO_2/SiO_2$ glasses. *Solid State Sci.* **2010**, *12*, 1570–1574. [CrossRef]
52. Shinozaki, K.; Honma, T.; Komatsu, T. Elastic properties and Vickers hardness of optically transparent glass-ceramics with fresnoite $Ba_2TiSi_2O_8$ nanocrystals. *Mater. Res. Bull.* **2011**, *46*, 922–928. [CrossRef]
53. Serbena, F.C.; Mathias, I.; Foerster, C.E.; Zanotto, E.D. Crystallization toughening of a model glass-ceramic. *Acta Mater.* **2015**, *86*, 216–228. [CrossRef]
54. Pinto, H.; Ito, L.; Crovace, M.; Ferreira, E.B.; Fauth, F.; Wroblewski, T.; Zanotto, E.; Pyzalla, A. Surface and bulk residual stresses in Li_2O-SiO_2 glass-ceramics. *J. Non-Cryst. Solids* **2007**, *353*, 2307–2317. [CrossRef]
55. Liu, X.; Zhou, J.; Zhou, S.; Yue, Y.; Qiu, J. Transparent glass-ceramics functionalized by dispersed crystals. *Prog. Mater. Sci.* **2018**, *97*, 38–96. [CrossRef]

nanomaterials

Article

UV-Excited Luminescence in Porous Organosilica Films with Various Organic Components

Md Rasadujjaman [1,2,*], Jinming Zhang [1], Dmitry A. Spassky [3,4], Sergej Naumov [5], Alexey S. Vishnevskiy [6], Konstantin A. Vorotilov [6], Jiang Yan [1], Jing Zhang [1,*] and Mikhail R. Baklanov [1,6,7]

1 Department of Microelectronics, North China University of Technology, Beijing 100144, China; jinming@naura.com (J.Z.); jiangyan@ncut.edu.cn (J.Y.); baklanovmr@gmail.com (M.R.B.)
2 Department of Physics, Mawlana Bhashani Science and Technology University, Santosh, Tangail 1902, Bangladesh
3 Skobeltsyn Institute of Nuclear Physics, Lomonosov Moscow State University, Moscow 119991, Russia; deris2002@mail.ru
4 Institute of Physics, University of Tartu, 50411 Tartu, Estonia
5 Leibniz Institute of Surface Engineering (IOM), 04318 Leipzig, Germany; sergej.naumov@iom-leipzig.de
6 Research and Education Center "Technological Center", MIREA—Russian Technological University (RTU MIREA), Moscow 119454, Russia; alexeysw@mail.ru (A.S.V.); vorotilov@live.ru (K.A.V.)
7 European Centre for Knowledge and Technology Transfer (EUROTEX), 1040 Brussels, Belgium
* Correspondence: rasadphy@mbstu.ac.bd (M.R.); zhangj@ncut.edu.cn (J.Z.)

Abstract: UV-induced photoluminescence of organosilica films with ethylene and benzene bridging groups in their matrix and terminal methyl groups on the pore wall surface was studied to reveal optically active defects and understand their origin and nature. The careful selection of the film's precursors and conditions of deposition and curing and analysis of chemical and structural properties led to the conclusion that luminescence sources are not associated with the presence of oxygen-deficient centers, as in the case of pure SiO_2. It is shown that the sources of luminescence are the carbon-containing components that are part of the low-k-matrix, as well as the carbon residues formed upon removal of the template and UV-induced destruction of organosilica samples. A good correlation between the energy of the photoluminescence peaks and the chemical composition is observed. This correlation is confirmed by the results obtained by the Density Functional theory. The photoluminescence intensity increases with porosity and internal surface area. The spectra become more complicated after annealing at 400 °C, although Fourier transform infrared spectroscopy does not show these changes. The appearance of additional bands is associated with the compaction of the low-k matrix and the segregation of template residues on the surface of the pore wall.

Keywords: low-k dielectrics; organosilica glass; interconnects; photoluminescence; oxygen deficient centers

Citation: Rasadujjaman, M.; Zhang, J.; Spassky, D.A.; Naumov, S.; Vishnevskiy, A.S.; Vorotilov, K.A.; Yan, J.; Zhang, J.; Baklanov, M.R. UV-Excited Luminescence in Porous Organosilica Films with Various Organic Components. *Nanomaterials* **2023**, *13*, 1419. https://doi.org/10.3390/nano13081419

Academic Editors: Yann Molard and Wojciech Pisarski

Received: 19 March 2023
Revised: 6 April 2023
Accepted: 14 April 2023
Published: 20 April 2023

1. Introduction

Porous organosilica films have many different applications, from catalysis and drug and gene delivery to microelectronics [1]. One of their most economically significant applications is with ultra-large-scale integration (ULSI) devices where porous low-dielectric constant (low-k) materials are used to reduce signal propagation delay in metallization wires [2]. However, the integration of porous low-k materials with metal wires faces numerous challenges, and the most important problems are related to the degradation of their dielectric properties and reliability. The most studied factors that degrade leakage current and reliability are related to structural modifications and adsorbed moisture in "plasma-damaged" low-k dielectrics and hydrocarbon residues formed during sacrificial porogen removal [3–15]. Photon irradiation increases the intrinsic defect density and creates trapped charges inside the low-k material, which can lead to reliability issues [16,17]. The knocking-off of atoms from the low-k material network can also occur during the ion sputtering process leading to the formation of Si vacancies such as EX centers or dangling carbon bonds, where the carbon-related defects contribute to higher leakage [18]. In addition, the formation of surface oxygen vacancies, probably due to the removal of terminal organic groups after Ar^+ sputtering, leads to the formation of sub-gap surface states at 5.0 and 7.2 eV [18]. Atomic defects such as non-bridging oxygen hole centers (NBOHC) and oxygen vacancies (E′centers) have been studied by electron spin resonance (ESR) spectroscopy, and the results of these studies of low-k materials are discussed and summarized in Refs. [17,19,20]. Recently, the UV-induced photoluminescence (PL) of a mesoporous organosilica low-k dielectric was studied [21], and it was concluded that the formation of oxygen-deficient centers (ODCs), ODC(I) (≡Si–Si≡) and ODC(II) (=Si:) centers, similar to those observed in pure SiO_2 [22,23] can explain the leakage current mechanism studied in Refs. [24,25]. However, it is not always easy to distinguish between the influence of atomic defects and residual carbon on the critical properties of organosilicate glass (OSG) low-k dielectrics.

So far, most defect studies have been performed with so-called methyl-terminated organosilica glasses: silica-like materials, where some oxygen bridging atoms in the silica matrix are replaced by two methyl groups (≡Si–O–Si≡ → ≡Si–CH_3 … CH_3–Si≡). This reduces the matrix density but makes the films sufficiently hydrophobic. These materials are mainly deposited by plasma-enhanced chemical vapor deposition (PECVD) [26]. However, the need to improve the mechanical properties and reliability of low-k dielectrics stimulated the extensive study of materials with different types of carbon bridges between silicon atoms [27–34]. Replacing the oxygen bridge with carbon allows for a significant improvement in mechanical properties due to the higher bending rigidity of the ≡Si–C–Si≡ bonds compared to the ≡Si–O–Si≡ bonds [31]. Evaporation-induced self-assembly (EISA) [35] using carbon-bridged alkoxysilane precursors has been shown to produce periodic mesoporous organosilica (PMO) with ordered porosity and hydrocarbon bridges in the film matrix are formed. Their properties, including the thermal and chemical resistance of different carbon bridges, have been extensively evaluated [36–38].

The present research aims to study the origin of ultra-violet (UV)-induced PL in porous organosilica films containing various organic components. The films deposited had a well-defined chemical composition and porosity. Using materials with different and controlled compositions and porosities (Figure 1), we hoped to understand the physical nature of the optically and electrically active defects.

Three different types of organosilica glasses were used in this research. The first type (1-1, 1-2, 2-2, 2-3) includes periodic mesoporous organosilica with ethylene and 1,4-benzene bridges in their matrix. The second type (2-1) film includes a recently proposed "hyperconnected structure" based on 1,3,5- and 1,3-benzene bridges [33]. The third type of film (3-1, 3-2) has a "classic" low-k structure with methyl terminal groups and random porosity [2,3]. The conditions of their preparation and properties are described in the experimental part and in the Supplementary Materials, and even more detailed information can be found in the original publications cited.

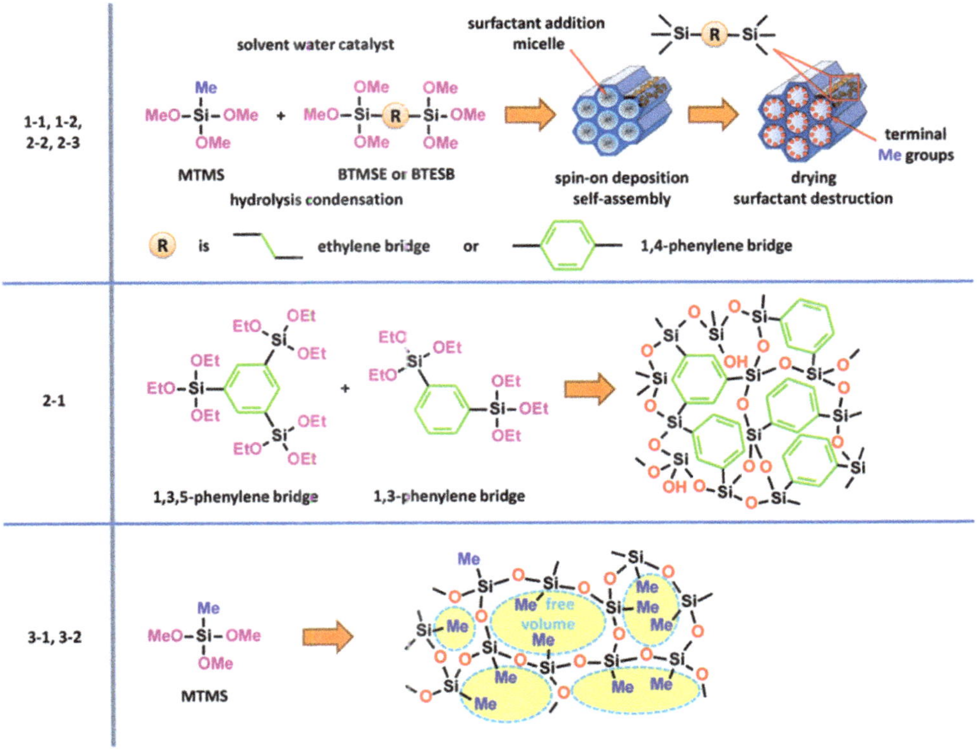

Figure 1. Schematic presentation of the samples used in this research where, Me is a methyl group (–OCH_3), OMe is a methoxy group, OEt is an ethoxy group (–OCH_2CH_3), MTMS is methyltrimethoxysilane, BTMSE is 1,2-bis(trimethoxysilyl)ethane, BTESB is 1,4-bis(triethoxysilyl)benzene. First type samples (1-1, 1-2, 2-2 and 2-3): Ethylene and 1,4-benzene bridged PMO materials with ordered porosity and methyl terminal groups on pore wall surface. Second type sample (2-1): Hybrid OSG films containing both 1,3,7-benzene and 1,3-benzene bridges. Third type samples (3-1, 3-2): methyl terminated OSG materials.

2. Materials and Methods

2.1. Material Preparation

i. The first type of sample includes ethylene-bridged PMO materials (Figure 1). The precursor solutions were prepared by co-hydrolysis of alkylenealkoxysilane—1,2-bis(triethoxysilyl)methane (BTMSE) with alkylalkoxysiloxane—methyltrimethoxysilane (MTMS) (Fluka, Buchs, Switzerland) under acidic conditions as described in detail in Refs. [39,40]. Due to the presence of both BTMSE and MTMS, the film contains both ethylene bridges and methyl terminal groups (Figure 1). This approach is used for low-k film preparation since the materials containing only a carbon bridge do not have sufficient hydrophobicity. Conversely, materials with only terminal methyl groups have insufficiently good mechanical properties due to the reduction in matrix connectivity. All films were deposited on Si (100) wafers, and film preparation was completed by a two-step annealing process: (1) soft-bake at 150 °C for 30 min on a hot plate to remove the solvent, (2) hard-bake at 400 °C for 30 min in a dry air oven. After deposition and curing, the wafers were cut for analysis by different techniques.

To prepare the films with different porosity but the same matrix composition, the matrix solution was divided into four parts, and different amounts of Brij®30 surfactant (Sigma-Aldrich, St. Louis, MO, USA) were added and stirred. The added surfactant acts

tion index (RI) were measured with a spectroscopic ellipsometer SE850 (Sentech, Berlin, Germany) (λ = 300–800 nm) using the Cauchy polynomial function. The porosity and pore size distribution of the films were characterized by atmospheric pressure EP [45]. Isopropyl alcohol (IPA) vapors (and heptane vapors, in some cases) diluted with dry N_2 carrier gas in a specially designed bubbler were used as an adsorptive. The open porosity of the films is calculated as the volume of adsorbed liquid adsorbate from RI values measured during IPA adsorption using a modified Lorentz–Lorenz equation [45]:

$$\frac{n_{eff}^2 - 1}{n_{eff}^2 + 2} = V\frac{n_{ads}^2 - 1}{n_{ads}^2 + 2} + (1 - V)\frac{n_s^2 - 1}{n_s^2 + 2} \quad (3)$$

where n_{eff}—is the measured RI of the porous film when the pores are gradually filled by adsorbate at different relative pressures, n_{ads}—is the RI of the liquid adsorbate, n_s—is the matrix RI, and V—is the volume of the condensed adsorbate. The calculation of the pore size distribution (PSD) is based on the analysis of adsorption isotherms. The analysis is based on the Kelvin equation, which describes the dependence of relative pressure (P/P_0) on meniscus curvature similar to the standard Barrett–Joyner–Halenda (BJH) procedure used in nitrogen adsorption porosimetry [46]. The micropore size uses the Dubinin–Radushkevich approach based on Polanyi's potential theory of adsorption. EP also allows us to calculate the so-called cumulative surface area. The specific surface area of each small group of pores dA_i are calculated from the corresponding pore volume and pore radius as $dA_i = cV_i/r_i$ (for cylindrical pores). A value of the cumulative surface area is obtained by assuming the values of dA_i over the whole pore system.

UV-induced luminescence of samples with ethylene bridge (1-1 and 1-2) and 1,3,5- and 1,3-benzene bridge (2-1) samples were measured on a FP-8300 spectrofluorometer (JASCO, Tokyo, Japan) using a continuous output Xe arc lamp with shielded lamp housing (150 W) and holographic concave grating in a modified Rowland mount monochromator. The radio-photometer system using monochromatic light was used to monitor the output intensity of the Xe lamp. The samples were mounted in a standard 10 mm rectangular cell holder SCE-846/D061161450 provided by JASCO. The wavelength accuracy and maximum resolution are 1 nm. The excitation and emission spectra are in the range in the energy range from 6.2 to 1.65 eV and the slit width is 5 nm–5 nm. Measurements were performed at room temperature and fully controlled using a Spectra Manager. The excitation and emission bandwidth were 5 nm at a scan speed of 1000 nm/min.

The methyl-terminated (3-1, 3-2) and 1,4-benzene-bridged samples (2-2 and 2-2) were measured at excitation energies of 6 to 10 eV using different systems. Luminescence and luminescence excitation spectra were measured with the photoluminescence end station of the FinEstBeAMS beamline at the 1.5 GeV storage ring of the MAX IV synchrotron facility [47]. The luminescence excitation spectra were measured with a spectral resolution of no less than 4 meV using fused silica and MgF_2 optical filters in the energy range of 4.5–7.0 eV and 5.5–11 eV, respectively. The samples were placed in an ARS closed-cycle helium cryostat equipped with a LakeShore 325 temperature controller, and the measurement temperature was equal to 7K. Before measurements, the samples were degassed at 350K in a vacuum of 10^{-9} mbar. An excitation flux curve obtained using a factory-calibrated AXUV-100G diode (OptoDiode Corp, Camarillo, CA, USA) was used to correct the excitation spectra. The luminescence spectra were recorded using a fiber-coupled Andor Shamrock SR-303i (Andor Technology Ltd., Belfast, UK) spectrometer equipped with a Hamamatsu H3259-01 photon counting head. The luminescence spectra were corrected for the spectral sensitivity of the registration channel. It is necessary to mention that by using photons with higher energy, we confirmed that we are not missing transitions that cannot be excited by photons with an energy of 6.2 eV. Cryogenic temperature reduces thermal noise and, therefore, should allow to detection of PL peaks of low intensity if they are present. The ability to generate detailed PLE spectra helps to better understand the origin of PL transitions. Using this system, we found a correspondence between the results obtained in these systems, which makes our conclusions more reliable.

It is necessary to mention that all other characteristics of the films, like dielectric constant, mechanical properties, surface roughness, pore ordering, plasma and VUV properties, have also been studied by using different instrumentations. These data are not discussed in this work but can be found in the references [36–40,48,49].

2.3. Calculation of the Energy Diagram of UV-Induced Processes

Density Functional Theory (DFT) calculations were carried out systematically, employing the PBE0 density functional [50–52]. The way in which the PBE0 functional is derived and the lack of empirical parameters fitted to specific properties make the PBE0 model a widely applicable method for quantum chemical calculations. The molecular geometries, energies and electronic structure of the molecules were studied at the PBE0/6-31G** level of theory, as implemented in the Jaguar 9.6 program [53]. This computational model has already been used successfully for calculations in our previous work [54,55]. Frequency calculations were performed at the same level of theory to obtain the total enthalpy (H) and Gibbs free energy (G) at a standard temperature of 298.15K using unscaled frequencies. The reaction enthalpies (ΔH) and Gibbs free energies of the reaction (ΔG) of the studied molecules were calculated as the difference in calculated H and G between the reactants and products, respectively. The energy of the excited states and oscillator strength (f) of transitions were calculated in the gas phase at the optimized ground state structure of model molecules using the time-dependent (TD) DFT method [56] at the PBE0/6-31G** level of theory. The excited states calculations were performed using the Full Linear Response (FLR) approximation [57], as implemented in the Jaguar 9.6 program. The number of excited states was set to 100 for two reasons: the first is that the initial guess might not accurately reflect the final states; the second is to ensure that the near-degeneracies are accounted for.

3. Results

3.1. Ethylene-Bridged PMO [39,40]

The FTIR spectra of fully cured ethylene-bridged PMO films (1-1, 1-2) are shown in Figure 2. The absorption band at 1250–1000 cm^{-1} corresponds to Si–O stretching modes. The soft baked (SB) at 150 °C films show the presence of hydroxyl and silanol groups (O–H, Si–OH) at 3700–3100 cm^{-1} and C=O group at 1750 cm^{-1}, which are removed after the hard bake (HB). The Si–OH concentration in SB films is higher in the films deposited without or with a small concentration of porogens. The possible reason is that the remaining porogen makes the films more hydrophobic. It can be seen that SB films still contain a significant amount of C–H_x (x = 2 or 3) groups in the wavenumber range (3000–2800 cm^{-1}) and C–H (1460 cm^{-1}). They mainly originate from the template (Brij®30), and this is the reason why their concentration increases with initial porogen concentration (Figure 2b). HB reduces those group concentrations (Figure 2d). The presence of Si–CH_3 terminal groups can be seen from the peak at ~1275 cm^{-1}, and Si–CH_3 groups also contribute to the intensity of the hydrocarbon peaks at ~2970 and ~2920 cm^{-1}.

Figure 3 shows the adsorption–desorption isotherms of heptane vapors measured by EP. The isotherms of all samples deposited with 0–30 wt% template do not have a hysteresis loop, indicating that the pores are cylindrical in shape. However, the samples deposited with a 50 wt% template have pronounced hysteresis loops typical for the formation of internal voids leading to "ink bottle"-like effects in isotherms [58].

A clear difference can be seen between the samples deposited with a BMTSE/MTMS ratios of 47/53 and 25/75. The isotherms in samples with a BTMSE/MTMS ratio of 25/75 have critical slopes at higher relative pressures P/P_0, indicating a larger pore size. They also have higher porosity than the samples deposited with a BTMSE/MTMS ratio of 47/53, and this difference becomes even more pronounced for the samples deposited with 30 and 50 wt% templates (Figure 3). A reasonable explanation is that the ethylene bridge increases the stiffness of the matrix and hinders the agglomeration of template molecules during matrix formation. However, no principal differences in the photoluminescence spectra are observed between 47/53 and 25/75 samples (Figure 4).

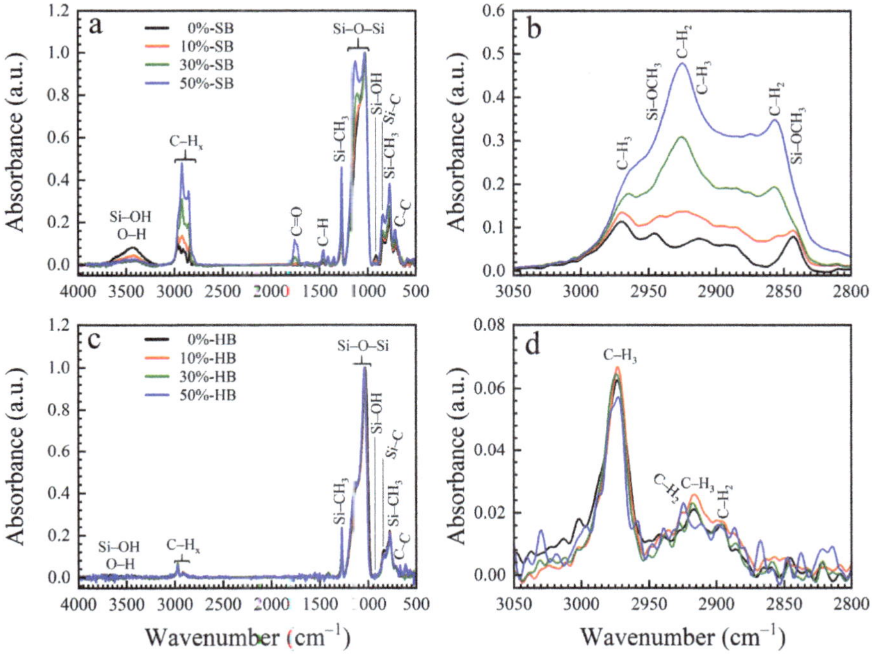

Figure 2. FTIR spectra of OSG low-k samples deposited with a BTMSE/MTMS precursor ratio of 25/75 and different porogen concentrations (0–50 wt%) after soft bake at 150 °C (SB) in the air (**a**,**b**) and hard bake at 400 °C (HB) in the air (**c**,**d**). FTIR spectra of the samples deposited with a BMTSE/MTMS ratio of 47/53 are qualitatively similar.

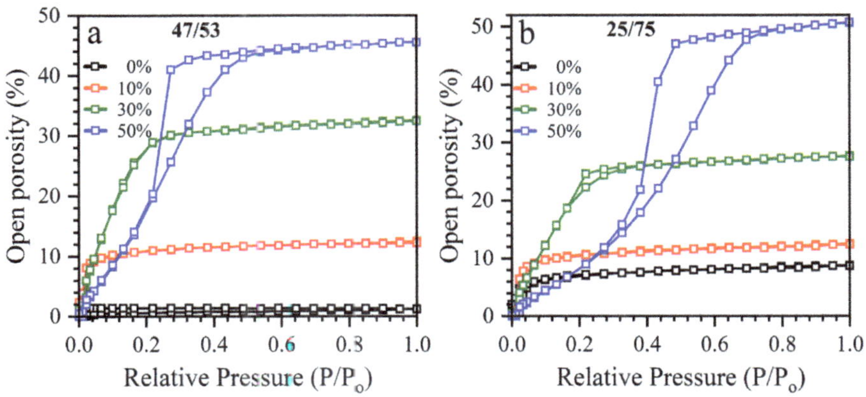

Figure 3. Adsorption–desorption isotherms of heptane vapor in the OSG low-k films deposited with BMTSE/MTMS ratio of 47/53 (**a**) and 25/75 (**b**) and different porogen concentrations (0–50 wt%).

The measured luminescence spectra clearly show three peaks located near 4.3, 3.3, and 2.9 eV. The position of these peaks is the same as in the paper [21], where the films of the same series (BTMSE/MTMS = 25/75 ratio, porosity 30%) were studied upon excitation by synchrotron radiation. The only differences relate to the relative intensities of the peaks, but it is likely a sample storage issue. The peaks interpretation in Ref. [21] is based on information reported for SiO_2. According to [22,23], the peaks at 4.3 and 3.2 eV are often observed in SiO_2 and can correspond to ODC(I) defects (≡Si–Si≡), and the peak at 2.9 eV was interpreted as emission from ODC(II) defect (=Si:).

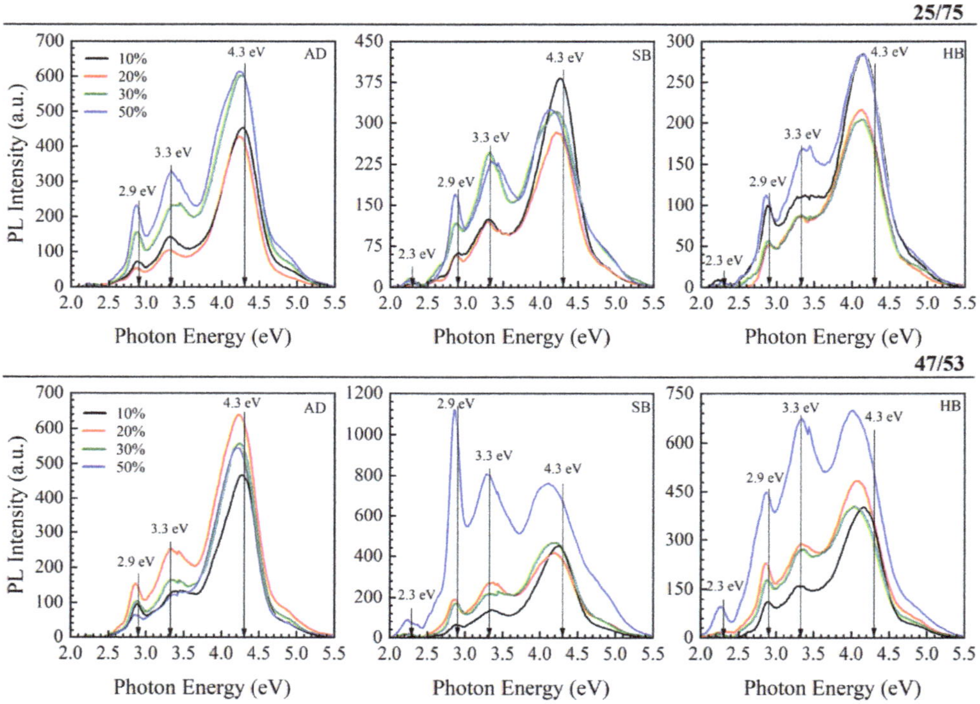

Figure 4. UV-induced room temperature luminescence of PMO low-k films containing both methyl terminal and ethylene bridging groups. The films have different porosity and were cured in air. The samples were deposited with a BTMSE/MTMS ratio of 25/75 (top line) and 47/53 (bottom line) with different porogen concentrations (10–50 wt%) as deposited (AD) after soft bake at 150 °C (SB) in air and hard bake at 400 °C (HB) in air, upon excitation with light of 6.2 eV. The data are reproduced from the paper "Effect of methyl terminal and ethylene bridging groups on porous organosilicate glass films: FTIR, ellipsometric porosimetry, luminescence dataset" by Md Rasadujjaman, J. Zhang, K. P. Mogilnikov, A. S. Vishnevskiy, J. Zhang, M. R. Baklanov (Data in Brief 35 (2021) 106895) with granted permission from ELSEVIER [40].

In our case, the positions of the 2.9 and 3.2 eV peaks are constant in all three types (AD, SB, HB) of films but the position of 4.3 eV changes during curing. This peak corresponds to 4.3 eV in as-deposited (AD) films and then decreases to ~4 eV in SB samples at 150 °C and HB samples at 400 °C. This shift is too significant to be assigned to only ODC(I). The shift of the 4.3 eV peak, in principle, can be interpreted as the result of the formation of additional peaks at $E < 4$ eV, which can be assigned to specific carbon-containing residues [59–63]. In the general case, the formation of (≡Si–Si≡) type defects should occur due to the destruction of certain bonds of Si atoms. It can be assumed that they can be formed due to the destruction of carbon bridges, which have the lowest thermal stability in this system. However, the –Si–$(CH_2)_2$–Si– bonds are sufficiently stable and certainly cannot be destructed in AD and SB films ($T < 200$ °C). Therefore, the assignment of this peak to ODC(I) (Ref. [21]) is doubtable. The peak at 2.9 eV in BTMSE/MTMS = 47/53 sample increases drastically during SB and remains quite high after HB. It also generates doubt on the assignment to ODC(II) since, according to Ref. [22], ODC(I) is typically much more abundant than ODC(II) because ODC(II) might be a product of a partial transformation of ODC(I).

Effect of Storage and ICP Oxygen Plasma

The evolution of PL intensity in a sample with a BTMSE/MTMS ratio of 45/53 (HB) is shown after storage in air in a clean room environment (Figure 5a). The PL intensity gradually increased every week and finally increased ~10 times after 1 month. Figure 5b shows that the first PL measurement of a sample stored after one month reduces the PL intensity by about 10%. The following measurements, one after another, with a break of a few minutes, slightly reduce the intensity.

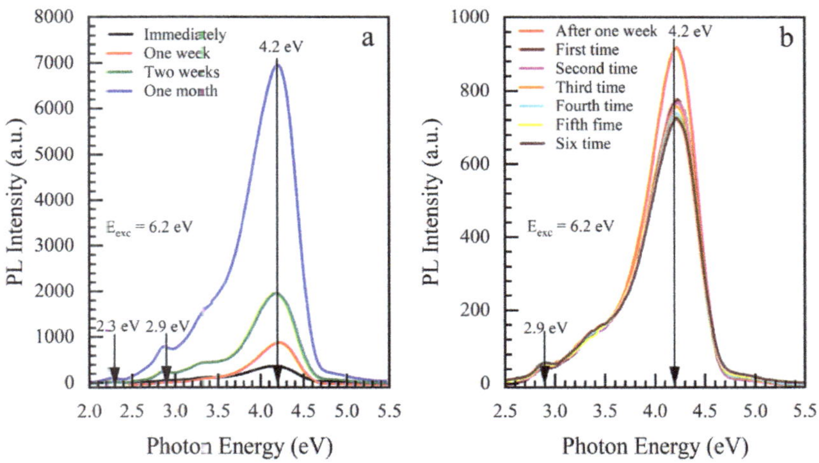

Figure 5. (**a**) Effect of storage in a clean room environment on the PL intensity for the OSG films deposited with a BTMSE/MTMS ratio of 47/53 and (**b**) PL measurements one after the another with a break of a few minutes.

The corresponding change in chemical composition during sample storage can be seen in FTIR spectra (Figure 6a). Air storage accumulates hydrocarbon residues from the clean room environment. Such an effect is well known for porous low-k materials: capillary forces enhance the adsorption and condensation of hydrocarbons and other residues. The observed decrease in PL intensity in air-stored samples after the first measurement may be related to the UV-induced desorption of part of the carbon residue.

When this sample is exposed to soft oxygen inductively coupled (ICP) plasma (Figure 6b), the concentration of the hydrocarbon-containing species is reduced. At the same time, also some loss of Si–CH_3 groups and hydrophilization is also observed (see Supplementary Materials Figure S1). The thickness of the low-k film before exposure to O_2 ICP plasma was equal to 466 nm; then, it reduced to 435 nm after 5 min and 399 nm after 10 min. Meanwhile, the corresponding reductions of Si–CH_3/Si–O–Si peaks ratio were from 0.173 to 0.163 and to 0.154, respectively. The changes of hydrocarbon residues peaks in the region 2800–3000 cm^{-1} were about three times larger. These observations suggest that the used plasma conditions were sufficiently soft to minimize plasma damage of the low-k matrix (a small change in Si–CH_3/Si–O–Si ratio), and the ICP-generated oxygen radicals are mainly consumed on the accumulated carbon residue.

Reducing the carbon concentration leads to a reduction in PL intensity at 4.2 eV (Figure 7). The change happening in the PL intensity during storage and exposure to O_2 ICP plasma clearly shows that the intensity of the peak near 4.1–4.2 eV is related to hydrocarbon residues accumulated during the template removal (pristine sample) and storage in air.

PL of hydrogenated carbon has been extensively studied in the past. According to these publications, in addition to the well-known broad PL bands at 2.1–2.33, 2.85–2.92 eV, and 3.17–3.22, other peaks located at 3.64–3.70, 3.93–4.01 and 4.34–4.56 eV in the UV region can also be found [64]. The observed change in PL spectra in the carbon-rich films deposited with high template concentration (50 wt%) and high BTMSE/MTMS ratio (Figure 4) and the increase in the intensity of certain peaks during thermal curing and exposure in O_2 ICP plasma most probably reflects the modification of carbon-containing components of the studied films.

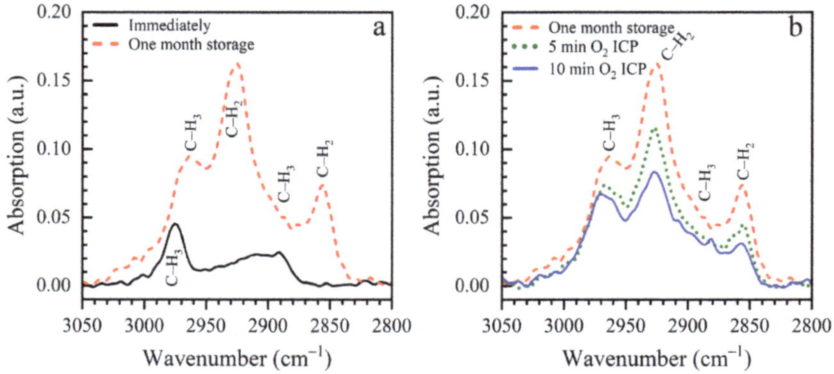

Figure 6. Accumulation of hydrocarbon residues from a clean room environment (**a**) and removal of these residues by ICP oxygen plasma (**b**).

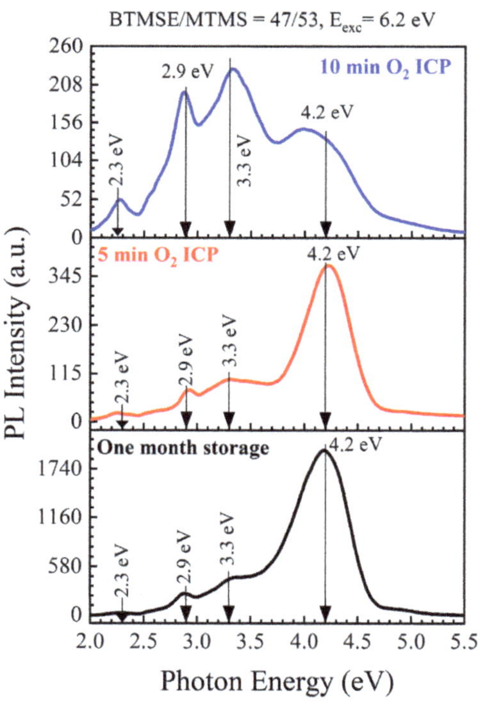

Figure 7. Reduction of 4.2 eV PL intensity during exposure to ICP oxygen plasma. The excitation energy of the photon was 6.2 eV.

3.2. Benzene-Bridged Organosilica Films

3.2.1. 1,3,5- and 1,3-Benzene-bridged Films [33,43,44]

The films with different molar ratios of 1,3,5-tris(triethoxysilyl)benzene to 1,3-bis(triethoxysilyl)benzene bridging organic groups (1:3 and 1:7) were deposited by spin-on coating followed by a soft bake in the air at 150 °C (SB) and hard bake in the air at 400 °C (HB). The concentrations of the non-ionic template (Brij®30) varied from 0 to 41 wt%. The chemical composition of the matrix of the films was evaluated and discussed, as well as refractive indices, mechanical properties, k-values, porosity and pore structure [43,44]. The films containing benzene bridging groups keep the pore size constant and equal to 0.81 nm while their porosity changes from 0 to 30%. The films containing a benzene bridge have a higher Young's modulus than plasma-enhanced chemical vapor deposition (PECVD) methyl-terminated low-k films with the same porosity [33]. The fabricated films show good stability after a long time of storage. FTIR spectra and porosity data generated by ellipsometric porosimetry are presented in the SI as Figures S2 and S3.

Figure 8 shows photoluminescence spectra for 1:3 and 1:7 samples. This means that the 1:3 sample has a 2.3 times higher concentration of 1,3,5-benzene rings than the 1:7 sample. The most important difference from ethylene-bridged films is that AD and SB films contain only 1 pronounced peak at around 3.9 eV. Looking at the PL spectra of samples AD and SB, one can conclude that the intensity increases with increasing porosity. The PL spectra of HB films are becoming more complicated, and new peaks near 2.3, 2.9 and 4.85 eV are observed.

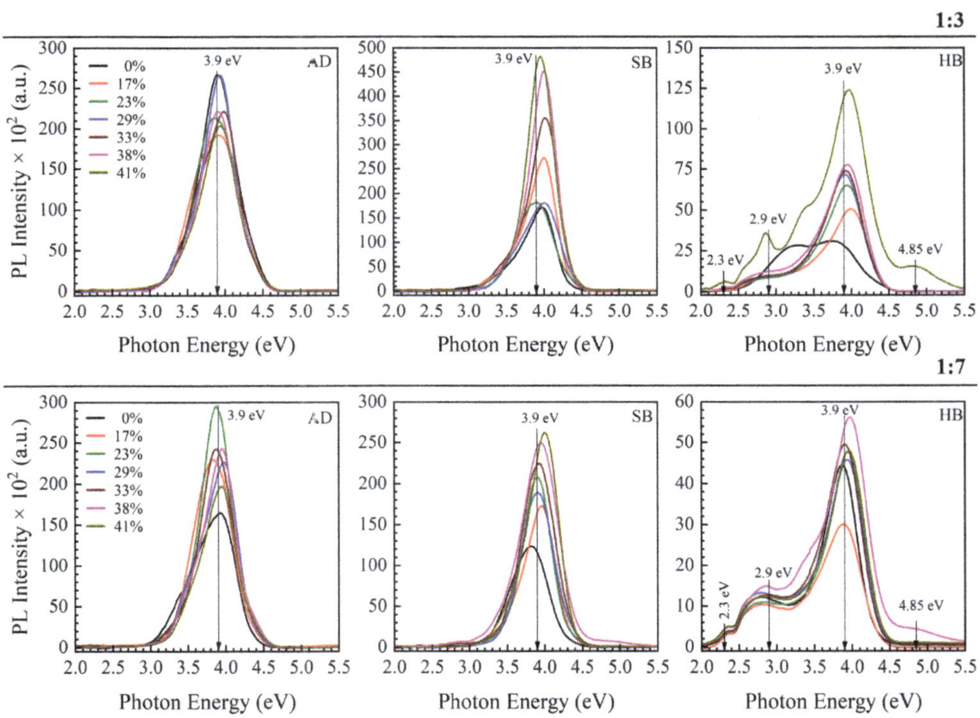

Figure 8. Photoluminescence spectra of OSG films containing 1,3,5- and 1,3-benzene bridges (samples 2-1 in Figure 1) and different porosity upon excitation with light of 6.2 eV. The porogen concentration varied from 0 to 41 wt%. The ratio of 1,3,5- and 1,3-benzene bridges is equal to 1:3 on the top line and 1:7 on the bottom lines. AD, SB and HB have the same meaning as in the samples with ethylene bridge (Figure 4).

3.2.2. 1,4-Benzene-bridged Films [42]

Two different types of 1,4-benzene-bridged films were prepared. The first one (1,4-BB) was prepared from pure BTESB (1,4-bis(triethoxysilyl)-benzene) without porogen and without methyl terminal precursors and had an intrinsic porosity of about 10%. The second type of film (1,4-BB-p) was also prepared with BTESB, but 30 wt% porogen was added into the precursor. As a result, the film has EP measured porosity of 30%. FTIR and EP data can be found in the SI, as Figures S4 and S5. All of these samples were fully cured: they went through the SB and HB processes. PL spectra of 1,4-BB film were measured at the photoluminescence station of the FinEstBeAMS beamline at the 1.5 GeV storage ring of the MAX IV synchrotron facility [47]. The measurements were done at 7K after outgassing in an ultrahigh vacuum (10^{-9} mbar) at 350K. PL was excited by UV photons with energy from 4 to 10 nm (PLE). PL spectra of non-porous films have clearly defined bands at 3.67 eV (at excitation by 5.6 eV) and 3.69 eV (at excitation by 7.3 eV) (Figure 9a). The UV damage threshold in OSG low-k films close to 6.0–6.2 eV [19,65].

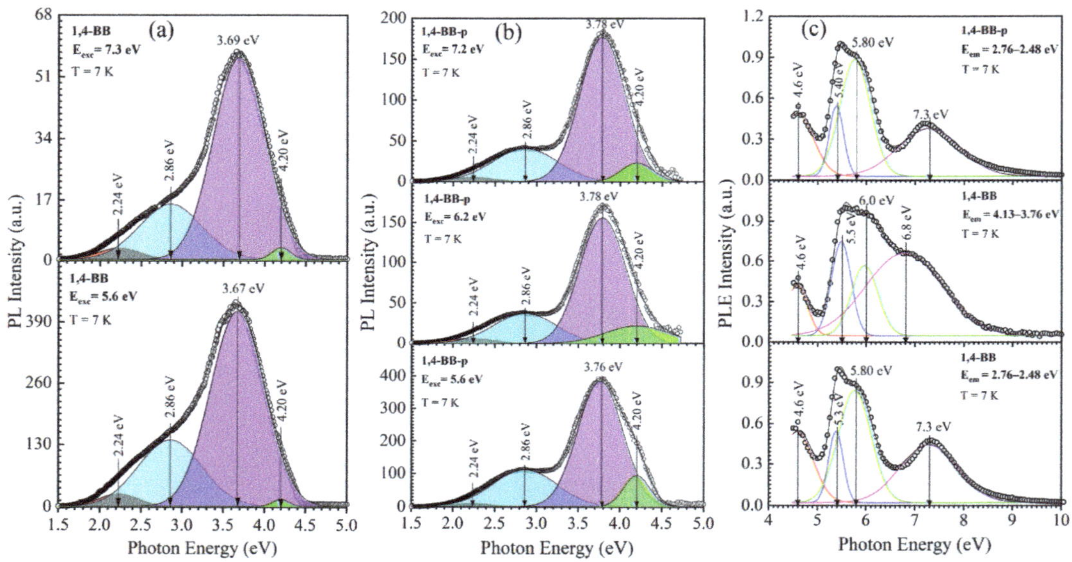

Figure 9. PL spectra of non-porous hard baked OSG film with 1,4-benzene (1,4-BB) bridge upon excitation by 5.6 and 7.3 eV photons (**a**) and porous films with 1,4-benzene bridge (1,4-BB-p) upon excitation by 5.6, 6.2 and 7.2 eV (**b**) and PL excitation spectra (PLE) for 1,4-BB film upon detection 2.76–2.48 eV and 4.13–3.76 eV and for 1,4-BB-p film upon detection 2.76–2.48 eV (**c**).

The results presented in Figure 9a show that PL intensity decreases approximately five times after exposure to 7.3 eV in comparison with the 5.6 eV exposed sample. However, the band position remains the same. A possible explanation is that 7.3 eV photons remove (destruct) part of the component responsible for PL emission at 3.67 eV. Similar conclusions can be made about the porous benzene-bridged films (Figure 9b). These films were irradiated by 5.6, 6.2 and 7.2 eV photons, but the difference in PL bands position is not remarkable. However, the emission intensity drops very much when comparing PL intensity after 6.2 and 7.2 eV excitation in comparison with 5.6 eV. Deconvolution shows the presence of a 2.86 eV peak, and there are also traceable emissions with wavelengths 2.24 and 4.2 eV. The introduction of porogen changes the main emission to 3.76–3.78 eV, and the presence of 4.20 eV is becoming more pronounced (Figure 9b). It supports our previous conclusion that emission at 3.9–4.2 eV is related to hydrocarbon residue: removal of porogen always tends to leave a certain amount of porogen residues [8–10]. The similarity of the PL spectra of these samples to the PL spectra of HB samples with 1,3,5/1,3-benzene bridges (Figure 8) shows that the key role is played by the presence of the benzene bridge rather than the type of their bond with Si atoms. Interesting information can also be obtained from PLE spectra (Figure 9c). The most important excitation bands in all cases correspond to 5.8–6.0 eV photons, but the 2.78-2.48 eV emission peaks also have excitation bands at 7.3 eV.

3.3. Methyl-Terminated Organosilica Films [66]

The films with only methyl terminal groups were fabricated using pure MTMS to prepare dense films and also with 30 wt% added porogen (MTMS-p). All of these samples were fully cured: they went through the SB and HB processes. FTIR spectra are shown in the SI in Figure S6. The measured porosities were equal to 7.5% and 33.1% (SI, Figure S7). The PL measurements were also done on the system used for 1,4-BB films. MTMS films at 7.0 eV photoluminescence excitation (PLE) showed only one peak with emission at 2.83 eV (Figure 10). The samples excited by UV light with an energy of 6.2 eV show an additional peak at 4.35 eV. The last peak is much more pronounced in the films prepared with a porogen, which also supports our previous conclusion that this peak can be related to the presence of porogen residue. MTMS-p also shows the presence of emission bands at 4.95 and 5.16 eV.

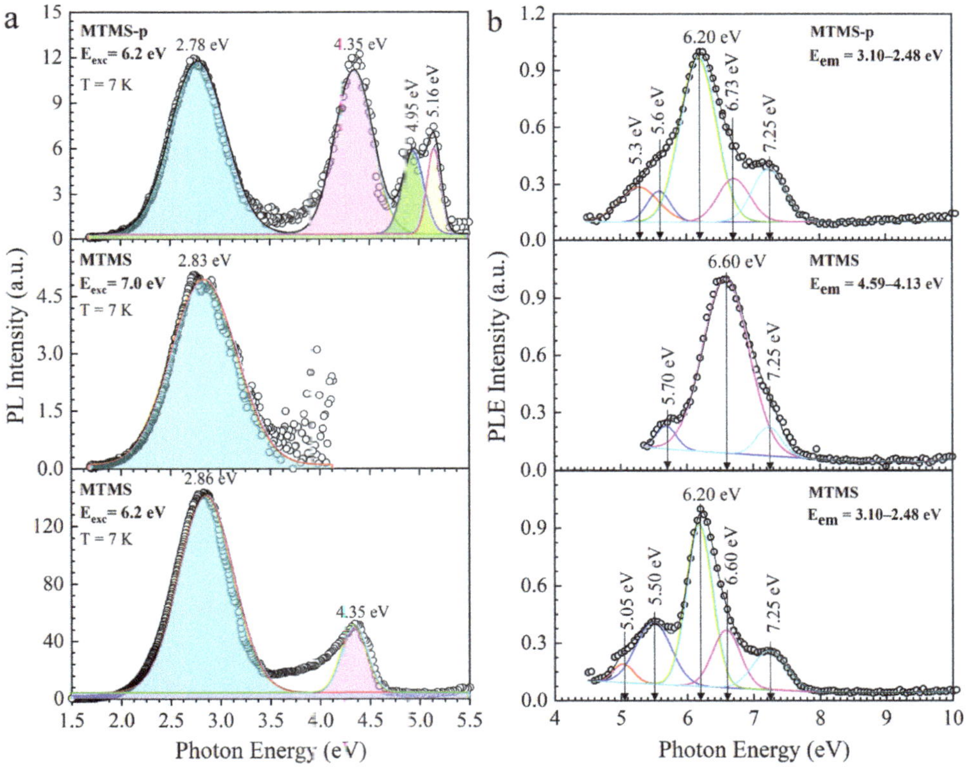

Figure 10. PL spectra of non-porous hard-baked OSG film with methyl terminal group (MTMS) upon excitation by 6.2 and 7.0 eV photons and porous methyl-terminated film (MTMS-p) upon excitation by 6.2 eV (**a**) and PL excitation spectra of MTMS films upon detection 3.10–2.48 and 4.59–4.13 eV and MTMS-p films upon detection 3.10–2.48 eV (**b**).

Dense MTMS samples excited with 7.0 eV photons have much lower PL intensity than the sample excited with 6.2 eV photons. The intensity of the 2,83 PL band has significantly reduced, and the band with an energy of 4.35 eV almost disappeared. This also suggests that the light of 7.0 eV is removing the PL component. It should be noted that 6.2 eV UV photons strongly reduce PL intensity in 1.4 BB films but not in the MTMS film. The most likely explanation is the higher UV resistance of the methyl terminal groups in comparison with the carbon bridges of PMO materials [36,37].

The excitation spectra (PLE) are shown in Figure 10b. Photoluminescence at 3.10–2.48 eV in both dense and porous samples have very pronounced excitation bands at 6.2 eV and also shoulders corresponding to 5.5 and 7.25 eV. The 4.59–4.13 emission bands in the dense MTMS film have a selective excitation band at 6.8 eV. The MTMS sample excited by 7.0 eV UV photons shows no emission band at 4.35 eV, and, therefore, it is reasonable to assume that this component is destroyed or removed during the exposure.

4. Discussion

The presented results allow us to discuss the origin and nature of the observed luminescence bands. The main results are summarized in Table 1. A clear correlation of PL bands with types of organic groups is observed. 1,4-benzene-bridged films (2-2) show the presence of 3.68 and 3.78 eV PL peaks. Similar emissions at 3.9 eV have demonstrated the films containing both 1,3,5- and 1,3-benzene bridges. It suggests that these emission bands are related to benzene bridges. However, HB (400 °C) films show the appearance of small intensity PL peaks near 2.9 and 4.2 eV. It is reasonable to assume that the appearance of these peaks is associated with the compaction of the film matrix, which occurs during calcination at 400 °C and initiates the segregation of unassembled porogen fragments and other residues from the low-k matrix to the surface of the pore walls [8,41].

Table 1. General characteristics of used OSG films and PL energy as a function of the carbon groups. For example, peaks at 3.78–3.9 eV are only observed in films with benzene bridges, so they can be attributed to the benzene bridge. A similar approach is used for other groups. The brackets show the low-intensity peaks, which appear mainly after HB.

Sample Number	Sample ID	Terminal Group ($-CH_3$)	Bridging Group	Porogen	Characteristics PL Peaks (eV)			
1-1	BTMSE/MTMS = 47/53	YES	Ethylene	YES	-	3.3	2.9	4.2
1-2	BTMSE/MTMS = 25/75	YES	Ethylene	YES	-	3.3	2.9	4.2
2-1	1,3,5/1,3-BB	NO	Benzene	YES	3.9	-	(2.90)	(4.85)
2-2	1,4-BB	NO	Benzene	NO	3.68	-	(2.86)	-
2-3	1,4-BB-p	NO	Benzene	YES	3.78	-	(2.86)	4.2
3-1	MTMS	YES	NO	NO	-	-	2.85	(4.35)
3-2	MTMS-p	YES	NO	YES	-	-	2.78	4.35
Assignment of PL peaks					Benzene bridge	Ethylene bridge	$-CH_3$ terminal	$C-H_x$

The samples with only terminal methyl groups show a PL band at 2.8–2.9 eV. Dense MTMS films only show the presence of this peak when PL is excited by 7.0 eV photons but also have a small peak at 4.2 eV when PL is excited by 6.2 eV photons. The porous MTMS-p films deposited with porogen show an increase in relative intensity of 4.25 eV peak compared to dense materials. The peaks with even higher energy are observed in this sample. An important observation (Table 1) is that all samples deposited with porogen have a strong peak at 4.2–4.3 eV, and this support our conclusion (Section 2.1) that it is related to porogen residues and other hydrocarbon residues.

Ethylene-bridged samples give emissions at 2.9, 3.3 and 4.2 eV. We already showed that PL peaks at 2.9 and 4.2 eV correspond to methyl terminal groups that are present in these films and porogen residues because the sample was deposited with a porogen. Therefore, the 3.3 eV peak reflects the presence of an ethylene bridge.

Diagrams of the Energy Levels Calculated by Using Density Functional Theory

Figure 11 shows energy level (Jablonski) diagrams calculated for model molecules reflecting the studied materials.

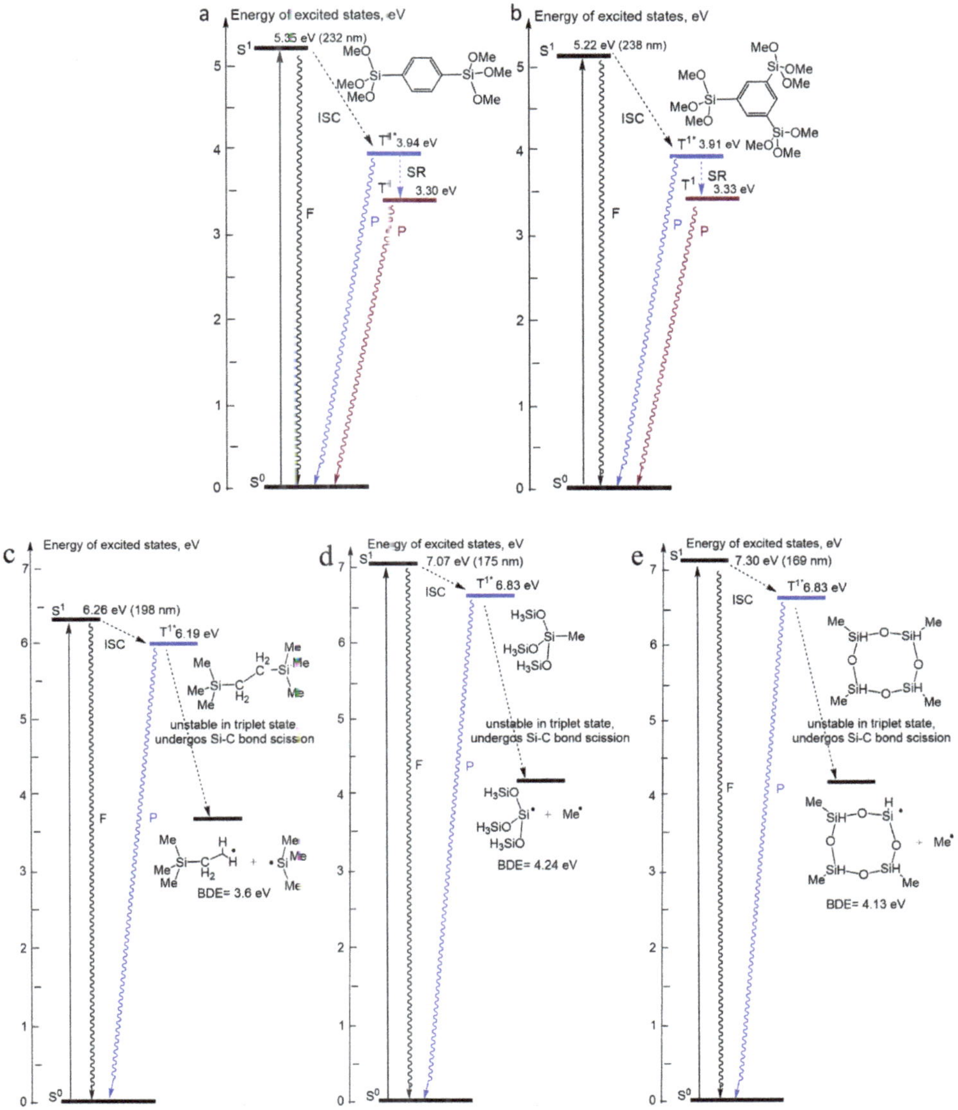

Figure 11. Calculated Jablonski diagram for model molecules containing: 1,4-benzene bridge (**a**), 1,3,5-benzene bridge (**b**), ethylene bridges (c), methyl-terminated (**d**,**e**). (F) is fluorescence, (P) is phosphorescence, ISC—intersystem crossing, SR—structural relaxation.

Figure 11a,b show the energy levels for 1,4-benzene and 1,3,5-benzene bridges. After excitation to the first excited singlet S^1 state, the molecule undergoes ISC to the excited triplet T^{1*} state. The T^{1*} state undergoes further structural relaxation to a more stable T^1 state due to the adjustment of molecular geometry to change the electron distribution upon excitation. Due to the large energy gap between ground S^0 and excited S^1 states, the deactivation of energy from the S^1 state may be observed by fluorescence. Otherwise, when ISC occurs, deactivation of energy by the phosphorescence of both T^{1*} and T^1 can be expected. Interestingly, the energy levels of all transitions in the 1,4-benzene and 1,3,5-benzene bridges are very similar despite the difference in bond structure. UV-induced excitation that initiates the singlet–singlet (S^0–S^1) transition has energies of 5.35 and 5.22 eV, which is very close to the maximum PLE energy of 5.4–5.5 eV measured for 1,4-benzene-bridged OSG materials. (Figure 9c). The energy of the PL photons measured for these films was equal to 3.7 ± 0.1 eV and 3.9 eV for 1,4- and 1,3,5-benzene-bridged OSG, respectively. The calculated energies for the triplet—singlet (T^{1*}–S^0) transition have very close values equal to 3.94 and 3.91 eV. Taking into account that model molecules were used for calculations, one can conclude that the agreement between the calculated and measured PL and PLE energies is perfect. Therefore, the energy diagrams presented in Figure 11a,b most likely correctly reflect the electronic transitions leading to PL.

Figure 11c shows the energy diagram for a molecule having both ethylene bridge and methyl terminal groups. After excitation to the first excited singlet S^1 state, molecules undergo ISC into the excited triplet T^{1*} state. However, the T^{1*} state is very unstable due to the high energy and undergoes Si–O bond cleavage during optimization instead of relaxing into a more stable T^1 state. Due to the large energy gap between ground S^0 and excited S^1 states, deactivation of the energy from the S^1 state through fluorescence may be observed. Otherwise, if ISC into T^{1*} state occurs, deactivation of the energy from T^{1*} through phosphorescence may be expected. The calculated energy for the singlet–singlet (S^0–S^1) transition for this molecule (6.26 eV) is quite close to the measured PLE energy for the sample 1-2 (BTMSE/MTMS 47/53), equal to 6.2 eV (Figure S8). However, we do not observe the emission of 6.19 eV corresponding to transition T^{1*}–S^0. Moreover, after the relaxation of T^{1*} to the T^1 state, this molecule dissociates, forming the radicals shown in Figure 11c. Nevertheless, the bond dissociation energy (BDE) is equal to 3.6 eV, which is close to PL observed with this compound ~3.3 eV. A simulation by Density Functional theory makes this radiative transition doubtful because of energetically preferential molecular destruction from the triplet state. It can be assumed that the destruction of these molecules occurs in parallel with UV radiation, and therefore the intensity of the characteristic 3.3 eV emission in the films with an ethylene bridge has a relatively low intensity (Figure 4).

Figure 11d,e show the energy diagrams for two different molecular structures representing methyl-terminated OSG materials. The first one reflects a model SiO_2-like structure where one Si bond is terminated by the methyl radical. In the second structure, we selected tetramethylcyclotetrasiloxane as a model to take into account the possibility of the second (hydrogen) terminal group being bonded to silicon. However, the energy diagrams in these two cases were very similar: 7.07 eV for the S^0–S^1 transition and 6.83 eV for the T^{1*} state. Only a small difference can be seen in BDE energy: 4.24 eV and 4.13 eV, respectively. It is obvious that the calculated values are quite different than the measured ones (6.2 eV for PLE and 2.8 eV for PL), although the PLE spectrum contains a band at 7.25 eV able to provide a S^0–S^1 transition in these molecules. The most important feature is that according to the DFT calculations, molecules in the T^{1*} state become unstable and dissociate to form radicals, as shown in Figure 11d,e. The most reasonable assumption that can explain the observed PL bands 2.78 eV in samples 3-1 and 3-2, and also the PL bands of 2.9 eV in samples 1-1 and 1.2 is PL of dissociation by-products. This mechanism needs further investigation. However, PL measurements of the MTMS samples were carried out at 7K, while the DFT calculations correspond to room temperature. Our estimations showed that temperature has little impact on the BDE, while the change in the Gibbs free energy of the reaction is significant. The dissociation of the Si–CH_3 bond in S^0 ground state DG(298K) = 78 kcal/mol and DG(7K) = 90 kcal/mol. For the dissociation from the triplet state DG(298K) = –104 kcal/mol and DG(7K) = –86 kcal/mol. Therefore, the probability of the Si–CH_3 dissociation reaction at 7K is much lower and therefore, parallel UV emission can also be expected. Similar phenomena were reported in the papers [67,68]. The calculated energy of the emission is about 4.2 eV, which overlaps with the peak attributed to carbon residue radiation. For this reason, these peaks are observed in all samples deposited with porogen and containing CH_3 terminal groups.

Therefore, the characteristic PL emissions of mesoporous organosilica films correspond to the introduced carbon fragments. It is most likely that they do not include emissions from oxygen-deficient centers (ODC), as was concluded in a recent publication [21]. This conclusion is based on our sample preparation strategy: the samples were not exposed to high-energy impacts capable of generating oxygen vacancies (no UV curing, no exposure to ion and plasma radiation). The curing temperature was not higher than 430 °C, and it is too low for the formation of oxygen-deficient centers [22]. It should be noted that our conclusion is consistent with the results of the ESR studies of various OSG low-k materials [17,32,69–73], which show the presence of carbon dangling bond-related signals in low-k films deposited using sacrificial porogen. A more detailed discussion of these defects and their effect on electrical properties can be found in the review [19] and the corresponding references cited in this paper under the numbers (306, 307, 316–324). The presence of carbon in the film in various forms ranging from the terminal and bridging carbon groups to clusters of elemental carbon originating from porogen or template residues, can give rise to deep energy levels in the insulator bandgap that causes low-field leakage currents [9,10,74].

Finally, similar conclusions were drawn based on the results of studying the PL in SiC_xO_y films deposited using thermal CVD [70]. Films deposited in this way should have properties similar to our MTMS (3-1, 3-2) films, which contain random porosity and methyl terminal groups. Using a parallel study of PL and ESR, the authors concluded that typical structural defects in the oxides, e.g., Si-related neutral oxygen vacancies or non-bridging oxygen hole centers, cannot be considered the dominant mechanism for the white luminescence from SiC_xO_y. It was concluded that PL from SiC_xO_y thin films can result from the generation of carriers due to electronic transitions associated with the C–Si/C–Si–O bonds during optical absorption, followed by recombination of these carriers between energy bands and in their tail states associated with Si–O–C/Si–C bonds. Although the detailed mechanism may differ from ours, the key importance of the Si–C and Si–O–C bonds is also emphasized.

5. Conclusions

The UV-induced photoluminescence of organosilica films with various combinations of ethylene and benzene bridging groups in their matrix and terminal methyl groups on the pore wall surface was studied to reveal optically active defects and understand their origin and nature. The careful selection of the films' precursors and conditions of deposition and curing, excluding breakage of chemical bonds and analysis of chemical and structural properties, led to the conclusion that luminescence sources are not associated with the presence of oxygen-deficient centers, as in the case in pure SiO_2, and has also been predicted for low-k organosilica films [21–23]. It is shown that the sources of luminescence are the carbon-containing components that are part of the low-k matrix (3 different types of benzene bridges), as well as the carbon residues formed upon removal of the template and destruction of organosilica samples containing ethylene bridge and methyl terminal groups. A fairly good correlation between the energy of the photoluminescence peaks and the chemical composition is observed (Table 1). This correlation is also confirmed by the results of density functional theory calculations. The spectra become more complicated after annealing at 400 °C, although Fourier transform infrared spectroscopy does not show these changes. The appearance of new peaks after annealing at 400 °C is associated with the segregation of hydrocarbon residues from the low-k matrix on the pore wall surface [8,41]. Therefore, photoluminescence spectroscopy can potentially be used as effective instrumentation to study the modification of these films. The use of photoluminescence spectroscopy can be important to understand the optical and electrical properties and reliability of integrated low-k dielectrics.

Photoluminescence intensity depends on the internal surface area. The correlation between the photoluminescence intensity of the 4.3 eV peak and the measured surface area calculated for the films measured immediately after full curing can be seen in SI, Figure S9. Further investigations of luminescence sources are planned, including their correlation with dielectric properties, the reliability of low-k dielectrics, and their evolution during various technological treatments, including the use of plasma, UV light and ion irradiation.

Supplementary Materials: The following supporting information can be downloaded at: https://www.mdpi.com/article/10.3390/nano13081419/s1, Table S1: Physical properties of organosilicate glass (OSG) films: first type: both ethylene bridges and methyl terminal groups with different precursor ratios (BTMSE/MTMS = 47/53 and 25/75); second type: a mixture of 1,3,5- and 1,3-benzene groups with a ratio of 1,3,5/1,3 = 1:3, 1,3,5/1,3 = 1:7 and 1,4-benzene bridges without porogen (1,4-BB) and with 30 wt% porogen (1,4-BB-p); third type: contains only methyl terminals without bridging groups (MTMS—without porogen, MTMS-p—with porogen). All films are completely cured in the air; Figure S1: Effect of chemical composition during storage of film (BTMSE/MTMS = 47/53, 20 wt% Brij®30) in a clean room environment and after soft oxygen ICP plasma. Loss of Si–CH_3 groups and hydrophilization was observed for films by using soft oxygen ICP plasma (remove the chemical residue). The thickness of low-k film before exposure to O_2 ICP plasma was equal to 466 nm, then it reduced to 435 nm after 5 min and 399 nm after 10 min. Meanwhile, the corresponding reductions of Si–CH_3/Si–O–Si peaks ratio were from 0.173 to 0.163 and to 0.154, respectively. The changes in hydrocarbon residue peaks in the region 2800–3000 cm^{-1} were about three times larger. These observations suggest that the plasma conditions were sufficiently soft to minimize plasma damage of the low-k matrix (small change Si–CH_3/Si–O–Si ratio), and the ICP-generated oxygen radicals are mainly consumed on the accumulated carbon residue. The equivalent depth of the CH_3 depleted region was estimated by using the approach described in the paper (F. Leroy et al. Journal of Physics D Applied Physics 48(43):435202) and was equal to ≈27 nm after 5 min and ≈45 nm after 10 min plasma exposure; Figure S2: FTIR spectra of 1,3,5- and 1,3-benzene-bridged ratio of 1:3 (bottom) and 1:7 (top) organosilica films with different porogen concentration s (0–41 wt%), soft bake at 200 °C and hard bake at 400 °C in air. The spectra are generally quite similar, but the films containing less 1,3 benzene bridges (1:3) are more hydrophilic; Figure S3: Adsorption–desorption isotherms for determining the value of open porosity, generated by ellipsometric porosimetry, of 1,3,5- and 1,3-benzene bridges ratio of 1:3 (a) and 1:7 (b) organosilica films with different porogen concentrations (17–41 wt%), soft bake at 150 °C and hard bake at 400 °C in air. The films containing less 1,3 benzene bridges (1:3) always have larger pore size at the same porogen concentration; Figure S4: FTIR spectra of pure 1,4-benzene-bridged organosilica films: dense (1,4-BB) and with 30 wt% porogen content 1,4-benzene-bridged (1,4-BB-p) organosilica films, both the films were completely cured in air. The porous 1,4-BB-p film contains a larger concentration of adsorbed water; Figure S5: Adsorption–desorption isotherms for determining the value of open porosity (a) and pore size distribution (b), generated by ellipsometric porosimetry, of pure 1,4-benzene-bridged organosilica films: dense (1,4-BB) and with 30 wt% porogen content 1,4-benzene-bridged (1,4-BB-p) organosilica films, both the films were cured completely in air; Figure S6: FTIR spectra of pure methyl-terminated organosilicate: dense (MTMS) and with 30 wt% porogen content methyl-terminated (MTMS-p), both the films were cured completely in air. The films with methyl terminal groups are much more hydrophobic than the films with benzene bridges, as expected; Figure S7: Adsorption–desorption isotherms for determining the value of open porosity (a) and pore size distribution (b), generated by ellipsometric porosimetry, of pure methyl-terminated organosilica films: dense (MTMS) and with 30 wt% porogen content methyl terminated (MTMS-p) organosilica films, both the films were cured completely in the air; Figure S8: UV-induced PL spectra measured at the different excitation energy for the ethylene-bridged films (BTMSE/MTMS = 47/53); Figure S9: Correlation between 4.3 eV UV-induced (upon excitation with light of 6.2 eV) room temperature PL emission with the measured surface area in hard-baked OSG low-k films containing both methyl terminal and ethylene bridging groups (BTMSE/MTMS ratio of 25/75) with different porosity. However, the PL bands at 2.9, 3.3 and 4.85 eV have the highest intensity in highly porous films. It can be assumed that the reason is that the terminal methyl groups and hydrocarbon residues are mainly located on the pore wall surface. However, this assumption needs further study.

Author Contributions: Conceptualization, M.R.B.; data curation, M.R., J.Z. (Jing Zhang), D.A.S., S.N. and A.S.V.; formal analysis, M.R., J.Z. (Jinming Zhang), D.A.S., S.N. and A.S.V.; funding acquisition, A.S.V., K.A.V., J.Z. (Jing Zhang) and M.R.B.; investigation, M.R.B.; methodology, M.R., J.Z.

(Jing Zhang), D.A.S., S.N. and A.S.V.; project administration, J.Z. (Jing Zhang) and M.R.B.; resources, J.Z. (Jing Zhang); supervision, K.A.V., J.Y., J.Z. (Jing Zhang) and M.R.B.; validation, M.R., J.Z. (Jinming Zhang), D.A.S., S.N. and A.S.V. and M.R.B.; visualization, M.R.; writing: original draft, M.R. and M.R.B.; writing: review & editing, M.R., A.S.V. and M.R.B. All authors have read and agreed to the published version of the manuscript.

Funding: This research was funded by Russian Science Foundation, grant number 23–79–30016; R&D Program of Beijing Municipal Education Commission, grant number KZ202210009014. The APC was funded by R&D Program of the Beijing Municipal Education Commission. We acknowledge MAX IV Laboratory for time on Beamline FinEstBeAMS under Proposal 20200327. Research conducted at MAX IV, a Swedish national user facility, is supported by the Swedish Research council under contract 2018-07152, the Swedish Governmental Agency for Innovation Systems under contract 2018-04969, and Formas under contract 2019-02496.

Institutional Review Board Statement: The study do not require ethical approval.

Informed Consent Statement: Not applicable.

Data Availability Statement: The data presented in this study are available on request from the corresponding author.

Acknowledgments: It is our pleasure to thank Dmitry Seregin, Yingjie Wang, Chunhui Liu, Haoyu Xu and Xuesong Wang for the help in sample preparation and FTIR analysis.

Conflicts of Interest: The authors declare no conflict of interest.

References

1. Van Der Voort, P.; Esquivel, D.; De Canck, E.; Goethals, F.; Van Driessche, I.; Romero-Salguero, F.J. Periodic Mesoporous Organosilicas: From simple to complex bridges; a comprehensive overview of functions, morphologies and applications. *Chem. Soc. Rev.* **2013**, *42*, 3913–3955. [CrossRef] [PubMed]
2. Baklanov, M.R.; Ho, P.S.; Zschech, E. (Eds.) *Advanced Interconnects for ULSI Technology*; John Wiley & Sons, Ltd.: Chichester, UK, 2012; ISBN 9781119963677.
3. Kondoh, E. Structural Change in Porous Silica Thin Film after Plasma Treatment. *Electrochem. Solid-State Lett.* **1999**, *1*, 224. [CrossRef]
4. Devine, R.A.B.; Ball, D.; Rowe, J.D.; Tringe, J.W. Irradiation and Humidity Effects on the Leakage Current in $SiO_2(CH_3)_y$-Based Intermetal Dielectric Films. *J. Electrochem. Soc.* **2003**, *150*, F151. [CrossRef]
5. Jousseaume, V.; Zenasni, A.; Favennec, L.; Gerbaud, G.; Bardet, M.; Simon, J.P.; Humbert, A. Comparison between E-beam and Ultraviolet Curing to Perform Porous a-SiOC:H. *J. Electrochem. Soc.* **2007**, *154*, G103. [CrossRef]
6. Gates, S.M.; Neumayer, D.A.; Sherwood, M.H.; Grill, A.; Wang, X.; Sankarapandian, M. Preparation and structure of porous dielectrics by plasma enhanced chemical vapor deposition. *J. Appl. Phys.* **2007**, *101*, 094103. [CrossRef]
7. Gerbaud, G.; Hediger, S.; Bardet, M.; Favennec, L.; Zenasni, A.; Beynet, J.; Gourhant, O.; Jousseaume, V. Spin-coated and PECVD low dielectric constant porous organosilicate films studied by 1D and 2D solid-state NMR. *Phys. Chem. Chem. Phys.* **2009**, *11*, 9729. [CrossRef] [PubMed]
8. Marsik, P.; Verdonck, P.; De Roest, D.; Baklanov, M.R. Porogen residues detection in optical properties of low-k dielectrics cured by ultraviolet radiation. *Thin Solid Films* **2010**, *518*, 4266–4272. [CrossRef]
9. Baklanov, M.R.; Zhao, L.; Besien, E.V.; Pantouvaki, M. Effect of porogen residue on electrical characteristics of ultra low-k materials. *Microelectron. Eng.* **2011**, *88*, 990–993. [CrossRef]
10. Van Besien, E.; Pantouvaki, M.; Zhao, L.; De Roest, D.; Baklanov, M.R.; Tőkei, Z.; Beyer, G. Influence of porosity on electrical properties of low-k dielectrics. *Microelectron. Eng.* **2012**, *92*, 59–61. [CrossRef]
11. Kikkawa, T.; Kuroki, S.; Sakamoto, S.; Kohmura, K.; Tanaka, H.; Hata, N. Influence of Humidity on Electrical Characteristics of Self-Assembled Porous Silica Low-k Films. *J. Electrochem. Soc.* **2005**, *152*, G560. [CrossRef]
12. Li, Y.; Ciofi, I.; Carbonell, L.; Heylen, N.; Van Aelst, J.; Baklanov, M.R.; Groeseneken, G.; Maex, K.; Tkei, Z. Influence of absorbed water components on SiOCH low-k reliability. *J. Appl. Phys.* **2008**, *104*, 034113. [CrossRef]
13. Biswas, N.; Lubguban, J.A.; Gangopadhyay, S. Electric field and temperature-induced removal of moisture in nanoporous organosilicate films. *Appl. Phys. Lett.* **2004**, *84*, 4254–4256. [CrossRef]
14. Ogawa, E.T.; Aubel, O. Electrical Breakdown in Advanced Interconnect Dielectrics. In *Advanced Interconnects for ULSI Technology*; John Wiley & Sons, Ltd.: Chichester, UK, 2012; pp. 369–434, ISBN 9780470662540.
15. Wu, C.; Li, Y.; Baklanov, M.R.; Croes, K. Electrical Reliability Challenges of Advanced Low-k Dielectrics. *ECS J. Solid State Sci. Technol.* **2015**, *4*, N3065–N3070. [CrossRef]
16. Sinha, H.; Ren, H.; Nichols, M.T.; Lauer, J.L.; Tomoyasu, M.; Russell, N.M.; Jiang, G.; Antonelli, G.A.; Fuller, N.C.; Engelmann, S.U.; et al. The effects of vacuum ultraviolet radiation on low- k dielectric films. *J. Appl. Phys.* **2012**, *112*, 111101. [CrossRef]

17. Afanas'ev, V.V.; Nguyen, A.P.D.; Houssa, M.; Stesmans, A.; Tőkei, Z.; Baklanov, M.R. High-resolution electron spin resonance analysis of ion bombardment induced defects in advanced low-κ insulators (κ = 2.0-2.5). *Appl. Phys. Lett.* **2013**, *102*, 172908. [CrossRef]
18. King, S.W.; French, B.; Mays, E. Detection of defect states in low-k dielectrics using reflection electron energy loss spectroscopy. *J. Appl. Phys.* **2013**, *113*, 044109. [CrossRef]
19. Baklanov, M.R.; Jousseaume, V.; Rakhimova, T.V.; Lopaev, D.V.; Mankelevich, Y.A.; Afanas'ev, V.V.; Shohet, J.L.; King, S.W.; Ryan, E.T. Impact of VUV photons on SiO_2 and organosilicate low-k dielectrics: General behavior, practical applications, and atomic models. *Appl. Phys. Rev.* **2019**, *6*, 011301. [CrossRef]
20. Mutch, M.J.; Pomorski, T.; Bittel, B.C.; Cochrane, C.J.; Lenahan, P.M.; Liu, X.; Nemanich, R.J.; Brockman, J.; French, M.; Kuhn, M.; et al. Band diagram for low-k/Cu interconnects: The starting point for understanding back-end-of-line (BEOL) electrical reliability. *Microelectron. Reliab.* **2016**, *63*, 201–213. [CrossRef]
21. Pustovarov, V.A.; Zatsepin, A.F.; Biryukov, D.Y.; Aliev, V.S.; Iskhakzay, R.M.K.; Gritsenko, V.A. Synchrotron-Excited Luminescence and Converting of Defects and Quantum Dots in Modified Silica Films. *J. Non-Cryst. Solids* **2023**, *602*, 122077. [CrossRef]
22. Skuja, L. Optically active oxygen-deficiency-related centers in amorphous silicon dioxide. *J. Non-Cryst. Solids* **1998**, *239*, 16–48. [CrossRef]
23. Salh, R. Defect Related Luminescence in Silicon Dioxide Network: A Review. In *Crystalline Silicon-Properties and Uses*; InTech: Houston, TX, USA, 2011; pp. 135–172.
24. Gismatulin, A.A.; Gritsenko, V.A.; Seregin, D.S.; Vorotilov, K.A.; Baklanov, M.R. Charge transport mechanism in periodic mesoporous organosilica low- k dielectric. *Appl. Phys. Lett.* **2019**, *115*, 082904. [CrossRef]
25. Perevalov, T.V.; Gismatulin, A.A.; Seregin, D.S.; Wang, Y.; Xu, H.; Kruchinin, V.N.; Spesivcev, E.V.; Gritsenko, V.A.; Nasyrov, K.A.; Prosvirin, I.P.; et al. Critical properties and charge transport in ethylene bridged organosilica low-κ dielectrics. *J. Appl. Phys.* **2020**, *127*, 195105. [CrossRef]
26. Grill, A.; Gates, S.M.; Ryan, T.E.; Nguyen, S.V.; Priyadarshini, D. Progress in the development and understanding of advanced low k and ultralow k dielectrics for very large-scale integrated interconnects—State of the art. *Appl. Phys. Rev.* **2014**, *1*, 011306. [CrossRef]
27. Dag, Ö.; Yoshina-Ishii, C.; Asefa, T.; MacLachlan, M.J.; Grondey, H.; Coombs, N.; Ozin, G.A. Oriented Periodic Mesoporous Organosilica (PMO) Film with Organic Functionality Inside the Channel Walls. *Adv. Funct. Mater.* **2001**, *11*, 213–217. [CrossRef]
28. Dubois, G.; Volksen, W.; Magbitang, T.; Sherwood, M.H.; Miller, R.D.; Gage, D.M.; Dauskardt, R.H. Superior mechanical properties of dense and porous organic/inorganic hybrid thin films. *J. Sol.-Gel. Sci. Technol.* **2008**, *48*, 187–193. [CrossRef]
29. Hoffmann, F.; Cornelius, M.; Morell, J.; Fröba, M. Silica-Based Mesoporous Organic–Inorganic Hybrid Materials. *Angew. Chemie Int. Ed.* **2006**, *45*, 3216–3251. [CrossRef]
30. Dubois, G.; Volksen, W.; Magbitang, T.; Miller, R.D.; Gage, D.M.; Dauskardt, R.H. Molecular Network Reinforcement of Sol–Gel Glasses. *Adv. Mater.* **2007**, *19*, 3989–3994. [CrossRef]
31. Li, H.; Knaup, J.M.; Kaxiras, E.; Vlassak, J.J. Stiffening of organosilicate glasses by organic cross-linking. *Acta Mater.* **2011**, *59*, 44–52. [CrossRef]
32. Vanstreels, K.; Wu, C.; Baklanov, M.R. Mechanical Stability of Porous Low-k Dielectrics. *ECS J. Solid State Sci. Technol.* **2015**, *4*, N3058–N3064. [CrossRef]
33. Burg, J.A.; Oliver, M.S.; Frot, T.J.; Sherwood, M.; Lee, V.; Dubois, G.; Dauskardt, R.H. Hyperconnected molecular glass network architectures with exceptional elastic properties. *Nat. Commun.* **2017**, *8*, 1019. [CrossRef]
34. Goethals, F.; Ciofi, I.; Madia, O.; Vanstreels, K.; Baklanov, M.R.; Detavernier, C.; Van Der Voort, P.; Van Driessche, I. Ultra-low-k cyclic carbon-bridged PMO films with a high chemical resistance. *J. Mater. Chem.* **2012**, *22*, 8281. [CrossRef]
35. Lu, Y.; Fan, H.; Doke, N.; Loy, D.A.; Assink, R.A.; LaVan, D.A.; Brinker, C.J. Evaporation-Induced Self-Assembly of Hybrid Bridged Silsesquioxane Film and Particulate Mesophases with Integral Organic Functionality. *J. Am. Chem. Soc.* **2000**, *122*, 5258–5261. [CrossRef]
36. Redzheb, M.; Prager, L.; Naumov, S.; Krishtab, M.; Armini, S.; Van Der Voort, P.; Baklanov, M.R. Effect of the C-bridge length on the ultraviolet-resistance of oxycarbosilane low-k films. *Appl. Phys. Lett.* **2016**, *108*, 012902. [CrossRef]
37. Seregin, D.S.; Naumov, S.; Chang, W.-Y.; Wu, Y.-H.; Wang, Y.; Kotova, N.M.; Vishnevskiy, A.S.; Wei, S.; Zhang, J.; Vorotilov, K.A.; et al. Effect of the C-bridge on UV properties of organosilicate films. *Thin Solid Films* **2019**, *685*, 329–334. [CrossRef]
38. Vishnevskiy, A.S.; Naumov, S.; Seregin, D.S.; Wu, Y.H.; Chuang, W.T.; Rasadujjaman, M.; Zhang, J.; Leu, J.; Vorotilov, K.A.; Baklanov, M.R. Effects of methyl terminal and carbon bridging groups ratio on critical properties of porous organosilicate-glass films. *Materials* **2020**, *13*, 4484. [CrossRef]
39. Rasadujjaman, M.; Wang, Y.; Zhang, L.; Naumov, S.; Attallah, A.G.; Liedke, M.O.; Koehler, N.; Redzheb, M.; Vishnevskiy, A.S.; Seregin, D.S.; et al. A detailed ellipsometric porosimetry and positron annihilation spectroscopy study of porous organosilicate-glass films with various ratios of methyl terminal and ethylene bridging groups. *Microporous Mesoporous Mater.* **2020**, *306*, 110434. [CrossRef]
40. Rasadujjaman, M.; Zhang, J.; Mogilnikov, K.P.; Vishnevskiy, A.S.; Zhang, J.; Baklanov, M.R. Effect of methyl terminal and ethylene bridging groups on porous organosilicate glass films: FTIR, ellipsometric porosimetry, luminescence dataset. *Data Br.* **2021**, *35*, 106895. [CrossRef]

41. Iacopi, F.; Travaly, Y.; Eyckens, B.; Waldfried, C.; Abell, T.; Guyer, E.P.; Gage, D.M.; Dauskardt, R.H.; Sajavaara, T.; Houthoofd, K.; et al. Short-ranged structural rearrangement and enhancement of mechanical properties of organosilicate glasses induced by ultraviolet radiation. *J. Appl. Phys.* **2006**, *99*, 053511 [CrossRef]
42. Rezvanov, A.A.; Vishnevskiy, A.S.; Seregin, D.S.; Schneider, D.; Lomov, A.A.; Vorotilov, K.A.; Baklanov, M.R. Benzene bridged hybrid organosilicate films with improved stiffness and small pore size. *Mater. Chem. Phys.* **2022**, *290*, 126571. [CrossRef]
43. Liu, C.; Lv, C.; Kohler, N.; Wang, X.; Lin, H.; He, Z.; Wu, Y.-H.; Leu, J.; Wei, S.; Zhang, J.; et al. Properties of organosilicate low- k films with 1,3- and 1,3,5-benzene bridges between Si atoms. *Jpn. J. Appl. Phys.* **2020**, *59*, SLLG01. [CrossRef]
44. Rasadujjaman, M.; Wang, X.; Wang, Y.; Zhang, J.; Arkhincheev, V.E.; Baklanov, M.R. Analytical Study of Porous Organosilicate Glass Films Prepared from Mixtures of 1,3,5- and 1,3-Alkoxysilylbenzenes. *Materials* **2021**, *14*, 1881. [CrossRef]
45. Baklanov, M.R.; Mogilnikov, K.P.; Polovinkin, V.G.; Dultsev, F.N. Determination of pore size distribution in thin films by ellipsometric porosimetry. *J. Vac. Sci. Technol. B Microelectron. Nanom. Struct.* **2000**, *18*, 1385. [CrossRef]
46. Haul, R.S.J.; Gregg, K.S.W. Sing: Adsorption, Surface Area and Porosity. 2. Auflage, Academic Press, London 1982. 303 Seiten, Preis: $49.50. *Ber. Bunsenges. Phys. Chem.* **1982**, *86*, 957. [CrossRef]
47. Chernenko, K.; Kivimäki, A.; Pärna, R.; Wang, W.; Sankari, R.; Leandersson, M.; Tarawneh, H.; Pankratov, V.; Kook, M.; Kukk, E.; et al. Performance and characterization of the FinEstBeAMS beamline at the MAX IV Laboratory. *J. Synchrotron Radiat.* **2021**, *28*, 1620–1630. [CrossRef]
48. Sandeep, S.; Vishnevskiy, A.S.; Raetz, S.; Naumov, S.; Seregin, D.S.; Husiev, A.; Vorotilov, K.A.; Gusev, V.E.; Baklanov, M.R. In-Situ Imaging of a Light-Induced Modification Process in Organo-Silica Films via Time-Domain Brillouin Scattering. *Nanomaterials* **2022**, *12*, 1600. [CrossRef]
49. Zotovich, A.I.; Zyryanov, S.M.; Lopaev, D.V.; Rezvanov, A.A.; Attallah, A.G.; Liedke, M.O.; Butterling, M.; Bogdanova, M.A.; Vishnevskiy, A.S.; Seregin, D.S.; et al. Modification of Porous Ultralow- k Film by Vacuum Ultraviolet Emission. *ACS Appl. Electron. Mater.* **2022**, *4*, 2760–2776. [CrossRef]
50. Perdew, J.P.; Burke, K.; Ernzerhof, M. Generalized Gradient Approximation Made Simple. *Phys. Rev. Lett.* **1996**, *77*, 3865–3868. [CrossRef]
51. Perdew, J.P.; Burke, K.; Ernzerhof, M. Generalized Gradient Approximation Made Simple. *Phys. Rev. Lett.* **1997**, *78*, 1396, Erratum in *Phys. Rev. Lett.* **1996**, *77*, 3865–3868. [CrossRef]
52. Adamo, C.; Barone, V. Toward reliable density functional methods without adjustable parameters: The PBE0 model. *J. Chem. Phys.* **1999**, *110*, 6158–6170. [CrossRef]
53. *Jaguar*, version 9.6; Schrodinger, Inc.: New York, NY, USA, 2017.
54. Naumov, S.; Herzog, B.; Abel, B. Spectra and Photorelaxation of Hydroxyphenyl-benzotriazole-Type UV Absorbers: From Monomers to Nanoparticles. *J. Phys. Chem. A* **2020**, *124*, 625–632. [CrossRef] [PubMed]
55. Weilbeer, C.; Sickert, M.; Naumov, S.; Schneider, C. The Brønsted Acid-Catalyzed, Enantioselective Aza-Diels-Alder Reaction for the Direct Synthesis of Chiral Piperidones. *Chem.-A Eur. J.* **2017**, *23*, 513–518. [CrossRef]
56. Bauernschmitt, R.; Ahlrichs, R. Treatment of electronic excitations within the adiabatic approximation of time dependent density functional theory. *Chem. Phys. Lett.* **1996**, *256*, 454–464. [CrossRef]
57. Pollard, W.T.; Friesner, R.A. Efficient Fock matrix diagonalization by a Krylov-space method. *J. Chem. Phys.* **1993**, *99*, 6742–6750. [CrossRef]
58. Everett, D.H. Adsorption Hysteresis. In *The Solid-Gas Interface*; Flood, E.A., Ed.; Decker: New York, NY, USA, 1967; pp. 1055–1113.
59. Skuja, L.; Leimane, M.; Bite, I.; Millers, D.; Zolotarjovs, A.; Vitola, V.; Smits, K. Ultraviolet luminescence of polycyclic aromatic hydrocarbons in partially consolidated sol-gel silica glasses. *J. Non-Cryst. Solids* **2022**, *577*, 121325. [CrossRef]
60. Zhao, J.; Mao, D.S.; Lin, Z.X.; Jiang, B.Y.; Yu, Y.H.; Liu, X.H.; Wang, H.Z.; Yang, G.Q. Intense short-wavelength photoluminescence from thermal SiO_2 films co-implanted with Si and C ions. *Appl. Phys. Lett.* **1998**, *73*, 1838–1840. [CrossRef]
61. Tabassum, N.; Nikas, V.; Ford, B.; Huang, M.; Kaloyeros, A.E.; Gallis, S. Time-resolved analysis of the white photoluminescence from chemically synthesized SiCxOy thin films and nanowires. *Appl. Phys. Lett.* **2016**, *109*, 043104. [CrossRef]
62. Nikas, V.; Tabassum, N.; Ford, B.; Smith, L.; Kaloyeros, A.E.; Gallis, S. Strong visible light emission from silicon-oxycarbide nanowire arrays prepared by electron beam lithography and reactive ion etching. *J. Mater. Res.* **2015**, *30*, 3692–3699. [CrossRef]
63. Ford, B.; Tabassum, N.; Nikas, V.; Gallis, S. Strong Photoluminescence Enhancement of Silicon Oxycarbide through Defect Engineering. *Materials* **2017**, *10*, 446. [CrossRef]
64. Koós, M.; Pócsik, I. Photoluminescence in Hydrogenated Amorphous Carbon. In *Physics and Applications of Non-Crystalline Semiconductors in Optoelectronics*; Springer Netherlands: Dordrecht, The Netherlands, 1997; pp. 361–378.
65. Prager, L.; Marsik, P.; Wennrich, L.; Baklanov, M.R.; Naumov, S.; Pistol, L.; Schneider, D.; Gerlach, J.W.; Verdonck, P.; Buchmeiser, M.R. Effect of pressure on efficiency of UV curing of CVD-derived low-k material at different wavelengths. *Microelectron. Eng.* **2008**, *85*, 2094–2097. [CrossRef]
66. Liu, C.; Qi, Q.; Seregin, D.S.; Vishnevskiy, A.S.; Wang, Y.; Wei, S.; Zhang, J.; Vorotilov, K.A.; Dultsev, F.N.; Baklanov, M.R. Effect of terminal methyl groups concentration on properties of organosilicate glass low dielectric constant films. *Jpn. J. Appl. Phys.* **2018**, *57*, 07MC01. [CrossRef]
67. Knolle, W.; Feldman, V.I.; Janovský, I.; Naumov, S.; Mehnert, R.; Langguth, H.; Sukhov, F.F.; Orlov, A.Y. EPR study of methyl and ethyl acrylate radical cations and their transformations in low-temperature matrices. *J. Chem. Soc. Perkin Trans.* **2002**, *2*, 687–699. [CrossRef]

68. Naumov, S.; Knolle, W.; Naumov, S.; Pöppl, A.; Janovský, I. The Dynamical Behavior of the s-Trioxane Radical Cation—A Low-Temperature EPR and Theoretical Study. *Molecules* **2014**, *19*, 17305–17313. [CrossRef] [PubMed]
69. Afanas'ev, V.V.; Keunen, K.; Nguyen, A.P.D.; Jivanescu, M.; Stesmans, A.; Tokei, Z.; Baklanov, M.R.; Beyer, G.P. *Symposium O—Materials, Processes, and Reliability for Advanced Interconnects for Micro- and Nanoelectronics*; Cambridge University Press: Cambridge, UK, 2011; p. 1335.
70. Nikas, V.; Gallis, S.; Huang, M.; Kaloyeros, A.E.; Nguyen, A.P.D.; Stesmans, A.; Afanas'ev, V.V. The origin of white luminescence from silicon oxycarbide thin films. *Appl. Phys. Lett.* **2014**, *104*, 061906. [CrossRef]
71. Pomorski, T.A.; Bittel, B.C.; Lenahan, P.M.; Mays, E.; Ege, C.; Bielefeld, J.; Michalak, D.; King, S.W. Defect structure and electronic properties of SiOC:H films used for back end of line dielectrics. *J. Appl. Phys.* **2014**, *115*, 234508. [CrossRef]
72. Mutch, M.J.; Lenahan, P.M.; King, S.W. Defect chemistry and electronic transport in low-κ dielectrics studied with electrically detected magnetic resonance. *J. Appl. Phys.* **2016**, *119*, 094102. [CrossRef]
73. Bittel, B.C.; Lenahan, P.M.; King, S.W. Ultraviolet radiation effects on paramagnetic defects in low-κ dielectrics for ultralarge scale integrated circuit interconnects. *Appl. Phys. Lett.* **2010**, *97*, 063506. [CrossRef]
74. Krishtab, M.; Afanas'ev, V.; Stesmans, A.; De Gendt, S. Leakage current induced by surfactant residues in self-assembly based ultralow-k dielectric materials. *Appl. Phys. Lett.* **2017**, *111*, 032908. [CrossRef]

nanomaterials

MDPI

Communication

Enhanced Photoluminescence of Crystalline Alq_3 Micro-Rods Hybridized with Silver Nanowires

Misuk Kim [1,†], Jiyoun Kim [2,†], Seongcheol Ju [1], Hyeonwoo Kim [1], Incheol Jung [1], Jong Hoon Jung [1], Gil Sun Lee [3], Young Ki Hong [4,5,*], Dong Hyuk Park [2,*] and Kyu-Tae Lee [1,*]

1 Department of Physics, Inha University, Incheon 22212, Republic of Korea
2 Department of Chemical Engineering, Program in Biomedical Science and Engineering, Inha University, Incheon 22212, Republic of Korea
3 Department of General Education, Kookmin University, Seoul 02707, Republic of Korea
4 Department of Physics, Gyeongsang National University, Jinju 52828, Republic of Korea
5 Research Institute of Natural Science, Gyeongsang National University, Jinju 52828, Republic of Korea
* Correspondence: ykhong@gnu.ac.kr (Y.K.H.); donghyuk@inha.ac.kr (D.H.P.); ktlee@inha.ac.kr (K.-T.L.); Tel.: +82-55-772-1407 (Y.K.H.); +82-32-860-7496 (D.H.P.); +82-32-860-7653 (K.-T.L.)
† These authors contributed equally to this work.

Abstract: An enhancement of the local electric field at the metal/dielectric interface of hybrid materials due to the localized surface plasmon resonance (LSPR) phenomenon plays a particularly important role in versatile research fields resulting in a distinct modification of the electrical, as well as optical, properties of the hybrid material. In this paper, we succeeded in visually confirming the LSPR phenomenon in the crystalline tris(8-hydroxyquinoline) aluminum (Alq_3) micro-rod (MR) hybridized with silver (Ag) nanowire (NW) in the form of photoluminescence (PL) characteristics. Crystalline Alq_3 MRs were prepared by a self-assembly method under the mixed solution of protic and aprotic polar solvents, which could be easily applied to fabricate hybrid Alq_3/Ag structures. The hybridization between the crystalline Alq_3 MRs and Ag NWs was confirmed by the component analysis of the selected area electronic diffraction attached to high-resolution transmission electron microscope. Nanoscale and solid state PL experiments on the hybrid Alq_3/Ag structures using a lab-made laser confocal microscope exhibited a distinct enhancement of the PL intensity (approximately 26-fold), which also supported the LSPR effects between crystalline Alq_3 MRs and Ag NWs.

Keywords: Alq_3; crystalline micro-rod; Ag; nanowire; hybridization; surface plasmon resonance

Citation: Kim, M.; Kim, J.; Ju, S.; Kim, H.; Jung, I.; Jung, J.H.; Lee, G.S.; Hong, Y.K.; Park, D.H.; Lee, K.-T. Enhanced Photoluminescence of Crystalline Alq_3 Micro-Rods Hybridized with Silver Nanowires. *Nanomaterials* **2023**, *13*, 825. https://doi.org/10.3390/nano13050825

Academic Editor: Junying Zhang

Received: 21 January 2023
Revised: 21 February 2023
Accepted: 21 February 2023
Published: 23 February 2023

1. Introduction

Luminescence, the spontaneous emission of light, arises from the transition of electrons from the highest occupied molecular orbital (HOMO) to the lowest unoccupied molecular orbital (LUMO). Various types of luminescence have been identified, including chemiluminescence, where electron excitation is caused by the energy released in a chemical reaction, electroluminescence, which is a result of the passage of an electric current, and photoluminescence (PL), which is generated by the absorption of photons [1–3].

Significant efforts have been dedicated to enhancing the PL efficiency of light-emitting organic materials by exploiting surface plasmon polaritons (SPPs), which are hybridized excitations propagating at the interface between metals and dielectrics when collective electron oscillations (i.e., SPs) couple with photons in metallic nanostructures [4–8]. Matching the energy of the SPPs in the metallic nanostructures with that of the emitted photons of the organic luminescent materials allows resonance to occur, thus leading to a significantly enhanced PL efficiency at the resonant wavelengths [9,10]. The resonant wavelength can be easily tuned by altering the materials, the geometric parameters of the metallic nanostructures, and the surrounding media, etc. [11–14]. These localized surface plasmon resonance

(LSPR) effects have been reported in various research fields, such as color change sensors for lights, DNA sensors, and color barcode nanowire (NW) [9–11,13,15–18].

The optical and electrical properties of tris(8-hydroxyquinoline) aluminum (Alq_3), which is one of the most widely used organic small molecules in diverse opto-electronic devices [19–21], were increased by modifying the chemical structures or changing the crystal structures [22–24] Furthermore, it has also been reported that the LSPR phenomena of Alq_3 and metal nanostructures influence its optical properties [25]. After the report on a crystalline form of the organic small molecules was published, such as rubrene, exhibiting highly stable and increasing PL characteristics [26–29], a single crystal based on a benzene derivative showed efficient and planar optical waveguiding capabilities, i.e., the lateral propagation of normally incident photon energy, which is very different from the axial waveguiding commonly observed in optical fibers [30]. These reports would imply that the crystalline organic small molecular structure can be an excellent and potential candidate for dielectric counterparts in hybridization, which is essential for the LSPR procedure [18,31–33].

This paper reports a large enhancement of the LSPR phenomenon in crystalline Alq_3 micro-rod (MR) hybridized with silver (Ag) NWs, which are denoted as "hybrid Alq_3/Ag NWs-MR" hereafter. Crystalline Alq_3 MRs were fabricated by a self-assembly method with the aid of a surfactant, as well as an amphiphilic solvent. For the hybrid Alq_3/Ag NWs-MRs, surface-functionalized Ag NWs were attached on the surface of the Alq_3 MR. The formation and hybridization between the crystalline Alq_3 MRs and Ag NWs were investigated using elementary analyses and high-resolution electron microscopy experiments. The nanoscale and solid state optical properties of the pristine Alq_3 MRs (i.e., without Ag NWs) and hybrid Alq_3/Ag NW-MRs were also compared to support the LSPR effects in the developed hybrid systems.

2. Materials and Methods

2.1. Synthesis of Hybrid Alq_3/Ag NWs-MRs

Alq_3 ($C_{27}H_{18}AlN_3O_3$, purity 99.995%) and cetyltrimethylammonium bromide (CTAB; $C_{19}H_{42}BrN$, purity 99.0%) were purchased from Sigma Aldrich (St. Louis, MO, USA) and used without further purification. For the crystalline Alq_3 MRs, 8 mg of Alq_3 powder was dissolved in 1 mL of tetrahydrofuran (THF). Next, an aqueous CTAB solution was prepared with a concentration of 1 $mg \cdot mL^{-1}$ in which CTAB acted as a surfactant. Subsequently, an Alq_3 solution was injected dropwise into an aqueous CTAB solution, resulting in a homogeneously dispersed crystalline Alq_3 MRs solution due to the amphiphilic nature of the THF, as well as the CTAB surfactants. It should also be noted that only Alq_3 nanoparticles (NPs) were obtained without a surfactant [10,11,24,34].

For the hybrid Alq_3/Ag NWs-MRs, 0.5 wt% of the Ag NWs-dispersed aqueous solution was added into 0.5 mL of the Alq_3 MRs dispersed solution. Ag NWs were fabricated with polyol methods, in which the surface of the each Ag NW was encapsulated by a thin layer of polyvinylpyrrolidone [35–37]. Thus, surface-functionalized Ag NWs were compatibly attached onto the surfaces of the crystalline Alq_3 MRs. The length and diameter of the Ag NW were a few tens of micrometers and approximately 40 nm, respectively.

2.2. Formation of Crystalline Hybrid MRs

The surface morphology of the crystalline Alq_3 MRs was analyzed using field emission scanning electron microscope (FE-SEM; Hitachi, Tokyo, Japan, SU-8010) at an acceleration voltage of 15 kV. The crystalline structures of the pristine and hybrid MRs were analyzed using a high-resolution transmission electron microscope (HR-TEM; Tecnai G2, Fei) with an acceleration voltage of 200 kV and selected area electron diffraction (SAED). X-ray diffraction (XRD; X'Pert Powder Diffractometer, PANalytical) patterns were acquired at a voltage of 40 kV and current of 40 mA using Cu-Kα radiation (λ = 1.540 Å). The scan rate was 0.02 $degree \cdot s^{-1}$, and the 2θ range was $2-80°$. Luminescent color charge-coupled device (CCD) images were acquired using an AVT Marlin F-033C (λ_{ex} = 435 nm) instrument.

2.3. *Photoluminescences of Crystalline Hybrid MRs*

The PL spectra were acquired using a lab-made laser confocal microscope (LCM) instrument. The 405 nm line of an unpolarized diode laser was used to excite the samples. The crystalline Alq_3 MRs were placed on a glass substrate mounted on the XY stage of the LCM. An oil immersion objective lens (N.A. of 1.4) was used to focus the unpolarized laser light on the samples with a spot size of approximately 200 nm. The scattered light was collected using the same objective lens, and the excitation laser light was filtered out through a long-pass edge filter (Semrock, Rochester, NY, USA). The red-shifted PL signal was focused onto a multimode fiber (core size = 50 μm), which acted as a pinhole for confocal detection. The other end of the multimode fiber was connected to a photomultiplier tube for the acquisition of the PL image or the input slit of a 0.3 m long monochromator equipped with a cooled CCD for acquisition of the PL spectra. To compare the brightness (i.e., luminescence intensity) of the CCD images of the pristine and hybrid MRs, the irradiation time was fixed at 0.1 s. For a quantitative comparison, the incident laser power and acquisition time for each LCM PL spectrum were fixed at 5 μW and 1 s, respectively, in all the LCM PL measurements [38,39]. For statistical justification, more than 20 spectra for the pristine and hybrid sample were averaged by using data plotting software (Gnuplot Ver 5.4).

3. Results and Discussion

The shapes and surface morphologies of the pristine Alq_3 MRs and hybrid Alq_3/Ag NWs-MRs were confirmed through SEM experiments. The crystallinity and structural characteristics were verified using HR-TEM and SAED experiments. The pristine Alq_3 MRs exhibited a uniform and continuous one-dimensional array, with a mean length of approximately 10 μm, as shown in the side view SEM image in Figure 1a. We observed that the pristine Alq_3 MRs exhibited a hexagonal cross-section, with a diameter of 0.5−1 μm, as shown in the magnified SEM image in Figure 1b. From the crystalline lattice image of the HR-TEM shown in Figure 1c, we can conclude that the Alq_3 MRs were well grown in a crystalline form. The SAED pattern in Figure 1d also demonstrates the intrinsic hexagonal crystallinity of Alq_3.

To obtain more convincing evidence of the hybridization of the Alq_3 MRs and Ag NWs, the morphology of the hybrid Alq_3/Ag NWs-MRs was carefully investigated using SEM, as shown in Figure 2a,b and the HR-TEM in Figure 2c. The surface-functionalized Ag NWs were intertwined with the crystalline Alq_3 MRs, which can be well observed in Figure 2b,d. HR-TEM acquires images through the transmitted electrons, which facilitated the observation of the coexistence of two different materials with distinct electron transmissions characteristics. As shown in Figure 2c, the Ag part in the hybrid MR is relatively dark due to the low electron transmission, whereas the Alq_3 part with good transmission is brighter than the Ag NWs are. Figure 2d–g present the energy-dispersive X-ray spectroscopic (EDX) mapping experiments conducted in the red, dashed box in Figure 2c, which shows the primary element distribution. The red color in Figure 2e, the yellow in Figure 2f, and the blue in Figure 2g denote aluminum (Al), carbon (C), and Ag, respectively. Three strands of the Alq_3 MRs can be identified in Figure 2e,f. In addition, Al and C were homogeneously distributed in each strand of the MR. The blue color distribution of Ag matches well with the shape of the Ag NW attached on the crystalline Alq_3 MRs, as shown in Figure 2g.

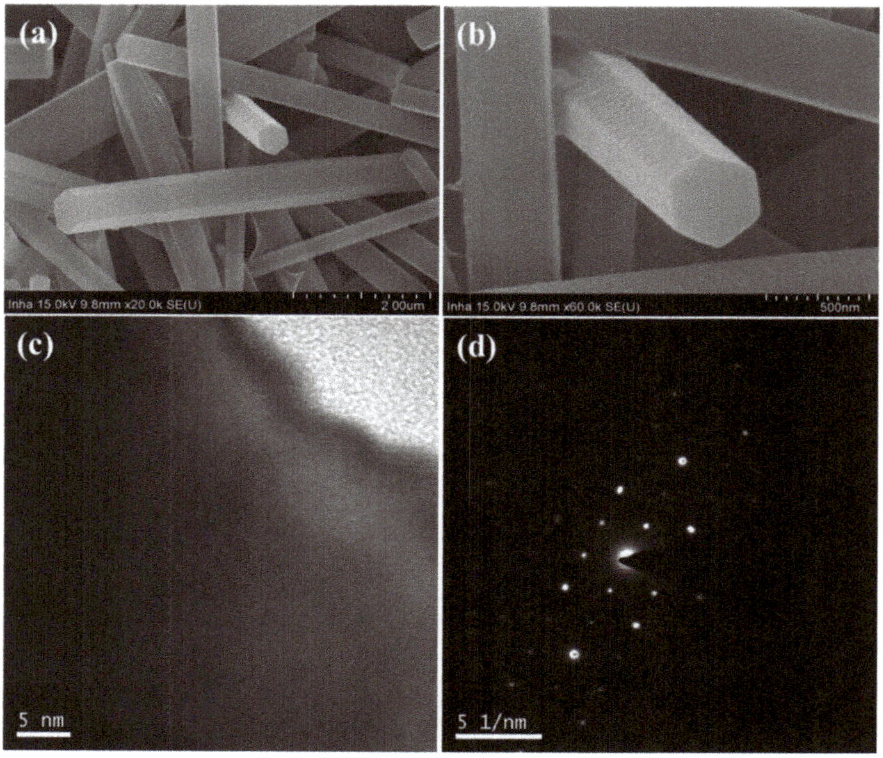

Figure 1. (**a**,**b**) show SEM images of the pristine Alq_3 MRs with different magnifications; (**c**) HR-TEM image of the single crystalline Alq_3 MR. (**d**) Corresponding SAED pattern of (**c**).

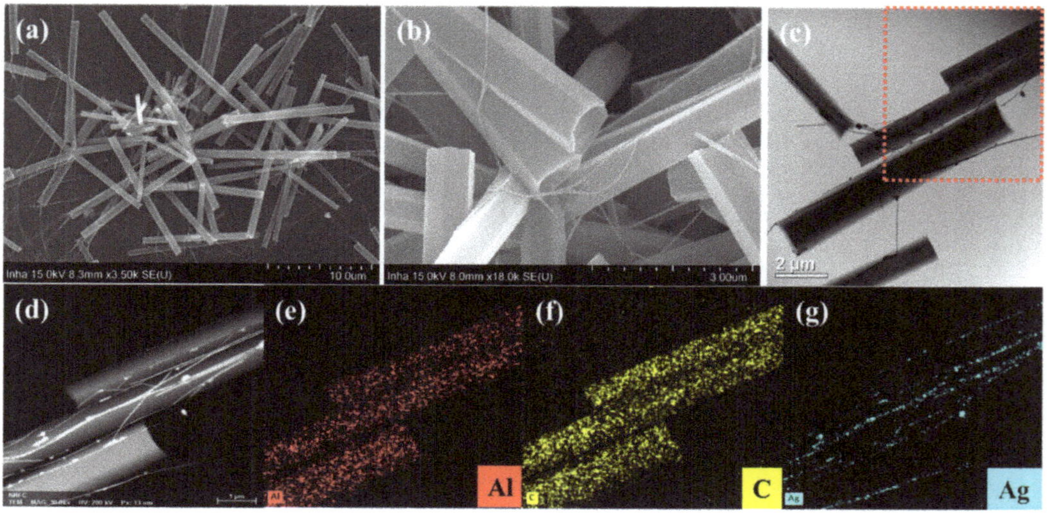

Figure 2. (**a**,**b**) show SEM images of the hybrid Alq_3/Ag NWs-MRs. (**c**) HR-TEM image of a single hybrid Alq_3/Ag NWs-MR. (**d**–**g**) EDX mapping images in the red dashed box of (**c**), exhibiting Al (**e**), C (**f**), and Ag (**g**), respectively.

Figure 3a shows the normalized ultraviolet (UV)-visible absorption spectra of the Ag NWs, pristine Alq_3 MRs, and hybrid Alq_3/Ag NWs-MRs. In the absorption spectra of the Ag NWs and pristine Alq_3 MRs, absorption peaks were observed at 355 and 371 nm, respectively.

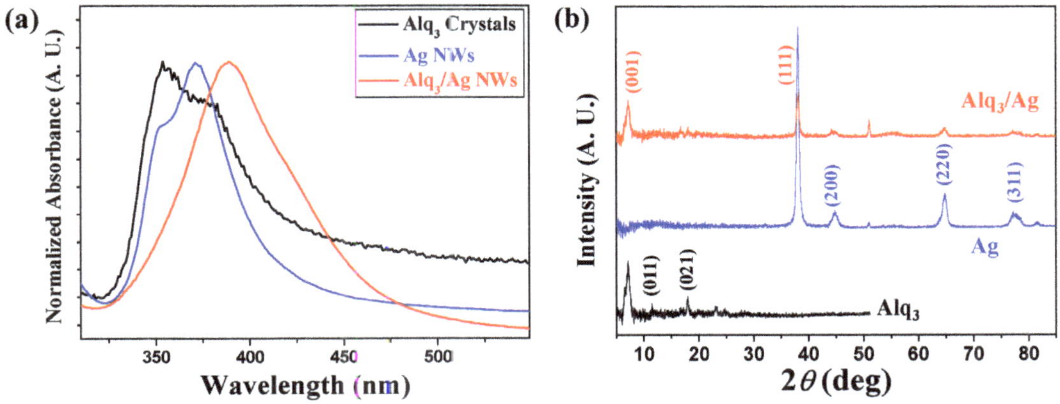

Figure 3. Comparison of the (**a**) UV absorption spectra and (**b**) XRD patterns of the pristine Alq_3 MRs (black), Ag NWs (blue), and hybrid Alq_3/Ag NWs-MRs (red), respectively.

It should be noted that there exists a relatively large overlap in the absorption spectrum of the pristine Alq_3 MRs with that of the Ag NWs, which are beneficial for energy transfer between them. The absorption at 355 nm due to the π-π^* transition plays an important role in the bright green emission of Alq_3, which corresponds to the plasmon band energy of Ag (E_g = 3.34 eV). Owing to the good matching of the energy levels, the interaction between Alq_3 and Ag contributes to the LSPR phenomenon. The spectrum of the hybrid Alq_3/Ag NWs-MRs showed an absorption peak at 389 nm. Compared with the intrinsic absorption of Ag NWs and Alq_3 MRs, the spectrum of the hybrid Alq_3/Ag NWs-MRs exhibited a relatively red shift, and a broader full width at half maximum because of the strong interaction between Alq_3 and Ag during LSPR coupling [5,9,40,41].

The XRD analysis shown in Figure 3b was performed as another method to determine the crystalline properties of the samples. Since the distances between the atoms and lattice structures are different depending on the materials, Miller indices were evaluated using Bragg's diffraction law and were compared with the measured 2θ values obtained using the XRD equipment. The black line in Figure 3b shows the XRD data of the pristine Alq_3 MRs. Alq_3 has d spacing values of 13.79, 7.68, and 4.97 Å, which were associated with the lattice planes of the (001), (011), and (021) directions, respectively. According to these XRD peaks, we can find that crystalline Alq_3 MRs grew with a typical pattern for α-Alq_3 [20]. In the case of Ag NWs, the XRD result shows that the d spacing values were 2.36, 2.05, 1.45, and 1.22 Å and their lattice planes were (111), (200), (220), and (311), respectively, as seen from the blue line of Figure 3b [37,42]. In the hybrid Alq_3/Ag NWs-MRs, we can find the XRD patterns of each material such as 2θ values of 6.4, 11.5, 17.8, 38, 44, and 64°, as shown in the red line of Figure 3b. These values of the hybrid Alq_3/Ag NWs-MRs coincide with individual XRD results of the Alq_3 and Ag, indicating the successful completion of hybridization in the crystalline form.

Figure 4a presents the PL spectrum of a single pristine Alq_3 MR. The maximum PL peak was observed at 511 nm, which is in good agreement with the relatively weak green light shown in Figure 4c. In Figure 4b, the PL intensities are compared with the hybrid Alq_3/Ag NWs-MR and pristine Alq_3 MR. The maximum PL peak of the hybrid Alq_3/Ag NWs-MRs was at 524 nm, which was slightly more red shifted compared to that of the pristine Alq_3 MR. However, the PL intensity of the hybrid Alq_3/Ag NWs-MR dramatically increased by approximately 26-fold. From the color CCD attached to LCM equipment, the luminescence images of the pristine Alq_3 MRs and hybrid Alq_3/Ag NWs-MR were directly obtained, as shown in Figure 4c,d, respectively. The luminescence image of the pristine Alq_3 MRs exhibited a weak green emission, whereas that of the hybrid Alq_3/Ag NWs-MRs was significantly brighter than the pristine ones were. When the plasmon activation energy of Ag NWs and the optical absorption energy of Alq_3 are well matched and harmonized, a plasmon resonance interaction occurs, and excited excitons are formed by the collective excitation of electrons [9–11,40,41]. Consequently, the emission efficiency of the hybrid Alq_3/Ag NWs-MRs was significantly increased, which could confirm the LSPR phenomenon through the improvement of the LCM PL intensities and brightness of the color CCD images. For quantitative comparison, the structural characteristics and optical features of the various organic/metal hybrid structures exhibiting a distinct PL intensity enhancement are listed in Table 1. Undoped poly (2-methoxy-5-(2′-ethylhexyloxy)-*p*-phenylene vinylene) (MEH-PPV) hybridized with gold (Au) NPs showed an approximate 30-fold enhancement of the PL intensity without a PL peak shift because the LSPR only results in the resonant energy transfer between MEH-PPV and Au [11]. However, PL peaks of the hybrid structures consisting of the doped polymeric nanostructures, e.g., electrochemically synthesized nanotubes (NTs) or NWs, were significantly increased and red shifted [4,5]. These results are attributed that the LSPR influences on both energy and charge transfer effect [9,10]. For the undoped organic small molecules, the enhancement of the PL intensity are only caused by the energy transfer effect. Therefore, it is very important to precisely control the crystalline nature of light-emitting organic small molecular materials, as well as metallic nanomaterials.

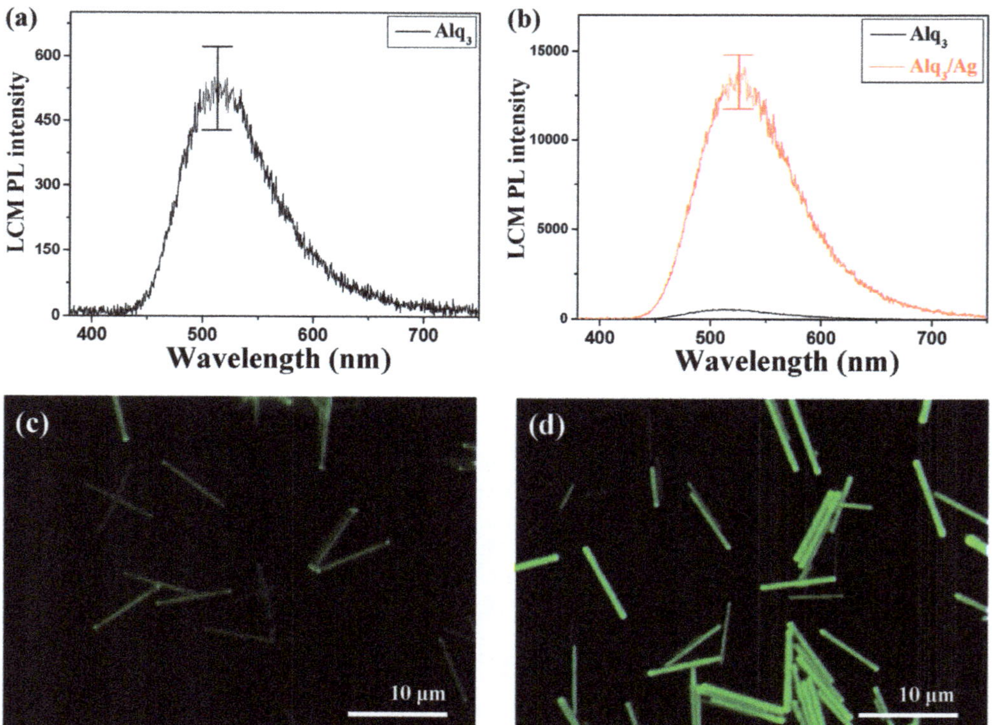

Figure 4. (**a**) PL spectrum of the pristine Alq_3 MRs. (**b**) Comparison of PL spectra of the pristine Alq_3 MRs and hybrid Alq_3/Ag NWs-MRs. (**c**,**d**) show the corresponding color CCD images of the pristine Alq_3 MRs (**a**) and hybrid Alq_3/Ag NWs-MRs (**b**), respectively.

Table 1. Comparison of the structural characteristics and typical PL features in the various hybrid organic/metal structures.

Organic Materials	Type	Metal	Type	PL Peak Shift	PL Enhancement (Fold)	Refs.
Alq_3	MR	Ag	NWs	X	~26	This Work
Alq_3	NW	Ag	NPs	X	4–5	[19]
Rubrene	1-dimensional MR	Au	NPs	X	~2	[18]
MEH-PPV	NP	Au	NPs	X	~30	[11]
Polythiophene	NT	Co/Cu/Ni	NT	O	40–70	[4]
Poly (3-methylthiophene)	NT	Ni	NT	O	~350	[5]

4. Conclusions

Ag and Alq_3, which are well matched with the plasmon excitation energy of metals and the band gap of organic semiconductors, were hybridized to improve the luminescent properties using the LSPR phenomenon. HR-TEM and SAED observations confirmed that Alq_3 grew in a crystalline form. The hybrid formation of the Ag NWs and crystalline Alq_3 MRs was confirmed through EDX and XRD analyses. The strength of the PL and color CCD images of the hybrid Alq_3/Ag NWs-MRs measured in the LCM PL equipment increased by approximately 26-fold compared to that of the pristine MRs, and they were significantly brighter. This indicated that strong LSPR coupling occurred. Future work would focus on the development of nanocomposite materials, where hybrid Alq_3/Ag NWs-MRs are embedded in dielectric matrices for the further enhancement of the photoluminescence by the LSPR and local field effects [43,44].

Author Contributions: Y.K.H., D.H.P. and K.-T.L. conceived and designed the experiments; M.K., J.K., S.J., H.K., I.J. and C.S.L. performed the experiments and analyzed the data; Y.K.H., D.H.P., J.H.J. and K.-T.L. contributed to the reagents, materials, and analytical tools. Y.K.H., D.H.P., J.H.J. and K.-T.L. contributed to the writing and reviewing of the manuscript. All authors have read and agreed to the published version of the manuscript.

Funding: This work was supported by INHA UNIVERSITY Research Grant.

Institutional Review Board Statement: Not applicable.

Informed Consent Statement: Not applicable.

Data Availability Statement: The data presented in this study are available upon request from the corresponding authors.

Conflicts of Interest: The authors declare no conflict of interest.

References

1. Schwartz, B.J. Conjugated Polymers As Molecular Materials: How Chain Conformation and Film Morphology Influence Energy Transfer and Interchain Interactions. *Annu. Rev. Phys. Chem.* **2003**, *54*, 141–172. [CrossRef] [PubMed]
2. Sun, R.; Wang, Y.; Wang, D.; Zheng, Q.; Kyllo, E.; Gustafson, T.; Epstein, A. High luminescent efficiency in light-emitting polymers due to effective exciton confinement. *Appl. Phys. Lett.* **2000**, *76*, 634–636. [CrossRef]
3. Kinkhabwala, A.; Yu, Z.; Fan, S.; Avlasevich, Y.; Müllen, K.; Moerner, W.E. Large single-molecule fluorescence enhancements produced by a bowtie nanoantenna. *Nat. Photonics* **2009**, *3*, 654–657. [CrossRef]
4. Joo, J.; Park, D.H.; Jeong, M.Y.; Lee, Y.B.; Kim, H.S.; Choi, W.J.; Park, Q.H.; Kim, H.J.; Kim, D.C.; Kim, J. Bright light emission of a single polythiophene nanotube strand with a nanometer-scale metal coating. *Adv. Mater.* **2007**, *19*, 2824–2829. [CrossRef]
5. Park, D.H.; Kim, H.S.; Jeong, M.Y.; Lee, Y.B.; Kim, H.J.; Kim, D.C.; Kim, J.; Joo, J. Significantly enhanced photoluminescence of doped polymer-metal hybrid nanotubes. *Adv. Funct. Mater.* **2008**, *18*, 2526–2534. [CrossRef]
6. Chiu, N.-F.; Lin, C.-W.; Lee, J.-H.; Kuan, C.-H.; Wu, K.-C.; Lee, C.-K. Enhanced luminescence of organic/metal nanostructure for grating coupler active long-range surface plasmonic device. *Appl. Phys. Lett.* **2007**, *91*, 083114. [CrossRef]
7. Park, H.-J.; Vak, D.; Noh, Y.-Y.; Lim, B.; Kim, D.-Y. Surface plasmon enhanced photoluminescence of conjugated polymers. *Appl. Phys. Lett.* **2007**, *90*, 161107. [CrossRef]

8. Kozhina, E.P.; Arzhanov, A.I.; Karimullin, K.R.; Bedin, S.A.; Andreev, S.N.; Naumov, A.V. Using Epi-Luminescence Microscopy to Visualize and Control the Distribution of Luminophores on a Highly-Developed Surface. *Bull. Russ. Acad. Sci. Phys.* **2021**, *85*, 1393–1399. [CrossRef]
9. Park, D.H.; Kim, M.S.; Joo, J. Hybrid nanostructures using π-conjugated polymers and nanoscale metals: Synthesis, characteristics, and optoelectronic applications. *Chem. Soc. Rev.* **2010**, *39*, 2439–2452. [CrossRef]
10. Hong, Y.K.; Park, D.H.; Lee, S.H.; Joo, J. Synthesis, characteristics, and applications of intrinsically light-emitting polymer nanostructures. *Adv. Polym. Sci.* **2013**, *259*, 201–244. [CrossRef]
11. Kim, M.S.; Park, D.H.; Cho, E.H.; Kim, K.H.; Park, Q.H.; Song, H.; Kim, D.C.; Kim, J.; Joo, J. Complex nanoparticle of light-emitting MEH-PPV with Au: Enhanced luminescence. *ACS Nano* **2009**, *3*, 1329–1334. [CrossRef]
12. Lee, K.C.; Lin, S.J.; Lin, C.H.; Tsai, C.S.; Lu, Y.J. Size effect of Ag nanoparticles on surface plasmon resonance. *Surf. Coat. Technol.* **2008**, *202*, 5339–5342. [CrossRef]
13. Park, D.H.; Hong, Y.K.; Cho, E.H.; Kim, M.S.; Kim, D.C.; Bang, J.; Kim, J.; Joo, J. Light-emitting color barcode nanowires using polymers: Nanoscale optical characteristics. *ACS Nano* **2010**, *4*, 5155–5162. [CrossRef]
14. Noh, H.S.; Cho, E.H.; Kim, H.M.; Han, Y.D.; Joo, J. Organic solar cells using plasmonics of Ag nanoprisms. *Org. Electron.* **2013**, *14*, 278–285. [CrossRef]
15. Park, D.H.; Kim, N.; Cui, C.; Hong, Y.K.; Kim, M.S.; Yang, D.H.; Kim, D.C.; Lee, H.; Kim, J.; Ahn, D.J.; et al. DNA detection using a light-emitting polymer single nanowire. *Chem. Commun.* **2011**, *47*, 7944–7946. [CrossRef]
16. Stebunov, Y.V.; Yakubovsky, D.I.; Fedyanin, D.Y.; Arsenin, A.V.; Volkov, V.S. Superior Sensitivity of Copper-Based Plasmonic Biosensors. *Langmuir* **2018**, *34*, 4681–4687. [CrossRef]
17. Mishra, S.K.; Zou, B.; Chiang, K.S. Surface-Plasmon-Resonance Refractive-Index Sensor with Cu-Coated Polymer Waveguide. *IEEE Photonics Technol. Lett.* **2016**, *28*, 1835–1838. [CrossRef]
18. Hwang, H.S.; Jo, S.G.; Lee, J.; Kim, J.; Joo, J. Enhanced Local and Nonlocal Photoluminescence of Organic Rubrene Microrods using Surface Plasmon of Gold Nanoparticles: Applications to Ultrasensitive and Remote Biosensing. *J. Phys. Chem. C* **2016**, *120*, 11612–11620. [CrossRef]
19. Khan, M.B.; Khan, Z.H. Ag-incorporated Alq_3 nanowires: Promising material for organic luminescent devices. *J. Lumines.* **2017**, *188*, 418–422. [CrossRef]
20. Hu, J.S.; Ji, H.X.; Cao, A.M.; Huang, Z.X.; Zhang, Y.; Wan, L.J.; Xia, A.D.; Yu, D.P.; Meng, X.M.; Lee, S.T. Facile solution synthesis of hexagonal Alq_3 nanorods and their field emission properties. *Chem. Commun.* **2007**, *29*, 3083–3085. [CrossRef]
21. Zlojutro, V.; Sun, Y.; Hudson, Z.M.; Wang, S. Triarylboron-functionalized 8-hydroxyquinolines and their aluminium(III) complexes. *Chem. Commun.* **2011**, *47*, 3837–3839. [CrossRef]
22. Chan, G.H.; Zhao, J.; Hicks, E.M.; Schatz, G.C.; Van Duyne, R.P. Plasmonic properties of copper nanoparticles fabricated by nanosphere lithography. *Nano Lett.* **2007**, *7*, 1947–1952. [CrossRef]
23. Chalana, S.R.; Ganesan, V.; Mahadevan Pillai, V.P. Surface plasmon resonance in nanostructured Ag incorporated ZnS films. *AIP Adv.* **2015**, *5*, 107207. [CrossRef]
24. Cui, C.; Park, D.H.; Kim, J.; Joo, J.; Ahn, D.J. Oligonucleotide assisted light-emitting Alq_3 microrods: Energy transfer effect with fluorescent dyes. *Chem. Commun.* **2013**, *49*, 5360–5362. [CrossRef]
25. Zeng, S.; Yu, X.; Law, W.C.; Zhang, Y.; Hu, R.; Dinh, X.Q.; Ho, H.P.; Yong, K.T. Size dependence of Au NP-enhanced surface plasmon resonance based on differential phase measurement. *Sens. Actuator B-Chem.* **2013**, *176*, 1128–1133. [CrossRef]
26. Park, D.H.; Jo, S.G.; Hong, Y.K.; Cui, C.; Lee, H.; Ahn, D.J.; Kim, J.; Joo, J. Highly bright and sharp light emission of a single nanoparticle of crystalline rubrene. *J. Mater. Chem.* **2011**, *21*, 8002–8007. [CrossRef]
27. Lee, J.W.; Jo, S.G.; Joo, J.; Kim, J.H.; Rho, H. Polarized photoluminescence characteristics of rubrene sheets with nanometer thickness. *Phys. Rev. B* **2012**, *86*, 075416. [CrossRef]
28. Park, H.J.; Kim, M.S.; Kim, J.; Joo, J. Fine control of photoluminescence and optical waveguiding characteristics of organic rubrene nanorods using focused electron-beam irradiation. *J. Phys. Chem. C* **2014**, *118*, 30179–30186. [CrossRef]
29. Park, C.J.; Kim, M.S.; Kim, J.; Joo, J. Electrically controlled photoluminescence efficiency of organic rubrene microplates. *Synth. Met.* **2015**, *199*, 394–399. [CrossRef]
30. Cho, E.; Choi, J.; Jo, S.; Park, D.-H.; Hong, Y.K.; Kim, D.; Lee, T.S. A Single-Benzene-Based Fluorophore: Optical Waveguiding in the Crystal Form. *ChemPlusChem* **2019**, *84*, 1130–1134. [CrossRef]
31. Cui, C.; Kim, S.; Ahn, D.J.; Joo, J.; Lee, G.S.; Park, D.H.; Kim, B.H. Unusual enhancement of fluorescence and Raman scattering of core-shell nanostructure of polydiacetylene and Ag nanoparticle. *Synth. Met.* **2018**, *236*, 19–23. [CrossRef]
32. Kim, T.J.; Kim, D.I.; Lee, Y.; Kim, J.; Joo, J. Enhanced blue photoluminescence and new crystallinity of Ag/organic rubrene core-shell nanoparticles through hydrothermal treatment. *Curr. Appl. Phys.* **2020**, *20*, 1201–1206. [CrossRef]
33. Dhakal, K.P.; Joo, J.; Kim, J. Molecular crystallization of rubrene thin films assisted by gold nanoparticles. *Curr. Appl. Phys.* **2022**, *39*, 304–310. [CrossRef]
34. Kim, M.; Kim, J.; Kim, H.; Jung, I.; Kwak, H.; Lee, G.S.; Na, Y.J.; Hong, Y.K.; Park, D.H.; Lee, K.T. Direct Visualization of UV-Light on Polymer Composite Films Consisting of Light Emitting Organic Micro Rods and Polydimethylsiloxane. *Polymers* **2022**, *14*, 1846. [CrossRef]
35. Gao, Y.; Jiang, P.; Liu, D.F.; Yuan, H.J.; Yan, X.Q.; Zhou, Z.P.; Wang, J.X.; Song, L.; Liu, L.F.; Zhou, W.Y.; et al. Synthesis, characterization and self-assembly of silver nanowires. *Chem. Phys. Lett.* **2003**, *380*, 146–149. [CrossRef]

36. Xia, Y.; Xiong, Y.; Lim, B.; Skrabalak, S.E. Shape-controlled synthesis of metal nanocrystals: Simple chemistry meets complex physics? *Angew. Chem. Int. Ed.* **2009**, *48*, 60–103. [CrossRef]
37. Kumar, D.; Singh, K.; Verma, V.; Bhatti, H.S. Microwave-assisted synthesis and characterization of silver nanowires by polyol process. *Appl. Nanosci.* **2015**, *5*, 881–890. [CrossRef]
38. Hong, Y.K.; Park, D.H.; Park, S.K.; Song, H.; Kim, D.C.; Kim, J.; Han, Y.H.; Park, O.K.; Lee, B.C.; Joo, J. Tuning and Enhancing Photoluminescence of Light-Emitting Polymer Nanotubes through Electron-Beam Irradiation. *Adv. Funct. Mater.* **2009**, *19*, 567–572. [CrossRef]
39. Hong, Y.K.; Park, D.H.; Jo, S.G.; Koo, M.H.; Kim, D.C.; Kim, J.; Kim, J.S.; Jang, S.Y.; Joo, J. Fine Characteristics Tailoring of Organic and Inorganic Nanowires Using Focused Electron-Beam Irradiation. *Angew. Chem. Int. Ed.* **2011**, *50*, 3734–3738. [CrossRef]
40. Chen, Y.; Munechika, K.; Ginger, D.S. Dependence of fluorescence intensity on the spectral overlap between fluorophores and plasmon resonant single silver nanoparticles. *Nano Lett.* **2007**, *7*, 690–696. [CrossRef]
41. Munechika, K.; Chen, Y.; Tillack, A.F.; Kulkarni, A.P.; Plante, I.J.L.; Munro, A.M.; Ginger, D.S. Spectral control of plasmonic emission enhancement from quantum dots near single silver nanoprisms. *Nano Lett.* **2010**, *10*, 2598–2603. [CrossRef]
42. Cheng, Z.; Liu, L.; Xu, S.; Lu, M.; Wang, X. Temperature dependence of electrical and thermal conduction in single silver nanowire. *Sci. Rep.* **2015**, *5*, 10718. [CrossRef] [PubMed]
43. Dolgaleva, K.; Boyd, R.W. Local-field effects in nanostructured photonic materials. *Adv. Opt. Photonics* **2012**, *4*, 1–77. [CrossRef]
44. Gladush, M.G.; Anikushina, T.A.; Gorshelev, A.A.; Plakhotnik, T.V.; Naumov, A.V. Dispersion of Lifetimes of Excited States of Single Molecules in Organic Matrices at Ultralow Temperatures. *J. Exp. Theor. Phys.* **2019**, *128*, 655–663. [CrossRef]

nanomaterials

Communication

Enhancing Photoluminescence of $CsPb(Cl_xBr_{1-x})_3$ Perovskite Nanocrystals by Fe^{2+} Doping

Chang Wu [1], Yan Li [2], Zhengyao Xia [1], Cheng Ji [1], Yuqian Tang [1], Jinlei Zhang [1], Chunlan Ma [1,*] and Ju Gao [1,3,*]

1 Jiangsu Key Laboratory of Micro and Nano Heat Fluid Flow Technology and Energy Application, School of Physical Science and Technology, Suzhou University of Science and Technology, Suzhou 215009, China
2 School of Environmental Science and Engineering, Suzhou University of Science and Technology, Suzhou 215009, China
3 School Optoelect Engn, Zaozhuang University, Zaozhuang 277160, China
* Correspondence: wlxmcl@mail.usts.edu.cn (C.M.); jugao@khu.hk (J.G.)

Abstract: The doping of impurity ions into perovskite lattices has been scrupulously developed as a promising method to stabilize the crystallographic structure and modulate the optoelectronic properties. However, the photoluminescence (PL) of Fe^{2+}-doped mixed halide perovskite NCs is still relatively unexplored. In this work, the Fe^{2+}-doped $CsPb(Cl_xBr_{1-x})_3$ nanocrystals (NCs) are prepared by a hot injection method. In addition, their optical absorption, photoluminescence (PL), PL lifetimes, and photostabilities are compared with those of undoped $CsPb(Br_{1-x}Cl_x)_3$ NCs. We find the Fe^{2+} doping results in the redshift of the absorption edge and PL. Moreover, the full width at half maximums (FWHMs) are decreased, PL quantum yields (QYs) are improved, and PL lifetimes are extended, suggesting the defect density is reduced by the Fe^{2+} doping. Moreover, the photostability is significantly improved after the Fe^{2+} doping. Therefore, this work reveals that Fe^{2+} doping is a very promising approach to modulate the optical properties of mixed halide perovskite NCs.

Keywords: photoluminescence; perovskite nanocrystals; Fe^{2+} doping

Citation: Wu, C.; Li, Y.; Xia, Z.; Ji, C.; Tang, Y.; Zhang, J.; Ma, C.; Gao, J. Enhancing Photoluminescence of $CsPb(Cl_xBr_{1-x})_3$ Perovskite Nanocrystals by Fe^{2+} Doping. *Nanomaterials* **2023**, *13*, 533. https://doi.org/10.3390/nano13030533

Academic Editor: Vincenzo Vaiano

Received: 30 December 2022
Revised: 19 January 2023
Accepted: 20 January 2023
Published: 28 January 2023

1. Introduction

Lead halide ABX_3 (X = Cl, Br, I) perovskites are very competitive semiconductors for solar cells, lighting, and photodetectors owing to their tremendous optical properties, such as a tunable fluorescent color across the whole visible range, strong absorption, high photoluminescence (PL) quantum yields (QYs), narrow PL bandwidths, and suppressed PL blinking [1–8]. Although the perovskites exhibit extraordinary potential and there are already many exciting progresses, the further commercialization of these promising semiconductors still confronts severe challenges. One is the inclusion of toxic Pb and another one is the instability against moisture, oxygen, heat, and electric current/irradiation [9–14]. In particular, for the mixed halide perovskites, the photo- or electric-induced ion migration is ineluctable, resulting in phase segregation [15–17]. Therefore, during the past years, many research works have been conducted to decrease the use of toxic Pb and improve the stability of perovskites [18–20]. To avoid the utilization of toxic lead, various lead-free perovskites, such as Sn^{2+}, Sn^{4+}, Mn^{2+}, and Cu^{2+}-based perovskites have been exploited [18–20]. Among them, tin-based perovskites have been the most explored [19]. However, the Sn^{2+} in perovskite is easily oxidized to Sn^{4+}, causing poor stability [21]. The PL QYs of tin-based perovskites and power conversion efficiency (PCE) of solar cells based on tin-based perovskites are very low [11].

Doping of other metal ions to partially replace Pb is another effective way to reduce the utilization of toxic lead, while maintaining or even enhancing the excellent optical and photoelectrical properties. The incorporation of appropriate impurity ions into host lattices has been exploited as a promising method to stabilize the crystallographic phases while modulating the optical, electronic, and magnetic properties of diverse semiconductors [22–28]. Regarding the halide perovskites, the partial substitution of Pb^{2+} by divalent metal ions, such as Cu^{2+}, Mg^{2+}, Fe^{2+}, Co^{2+}, Ni^{2+}, and Mn^{2+} at the B sites of the perovskite lattice have been demonstrated [29–36]. Klug et al revealed that the perovskite films retain an excellent photovoltaic performance if less than 3% Pb^{2+} ions are substituted by homovalent metal species due to the high tolerance of the perovskite lattices [31]. To date, many different research groups have recently demonstrated that Mn^{2+} ions can be doped into the B sites of perovskite lattices by using a facile approach [32–35]. Moreover, a broad PL peak at about 600 nm is induced by the $^4T_1$ to $^6A_1$ transition of Mn^{2+}, which can be applied for multicolor luminescence [35]. After anion exchange reactions between Mn-doped $CsPbCl_3$ and $CsPbBr_3$, fluorescence color gamut almost covering the entire visible spectrum are obtained [35]. Thanks to the high stability and wide color gamut, color converters for light-emitting diodes (LEDs) were constructed. Similarly, the doping of Ni^{2+} ions into all inorganic perovskite nanocrystals (NCs) can also modulate the PL. Sun et al have a general strategy for the synthesis of Ni-doped $CsPbCl_3$ NCs, which shows a strong single-color violet emission with a maximum PL QY of 96.5% [36].

Fe ions, as the earth-abundant elements, are eco-friendly and low-cost. The doping of Fe ions in perovskites has also attracted research interests. $CH_3NH_3PbCl_3$ single crystals with different concentrations of Fe^{2+}/Fe^{3+} doping were synthesized by Cheng et al [37]. In addition, the crystal structure, optical, and optoelectronic properties were investigated. They revealed that Fe^{2+} is prone to replacing Pb^{2+} and the optoelectronic properties are seriously deteriorated. On the contrary, Hu et al. reported that an appropriate amount of Fe^{2+} doping into the lattice of $CsPbCl_3$ NCs not only improved the homogeneity of the size of the NCs, but also enhanced the PL QY and average PL lifetimes [38]. Therefore, the impact of Fe^{2+} doping on the optoelectronic properties of perovskite NCs is still unclear.

In this work, the Fe^{2+}-doped $CsPb_{1-x}Fe_x(Br_{1-x}Cl_x)_3$ NCs are prepared by mixing $FeCl_2$ (x mmol) and $PbBr_2$ (1−x mmol) during the hot injection process. When x is not zero, Pb^{2+} at the B sites are partially replaced by Fe^{2+}; meanwhile, the X sites are also partially occupied by Cl-. Therefore, the PL properties of Fe^{2+}-doped mixed halide perovskite NCs are investigated. To our knowledge, the PL of Fe^{2+}-doped mixed halide perovskite NCs is studied for the first time. The morphology and size distribution of the Fe^{2+}-doped perovskite NCs are investigated by a transmission electron microscope (TEM). The optical absorption, PL, PL lifetimes, and photostabilities of the Fe^{2+}-doped perovskite NCs are measured, which are compared with those of the undoped $CsPb(Br_{1-x}Clx)_3$ NCs.

2. Materials and Methods

2.1. Synthesis of $CsPb_{1-x}Fe_x(Br_{1-x}Cl_x)_3$ (x = 0, 0.1, 0.2, and 0.3) NCs

Cs-oleate was prepared by dispersing Cs_2CO_3 powders (0.407 g, 1.25 mmol) into a mixture of 18 mL octadecene (ODE, Aladdin, 90%) and 1.74 mL oleic acid (OA, Aldrich, 90%), which was then heated to 150 °C under N_2 atmosphere until all Cs_2CO_3 powders were reacted. Then, $FeCl_2$ (x mmol, x = 0, 0.1, 0.2, and 0.3) and $PbBr_2$ (1−x mmol) were mixed with OLA (1.5 mL), oleylamine (OLA, 70%, 1.5 mL), trioctylphosphine (TOP, 90%, 1 mL), and ODE (10 mL) in a 100 mL 3-neck flask. The mixture was then degassed at 110 °C for 40 min. After that, the mixture was heated to 170 °C under N_2 atmosphere. After reaction for 15 s, the hot Cs-oleate precursor (1 mL) was quickly injected. Subsequently, an ice-water bath was used to cool down the reaction. The resulting solution was centrifuged at 5000 r/m for 5 min, the supernatant was discarded. To remove the residual reactants, the solids were redispersed in hexane and centrifuged again for 5 min. When x = 0.4, well-shaped perovskite nanocrystals cannot be obtained. The reaction was conducted under protection of N_2 and no other oxidant was added. Thus, the oxidation of feeding Fe^{2+} to Fe^{3+} is negligible. Moreover, Fe^{2+} is expected to replace Pb^{2+} as confirmed previously [37,38]. It is reasonable that the main valence of doping iron is bivalent.

2.2. Characterizations

Transmission electron microscope (TEM) images were captured on the JEM-2100F electron microscope (JOEL, Japan Electronics Co., Ltd, Tokyo, Japan). Elemental analysis was conducted using energy-dispersive X-ray spectroscopy (EDS) coupled on the TEM. The absorption spectra were measured by using a Shimazu UV2600 UV-Vis spectrophotometer, Shimazu Co., Ltd, Tokyo, Japan). The PL spectra were measured by a Maya 2000 Pro high sensitivity spectrometer (Ocean Optics Co., Ltd, Orlando, USA). The time-resolved PL decay curves were measured by a PicoHarp 300 time-correlated single photon counting (TCSPC) system (Pico Quant Co., Ltd, Berlin, Germany).

3. Results and Discussion

3.1. Structure

Figure 1a,b show the TEM and size distribution of the $CsPb(Br_{0.8}Cl_{0.2})_3$ NCs without the Fe^{2+} doping. The Fe^{2+}-doped perovskite NCs were prepared by a hot injection method. The feeding ratio between the $FeCl_2$ (x mmol) and $PbBr_2$ (1−x mmol) was adjusted to simultaneously control the doping concentration and halide composition. Figure 1c shows the TEM image of the $CsPb_{1-x}Fe_x(Br_{1-x}Cl_x)_3$ NCs when x = 0.2. It shows that the NCs are monodispersed with regular cubic shapes. In addition, a size distribution is obtained based on the TEM image. As shown in Figure 1d, the sizes of the $CsPb_{0.8}Fe_{0.2}(Br_{0.8}Cl_{0.2})_3$ NCs are about 9.3 nm with a relatively uniform distribution. The elemental distributions are measured by energy-dispersive X-ray spectroscopy (EDS) equipped on the TEM. The result shown in Figure 1e suggests the homogeneous presence of Fe, indicating the Fe^{2+} ions are successfully doped into NCs. The actual atomic ratio between Pb and Fe determined by the whole EDS mapping of Figure 1e is about 4.3:1 in Figure 1f. Both the shape and size distribution of the $CsPb(Br_{1-x}Cl_x)_3$ NCs are very similar to those of the $CsPb_{1-x}Fe_x(Br_{1-x}Cl_x)_3$ NCs. Therefore, we found the doping of the Fe^{2+} ion at a low concentration has a negligible influence on the growth of perovskite NCs.

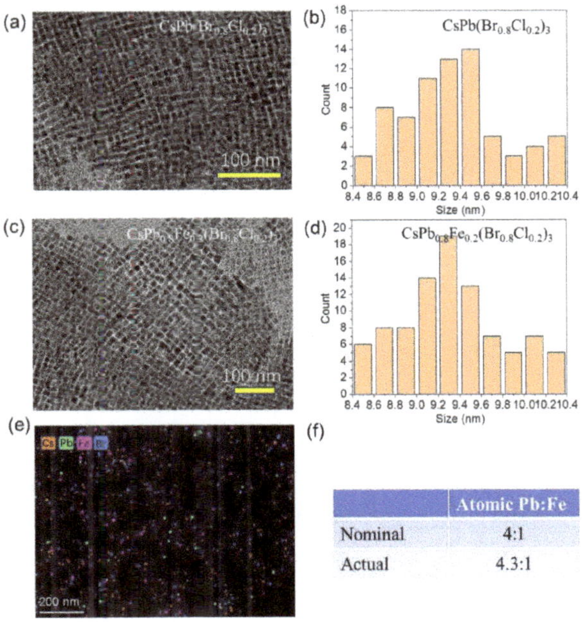

Figure 1. (**a**,**b**) TEM image (**a**) and size distribution (**b**) of the $CsPb(Br_{0.8}Cl_{0.2})_3$ NCs. (**c**,**d**) TEM image and size distribution of the $CsPb_{0.8}Fe_{0.2}(Br_{0.8}Cl_{0.2})_3$ NCs. (**e**) Elemental mappings of the $CsPb_{0.8}Fe_{0.2}(Br_{0.8}Cl_{0.2})_3$ NCs. (**f**) Atomic ratio between Pb and Fe of the $CsPb_{0.8}Fe_{0.2}(Br_{0.8}Cl_{0.2})_3$ NCs.

The absorption spectra of the $CsPb_{1-x}Fe_x(Br_{1-x}Cl_x)_3$ NCs are shown in Figure 2a. All the absorption spectra show a sharp edge with a negligible Urbach tail, indicating the low density of defect trapping. When x is not zero, Pb^{2+} at the B sites are partially replaced by Fe^{2+}; meanwhile, the X sites are also partially occupied by Cl^-. Typically, as the fraction of Cl^- increases, the bandgap of the perovskite increases and the PL peak position blueshifts [39]. The relationship between the absorption edge position and x value is presented in Figure 2b. Surprisingly, the absorption edge of the $CsPb_{0.9}Fe_{0.1}(Br_{0.9}Cl_{0.1})_3$ NCs redshift slightly compared to that of the $CsPbBr_3$. As x increases from 0 to 0.3, the absorption edge slightly redshifts first and then blueshifts. Mixed halide $CsPb(Br_{1-x}Cl_x)_3$ perovskite NCs without the Fe doping are also synthesized by mixing $PbCl_2$ (x mmol) and $PbBr_2$ (1−x mmol) during the preparation. The absorption edge positions of the $CsPb(Br_{1-x}Cl_x)_3$ NCs without the Fe^{2+} doping are added for comparison. The absorption edge position gradually blueshifts as x increases. All the absorption edges of the Fe^{2+}-doped NCs redshift compared to those of the corresponding perovskite NCs without the Fe^{2+} doping.

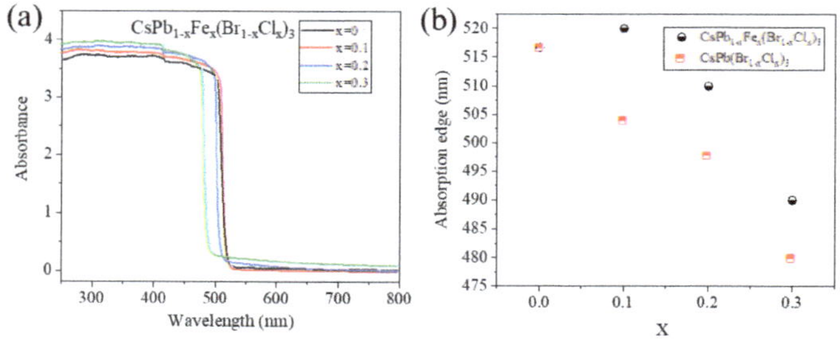

Figure 2. (**a**) Absorption spectra of the $CsPb_{1-x}Fe_x(Br_{1-x}Cl_x)_3$ NCs. (**b**) Absorption band edge positions of the $CsPb_{1-x}Fe_x(Br_{1-x}Cl_x)_3$ NCs at different x. Absorption band edge positions of the $CsPb_{1-x}Fe_x(Br_{1-x}Cl_x)_3$ are also plotted for comparison.

3.2. Photoluminescence Properties

Photos of the $CsPb_{1-x}Fe_x(Br_{1-x}Cl_x)_3$ NC solutions under natural daylight and UV light are shown in Figure 3a. As x changes from 0 to 0.3, the fluorescent color turns from green to cyan. The corresponding PL spectra of the $CsPb_{1-x}Fe_x(Br_{1-x}Cl_x)_3$ NCs with x = 0 to 0.3 are displayed in Figure 3b. For x = 0, the PL peak position of the $CsPbBr_3$ NCs appears at about 519 nm, which is in good agreement with previous work. Similar to the shift of the absorption edge, when x = 0.1, the PL peak position of the $CsPb_{1-x}Fe_x(Br_{1-x}Cl_x)_3$ NCs redshifts to 521 nm. As x increases to 0.2 and 0.3, the PL peak positions apparently blueshift. When x = 0.3, the peak position appears at 490 nm, explaining the cyan fluorescent color. The PL spectra of the $CsPb(Br_{1-x}Cl_x)_3$ NCs without the Fe^{2+} doping are shown in Figure 3c. The PL peak position monotonously blueshifts as the x value increases without any exception. The dependencies of the PL peak position on the x value for these two groups of perovskite NCs are plotted in Figure 3d. For the same x value, the PL peak position of the $CsPb_{1-x}Fe_x(Br_{1-x}Cl_x)_3$ NCs always redshifts compared to that of the $CsPb(Br_{1-x}Cl_x)_3$ NCs, which is again in good agreement with the shift of the absorption edge. Therefore, we can conclude that the Fe^{2+} doping results in the redshift of PL.

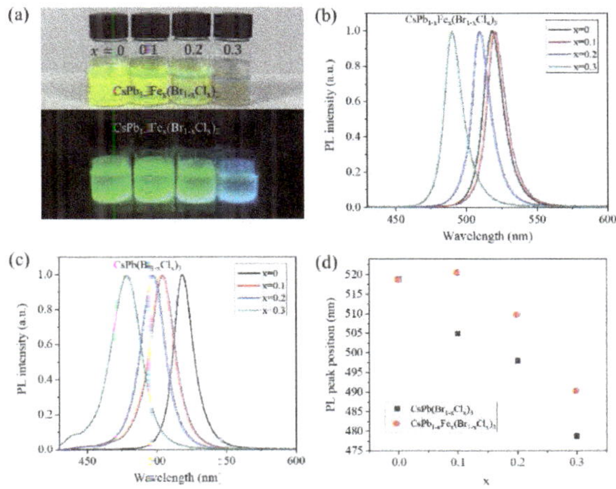

Figure 3. (**a**) Photographs of $CsPb_{1-x}Fe_x(Br_{1-x}Clx)_3$ NC solutions under natural light (top) and UV light (bottom). (**b**) Normalized PL spectra of $CsPb_{1-x}Fe_x(Br_{1-x}Cl_x)_3$ NC solutions. (**c**) Normalized PL spectra of $CsPb(Br_{1-x}Cl_x)_3$ NC solutions. (**d**) Comparison of PL peak positions of $CsPb_{1-x}Fe_x(Br_{1-x}Cl_x)_3$ and $CsPb(Br_{1-x}Cl_x)_3$ NC solutions at different x (x = 0, 0.1, 0.2, 0.3). The excitation wavelength kept as 405 nm.

Many different effects can be responsible for the PL redshift. For example, the large size means a narrowed bandgap due to the size effect, which can explain the PL redshift [39]. Moreover, photon reabsorption is another possible reason for the PL redshift [40,41]. As revealed previously [41], evident photon reabsorption prevailingly exists in various halide perovskite materials due to their high absorption coefficient and small Stokes shift. The Fe^{2+} doping may increase the aggregation of NCs, leading to the PL redshift. However, the TEM images do not show either an increased size or increased aggregation, excluding these factors. In addition, the element doping may expand the lattice spacings, causing the shrinkage of the bandgap. However, here, the doping element is Fe^{2+} with a radius of ~0.76 Å, which is smaller than the Pb^{2+} with a radius of ~1.33 Å. In fact, previous work has reported that the doping of Fe^{2+} in $CsPbCl_3$ NCs results in a slight blueshift of the PL peak position [38]. Furthermore, the surface defects generally induce shallow traps, leading to a narrowing of the bandgap. The Fe^{2+} doping may induce shallow traps in the perovskite lattices, which leads to the PL redshift. This kind of possibility is evaluated by our further investigation.

The PL spectra of the $CsPb_{1-x}Fe_x(Br_{1-x}Cl_x)_3$ NCs and $CsPb(Br_{1-x}Cl_x)_3$ NCs are separately compared in Figure 4a–c for x = 0.1 to 0.3. It can be clearly seen that, not only the PL peak position redshifts but also the bandwidth decreases in the Fe^{2+}-doped samples. The full widths at half maximums (FWHMs) are compared in Figure 4d. Especially when x = 0.3, the FWHM of the PL of the $CsPb(Br_{1-x}Cl_x)_3$ NCs is about 26 nm, which decreases to about 18 nm after the Fe^{2+} doping. The PL bandwidth depends on both intrinsic effects, such as electron–phonon interactions, shallow defect states, and extrinsic factors, such as size polydispersity. Herein, the size distribution is little changed by the Fe^{2+} doping. Therefore, the reduced FWHM is mainly attributed to the intrinsic effects. It is well supposed that both the doping of Fe^{2+} improves the crystallinity and diminishes the defect states. To verify this point, the PL QYs of the perovskite NCs before and after the Fe^{2+} doping were measured and compared. As shown in Figure 4e, the PL QYs are improved after different concentrations of the Fe^{2+} doping. Therefore, both the FWHMs and PL QYs suggest the density of defects is reduced by the appropriate Fe^{2+} doping.

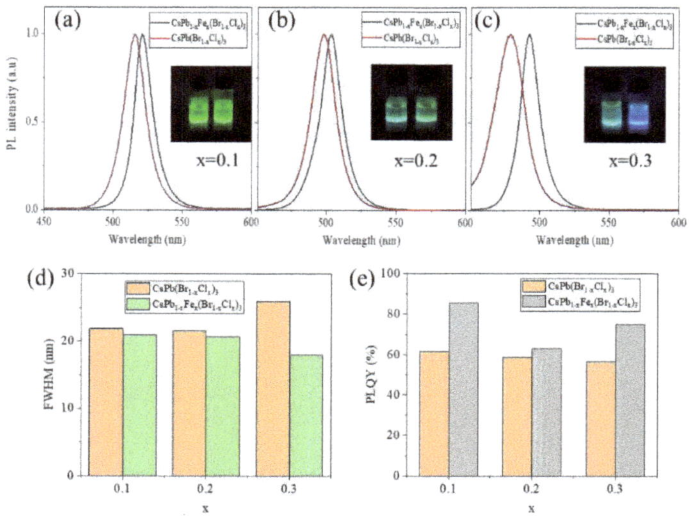

Figure 4. (**a**–**c**) Comparison of PL spectra of the $CsPb_{1-x}Fe_x(Br_{1-x}Cl_x)_3$ and $CsPb(Br_{1-x}Cl_x)_3$ NC solutions with x = 0.1 (**a**), 0.2 (**b**), and 0.3 (**c**). (**d**) FWHM of the two kinds of NCs with different x values. (**e**) PL QY of the two kinds of NCs with different x values.

To further understand the PL of the $CsPb_{1-x}Fe_x(Br_{1-x}Cl_x)_3$ NCs, the time-resolved PL decay spectra were measured. As shown in Figure 5a, all the time-resolved PL decays are fitted by an exponential function. The PL lifetimes are shown in Figure 5b. The PL lifetimes of the undoped $CsPb(Br_{1-x}Cl_x)_3$ NCs are also added for comparison. The PL lifetimes of the undoped $CsPb(Br_{1-x}Cl_x)_3$ monotonously decrease as x increases, which has been widely reported previously [39]. However, for the $CsPb_{1-x}Fe_x(Br_{1-x}Cl_x)_3$ NCs, the PL lifetimes increase to about 70 ns when x = 0.1 and 0.2, which are significantly longer than those of the undoped $CsPb(Br_{1-x}Cl_x)_3$. The time-resolved PL decay spectra indicate the PL lifetimes of the mixed halide perovskite NCs are increased by the Fe doping. The defect trapping usually leads to a short PL lifetime [42,43]. Thus, a longer PL lifetime implies a lower defect density [42,43]. Therefore, the PL lifetime further suggests the defect density is decreased after the Fe^{2+} doping, which is in good agreement with the decreased FWHMs and improved PL QYs. Back to the PL peak positions shown in Figure 3, the redshift of the emission cannot be attributed to the shallow traps caused by the Fe^{2+} doping.

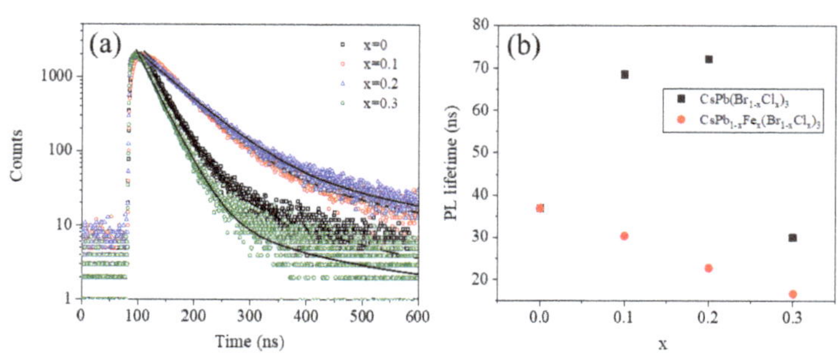

Figure 5. (**a**) Time-resolved PL spectra of the $CsPb_{1-x}Fe_x(Br_{1-x}Cl_x)_3$; (**b**) PL lifetime parameters of the $CsPb_{1-x}Fe_x(Br_{1-x}Cl_x)_3$ obtained from numerical fitting on (**a**). PL lifetimes of the $CsPb(Br_{1-x}Cl_x)_3$ are also plotted for comparison.

As is well known, the photostability of the $CsPb(Br_{1-x}Cl_x)_3$ NCs is generally worse than that of the $CsPbBrCl_3$ NCs due to the phase segregation [15–17]. After continuous irradiation, the mixed $CsPb(Br_{1-x}Cl_x)_3$ NCs transform to a separated Br rich phase and Cl rich phase, and the PL shows typical redshifts and quenches largely. As shown in Figure 6a, the PL intensity of the $CsPb(Br_{1-x}Cl_x)_3$ NCs quenches to about half of the original intensity after 1 W UV irradiation for 30 minutes. After the Fe^{2+} doping, we find the photostability is significantly improved. The PL intensity only decreases by ca.15% after UV irradiation for 30 minutes. The doping of the appropriate concentration of Fe^{2+} (~0.76 Å) with an ionic radius smaller than Pb^{2+} (~1.33 Å) can enhance the formation energies of perovskite lattices and, thus, essentially improve the structural stability [44,45]. Moreover, the stability usually depends on the crystal quality. The improved photostability is consistent with the decreased defect density.

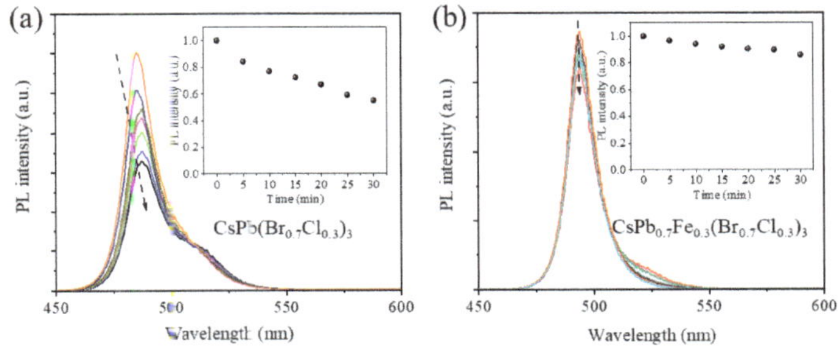

Figure 6. (**a**) PL spectra of $CsPb(Br_{1-x}Cl_x)_3$ NCs under continuous UV irradiation. (**b**) PL spectra of $CsPb_{1-x}Fe_x(Br_{1-x}Cl_x)_3$ NCs under continuous UV irradiation. Insets: PL intensity versus irradiation time.

4. Conclusions

In summary, the Fe^{2+}-doped perovskite NCs are prepared by a hot injection method. In addition, their optical properties, including absorption, PL, and PL lifetimes are compared with those of the undoped $CsPb(Br_{1-x}Cl_x)_3$ NCs. We find that Fe^{2+} doping results in the redshift of the absorption edge and PL. Moreover, the FWHMs are decreased and PL QYs are improved by the Fe^{2+} doping, suggesting the density of defects is reduced. The extended PL lifetimes further verify the defect density is decreased after the Fe^{2+} doping. Moreover, the photostability is significantly improved after the Fe doping. Therefore, this work reveals that Fe^{2+} doping is a very promising approach to modulate the optical properties of mixed halide perovskite NCs.

Author Contributions: Conceptualization, C.W. and Y.L.; methodology, Z.X., C.J. and Y.T.; writing draft, J.Z.; editing, C.M.; project administration, J.G. All authors have agreed to the published version of the manuscript.

Funding: This work is supported by the National Natural Science Foundation of China (No. 62004136), and Natural Science Foundation of the Jiangsu Higher Education Institutions of China (20KJB140019 and 19KJB150018). This work is supported by the Innovation and Entrepreneurship Training Program for College Students (202110332020Z). This work is supported by the Doctor of Entrepreneurship and Innovation in Jiangsu Province (No. (2020)30790).

Institutional Review Board Statement: Not applicable.

Data Availability Statement: All the experimental/calculation date that support the findings of this study are available from the corresponding authors upon reasonable request.

Conflicts of Interest: The authors declare no conflict of interest.

References

1. Jeon, N.J.; Na, H.; Jung, E.H.; Yang, T.-Y.; Lee, Y.G.; Kim, G.; Shin, H.-W.; Il Seok, S.; Lee, J.; Seo, J. A fluorene-terminated hole-transporting material for highly efficient and stable perovskite solar cells. *Nat. Energy* **2018**, *3*, 682–689. [CrossRef]
2. Akihiro, K.; Kenjiro, T.; Yasuo, S.; Tsutomu, M. Organometal halide perovskites as visible-light sensitizers for photovoltaic cells. *J. Am. Chem. Soc.* **2009**, *131*, 6050.
3. Kim, D.; Jung, H.J.; Park, I.J.; Larson, B.W.; Dunfield, S.P.; Xiao, C.; Kim, J.; Tong, J.; Boonmongkolras, P.; Ji, S.G.; et al. Efficient, stable silicon tandem cells enabled by anion-engineered wide-bandgap perovskites. *Science* **2020**, *368*, 155–160. [CrossRef] [PubMed]
4. Zhang, X.L.; Xu, B.; Zhang, J.B.; Gao, Y.; Zheng, Y.J.; Wang, K.; Sun, X.W. All-inorganic perovskite nanocrystals for high-efficiency light emitting diodes: Dual-phase $CsPbBr_3$-$CsPb_2Br_5$ composites. *Adv. Funct. Mater.* **2016**, *26*, 4595. [CrossRef]
5. Song, J.Z.; Li, J.H.; Li, X.M.; Xu, L.M.; Dong, Y.H.; Zeng, H.B. Quantum dot light-emitting diodes based on inorganic perovskite cesium lead halides ($CsPbX_3$). *Adv. Mater.* **2015**, *27*, 7162. [CrossRef]
6. Fang, H.J.; Li, J.W.; Ding, J.; Sun, Y.; Li, Q.; Sun, J.L.; Wang, L.D.; Yan, Q.F. An origami perovskite photodetector with spatial recognition ability. *ACS Appl. Mater. Interfaces* **2017**, *9*, 10921. [CrossRef]
7. Cao, F.R.; Tian, W.; Wang, M.; Li, L. Polarized ferroelectric field-enhanced self-powered perovskite photodetector. *ACS Photonics* **2018**, *5*, 3731. [CrossRef]
8. Ramasamy, P.; Lim, D.H.; Kim, B.; Lee, S.H.; Lee, M.S.; Lee, J.S. All-inorganic cesium lead halide perovskite nanocrystals for photodetector applications. *Chem. Commun.* **2016**, *52*, 2067. [CrossRef]
9. Hao, F.; Stoumpos, C.C.; Cao, D.H.; Chang, R.P.H.; Kanatzidis, M.G. Lead-free solid-state organic−inorganic halide perovskite solar cells. *Nat. Photonics* **2014**, *8*, 489–494. [CrossRef]
10. Volonakis, G.; Filip, M.R.; Haghighirad, A.A.; Sakai, N.; Wenger, B.; Snaith, H.J.; Giustino, F. Lead-free halide double perovskites via heterovalent substitution of noble metals. *J. Phys. Chem. Lett.* **2016**, *7*, 1254–1259. [CrossRef]
11. Savory, C.N.; Walsh, A.; Scanlon, D.O. Pb-free halide double perovskites dupport high-efficiency solar cells. *ACS Energy Lett.* **2016**, *1*, 949–955. [CrossRef] [PubMed]
12. Zhao, X.G.; Yang, J.H.; Fu, Y.H.; Yang, D.W.; Xu, Q.L.; Yu, L.P.; Wei, S.H.; Zhang, L.J. Lead-free inorganic halide perovskites for solar cells via cation-transmutation. *J. Am. Chem. Soc.* **2017**, *139*, 2630–2638. [CrossRef] [PubMed]
13. Wu, X.L.; Xiong, S.J.; Liu, Z.; Chen, J.; Shen, J.C.; Li, T.H.; Wu, P.H.; Chu, P.K. Green light stimulates terahertz emission from mesocrystal microspheres. *Nat. Nanotech.* **2011**, *6*, 103–106. [CrossRef] [PubMed]
14. Yang, B.; Chen, J.S.; Hong, F.; Mao, X.; Zheng, K.B.; Yang, S.Q.; Li, Y.J.; Pullerits, T.; Deng, W.Q.; Han, K.L. Lead-free, air-stable all-inorganic cesium bismuth halide perovskite nanocrystals. *Angew. Chem. Int. Ed.* **2017**, *56*, 12471. [CrossRef]
15. Unger, E.L.; Hoke, E.T.; Bailie, C.D.; Nguyen, W.H.; Bowring, A.R.; Heumuller, T.; Christoforo, M.G.; McGehee, M.D. Hysteresis and transient behavior in current-voltage measurements of hybrid-perovskite absorber solar cells. *Energy Environ. Sci.* **2014**, *7*, 3690–3698. [CrossRef]
16. Yuan, Y.; Huang, J. Ion migration in organometal trihalide perovskite and its impact on photovoltaic efficiency and stability. Acc. *Chem. Res.* **2016**, *49*, 286–293. [CrossRef]
17. Brennan, M.C.; Ruth, A.; Kamat, P.V.; Kuno, M. Photoinduced anion segregation in mixed halide perovskites. *Trends Chem.* **2020**, *2*, 282–301. [CrossRef]
18. Zhao, R.; Yang, C.; Wang, H.G.; Jiang, K.; Wu, H.; Shen, S.P.; Wang, L.; Sun, Y.; Jin, K.J.; Gao, J.; et al. Emergent multiferroism with magnetodielectric coupling in $EuTiO_3$ created by a negative pressure control of strong spin-phonon coupling. *Nat. Commun.* **2022**, *13*, 1–9.
19. Jellicoe, T.C.; Richter, J.M.; Glass, H.F.J.; Tabachnyk, M.; Brady, R.; Dutton, S.E.; Rao, A.; Friend, R.H.; Credgington, D.; Greenham, N.C.; et al. Synthesis and optical properties of lead-free cesium tin halide perovskite nanocrystals. *J. Am. Chem. Soc.* **2016**, *138*, 2941. [CrossRef]
20. Grandhi, G.K.; Matuhina, A.; Liu, M.N.; Annurakshita, S.; Ali-Loytty, H.; Bautista, G.; Vivo, P. Lead-free cesium titanium bro mide double perovskite nanocrystals. *Nanomaterials* **2021**, *11*, 1458. [CrossRef]
21. Noel, N.K.; Stranks, S.D.; Abate, A.; Wehrenfennig, C.; Guarnera, S.; Haghighirad, A.A.; Sadhanala, A.; Eperon, G.E.; Pathak, S.K.; Johnston, M.B.; et al. Lead-free organic–inorganic tin halide perovskites for photovoltaic applications. *Energy Environ. Sci.* **2014**, *7*, 3061–3068. [CrossRef]
22. Wang, F.; Han, Y.; Lim, C.S.; Lu, Y.H.; Wang, J.; Xu, J.; Chen, H.Y.; Zhang, C.; Hong, M.H.; Liu, X.G. Simultaneous phase and size control of upconversion nanocrystals through lanthanide doping. *Nature* **2010**, *463*, 1061. [CrossRef] [PubMed]
23. Dong, H.; Sun, L.D.; Wang, Y.F.; Ke, J.; Si, R.; Xiao, J.W.; Lyu, G.M.; Shi, S.; Yan, C.H. Efficient tailoring of upconversion selectivity by engineering local structure of lanthanides in Na_xREF_{3+x} nanocrystals. *J. Am. Chem. Soc.* **2015**, *137*, 6569. [CrossRef]
24. Feng, X.D.; Sayle, D.C.; Wang, Z.L.; Paras, M.S.; Santora, B.; Sutorik, A.C.; Sayle, T.X.T.; Yang, Y.; Ding, Y.; Wang, X.D.; et al. Converting ceria polyhedral nanoparticles into single-crystal nanospheres. *Science* **2006**, *312*, 1504. [CrossRef] [PubMed]
25. Begum, R.; Parida, M.R.; Abdelhady, A.L.; Murali, B.; Alyami, N.M.; Ahmed, G.H.; Hedhili, M.N.; Bakr, O.M.; Mohammed, O.F. Engineering interfacial charge transfer in CsPbBr3 perovskite nanocrystals by heterovalent doping. *J. Am. Chem. Soc.* **2017**, *139*, 731. [CrossRef]

26. Mao, K.H.; Zhang, J.L.; Guo, Z.J.; Liu, L.Z.; Ma, H.; Chin, Y.Y.; Lin, H.J.; Bao, S.S.; Xie, H.Q.; Yang, R.; et al. Constructing asymmetrical Ni-centered {NiN_2O_4} octahedra in layered metal−organic structures for near-room-temperature single-phase magnetoelectricity. *J. Am. Chem. Soc.* **2020**, *142*, 12841–12849. [CrossRef] [PubMed]
27. Guo, Z.J.; Mao, K.H.; Ma, G.D.; Li, G.A.; Wu, Q.F.; Chen, J.; Bao, S.S.; Yu, G.L.; Li, S.H.; Zhang, J.L.; et al. Light-Induced Tunable Ferroelectric Polarization in DipoleEmbedded Metal−Organic Framework. *Nano Lett.* **2022**, *22*, 10018–10024. [CrossRef]
28. Jiang, Y.C.; He, A.P.; Luo, K.; Zhang, J.L.; Liu, G.Z.; Zhao, R.; Zhang, Q.; Wang, Z.; Zhao, C.; Wang, L.; et al. Giant bipolar unidirectional photomagnetoresistance. *Proc. Natl. Acad. Sci. USA* **2022**, *27*, 119. [CrossRef] [PubMed]
29. Xia, Z.Y.; Li, Y.; Ji, C.; Jiang, Y.C.; Ma, C.L.; Gao, J.; Zhang, J.L. Slow relaxation behavior of a mononuclear Co(II) complex featuring long axial Co-O bond. *Nanomaterials.* **2022**, *12*, 707. [CrossRef] [PubMed]
30. Zhang, W.Z.; Li, X.; Peng, C.C.; Yang, F.; Lian, L.Y.; Guo, R.D.; Zhang, J.B.; Wang, L. $CsPb(Br/Cl)_3$ perovskite nanocrystals with bright blue emission synergistically modified by calcium halide and ammonium ion. *Nanomaterials* **2022**, *12*, 2026. [CrossRef]
31. Klug, M.T.; Osherov, A.; Haghighirad, A.A.; Stranks S.D.; Brown, P.R.; Bai, S.; Wang, J.T.W.; Dang, X.N.; Bulovic, V.; Snaith, H.J.; et al. Tailoring metal halide perovskites through metal substitution: Influence on photovoltaic and material properties. *Science* **2017**, *10*, 236–246. [CrossRef]
32. Parobek, D.; Roman, B.J.; Dong, Y.; Jin, H.; Lee, E.; Sheldon, M.; Son, D.H. Exciton-to-dopant energy transfer in Mn-doped cesium lead halide perovskite nanocrystals. *Nano Lett.* **2016** *16*, 7376. [CrossRef] [PubMed]
33. Liu, H.W.; Wu, Z.N.; Shao, J.R.; Yao, D.; Gao, H.; Liu. Y.; Yu, W.L.; Zhang, H.; Yang, B. $CsPb_xMn_{1-x}Cl_3$ perovskite quantum Dots with high Mn substitution ratio. *ACS Nano* **2017**, *11*, 2239. [CrossRef] [PubMed]
34. Liu, W.Y.; Lin, Q.L.; Li, H.B.; Wu, K.F.; Robel, I.; Pietryga, J.M.; Klimov, V.I. Mn^{2+}-doped lead halide perovskite nanocrystals with dual-color emission controlled by halide content. *J. Am. Chem. Soc.* **2016**, *138*, 14954. [CrossRef] [PubMed]
35. Ma, K.W.; Sheng, Y.H.; Wang, G.Z.; Zhang, X.W.; Di, Y.S.; Liu, C.H.; Yu, L.Y.; Dong, L.F.; Gan, Z.X. Stable and multicolor solid-state luminescence of Mn doped CsPb $(Cl/Br)_3$perovskite quantum dots and its application in light-emitting diodes. *J. Lumin.* **2022**, *243*, 118622. [CrossRef]
36. Yong, Z.J.; Guo, S.Q.; Ma, J.P.; Zhang, J.Y.; Li, Z.Y.; Chen, Y.M.; Zhang, B.B.; Zhou, Y.; Shu, J.; Gu, J.L.; et al. Doping-enhanced short-range order of perovskite nanocrystals for near-unity violet luminescence quantum yield. *J. Am. Chem. Soc.* **2018**, *140*, 9942–9951. [CrossRef]
37. Cheng, X.H.; Jing, L.; Yuan, Y.; Du, S.J.; Zhang, J.; Zhan, X.Y.; Ding, J.X.; Shi, G.D.T.H. Fe2+/Fe3+ doped into MAPbCl3 single crystal: Impact on crystal growth and optical and photoelectronic properties. *J. Phys. Chem. C* **2019**, *123*, 1669–1676. [CrossRef]
38. Hu, Y.; Zhang, X.Y.; Yang, C.Q.; Lia, J.; Wang, L. Fe^{2+} doped in $CsPbCl_3$ perovskite nanocrystals: Impact on the luminescence and magnetic properties. *RSC Adv.* **2019**, *9*, 33017–33022. [CrossRef]
39. Protesescu, L.; Yakunin, S.; Bodnarchuk, M.I.; Krieg, F.; Caputo, R.; Hendon, C.H.; Yang, R.X.; Walsh, A.; Kovalenko, M.V. Nanocrystals of cesium lead halide perovskites (CsPbX3, X = Cl, Br, and I): Novel optoelectronic materials showing bright emission with wide color gamut. *Nano Lett.* **2015**, *15*, 3692. [CrossRef]
40. Gan, Z.X.; Wen, X.M.; Chen, W.J.; Zhou, C.H.; Yang, S.; Cao, G.Y.; Ghiggino, K.P.; Zhang, H.; Jia, B.H. The dominant energy transport pathway in halide perovskites: Photon recycling or carrier diffusion. *Adv. Energy Mater.* **2019**, *9*, 1900185. [CrossRef]
41. Gan, Z.X.; Chen, W.J.; Yuan, L.; Cao, G.Y.; Zhou, C.H.; Huang, S.J.; Wen, X.M.; Jia, B.H. External stokes shift of perovskite nanocrystals enlarged by photon recycling. *Appl. Phys. Lett.* **2019**, *114*, 011906. [CrossRef]
42. Wenger, B.; Nayak, P.K.; Wen, X.M.; Kesava, S.V.; Noel, N.K.; Snaith, H.J. Consolidation of the optoelectronic properties of $CH_3NH_3PbBr_3$ perovskite single crystals. *Nat. Commun.* **2017**, *8*, 590. [CrossRef]
43. Sheng, Y.H.; Liu, C.H.; Yu, L.Y.; Yang, Y.Y.; Hu, F.R.; Sheng, C.; Di, Y.S.; Dong, L.F.; Gan, Z.X. Microsteganography on all inorganic perovskite micro-platelets by direct laser writing. *Nanoscale* **2021**, *13*, 14450. [CrossRef]
44. Zou, S.H.; Liu, Y.S.; Li, J.H.; Liu, C.P.; Feng, R.; Jiang, F.L.; Li, Y.X.; Song, J.Z.; Zeng, H.B.; Hong, M.C.; et al. Stabilizing cesium lead halide perovskite lattice through Mn (II) substitution for air-stable light-emitting diodes. *J. Am. Chem. Soc.* **2017**, *139*, 11443–11450. [CrossRef] [PubMed]
45. Cherevkov, S.; Azizov, R.; Sokolova, A.; Nautran, V.; Miruschenko, M.; Arefina, I.; Baranov, M.; Kurdyukov, D.; Stovpiaga, E.; Golubev, V.; et al. Interface chemical modification between all-inorganic perovskite nanocrystals and porous silica microspheres for composite materials with improved emission. *Nanomaterials* **2021**, *11*, 119. [CrossRef] [PubMed]

nanomaterials

MDPI

Article

Enhancing Photoluminescence Intensity and Spectral Bandwidth of Hybrid Nanofiber/Thin-Film Multilayer Tm^{3+}-Doped SiO_2–HfO_2

Nurul Izzati Zafirah Zulfikri [1], Abdel-Baset M. A. Ibrahim [1], Nur Amalina Mustaffa [1], Rozan Mohamad Yunus [2] and Suraya Ahmad Kamil [1,*]

1 Faculty of Applied Sciences, Universiti Teknologi MARA, 40450 Shah Alam, Selangor, Malaysia
2 Fuel Cell Institute, Universiti Kebangsaan Malaysia, 43600 Bangi, Selangor, Malaysia
* Correspondence: suraya_ak@uitm.edu.my

Abstract: Multilayering of optical thin films is widely used for a range of purposes in photonic technology, but the development of nanofiber structures that can outperform thin films and nanoparticles in optical applications cannot simply be disregarded. Hybrid structures composed of Tm^{3+}-doped SiO_2–HfO_2 in the form of nanofibers (NFs) and thin films (TFs) are deposited on a single substrate using the electrospinning and dip-coating methods, respectively. Ultrafine nanofiber strands with a diameter of 10–60 nm were fabricated in both single and multilayer samples. Enhanced photoluminescence emission intensity of about 10 times was attained at wavelengths of around 457, 512 and 634 nm under an excitation of 350 nm for NF-TF-NF* hybrid structures when compared with single-layered NF and TF structures. The arrangement of nanofibers and thin films in a multilayer structure influenced the luminescence intensity and spectral bandwidth. High transparency in the range of 75–95% transparency across the wavelength of 200–2000 nm was achieved, making it ideal for photonic application. Theoretical findings obtained through IMD software were compared with experimental results, and they were found to be in good agreement.

Keywords: nanofiber; thin film; photoluminescence; rare earth; optical materials; multilayering; thulium; silica-hafnia

Citation: Zulfikri, N.I.Z.; Ibrahim, A.-B.M.A.; Mustaffa, N.A.; Mohamad Yunus, R.; Ahmad Kamil, S. Enhancing Photoluminescence Intensity and Spectral Bandwidth of Hybrid Nanofiber/Thin-Film Multilayer Tm^{3+}-Doped SiO_2–HfO_2. *Nanomaterials* **2022**, *12*, 3739. https://doi.org/10.3390/nano12213739

Academic Editor: Zhixing Gan

Received: 28 September 2022
Accepted: 21 October 2022
Published: 25 October 2022

1. Introduction

Thin-film deposition is a very common technique used in a variety of applications, including telecommunications, solar cells, integrated circuits, semiconductor devices, wireless communications, photoconductors and light crystal displays, light emitting diodes, transistors and other emerging technologies [1]. To meet the demands of modern technology, thin films have undergone numerous modifications throughout the years. One of the most well-known techniques that produces notable results is layering. In fact, multilayer optical thin films have been widely exploited to obtain specific optical characteristics for certain applications in a variety of fields of optical technology. Numerous studies on thin-film stacking using a variety of compositions, designs and approaches have been conducted. The general concept of multilayering thin-film structures is based on the fact that optical interference is made of two layers of high and low refractive indexes that are arranged alternately onto a single substrate [2–4]. Moreover, a previous study conducted by Rahmani and Ardyanian [5] discovered that as the number of thin-film layers of ZnO and TiO_2 increases, the band gap of sample also increases, leading to a higher reflective index and absorbance. Thus, a multilayer structure is the best choice for photonic applications.

The optical applications of electrospun nanofibers have sparked much interest, because they can be used in sub-wavelength components for light generation, confinement, guiding and detection [6,7]. The excellent performance of nanofiber structures in terms of charge and energy transfer makes them preferable to nanoparticles and thin films due their unique physical properties; namely, high surface-to-volume ratio and high porosity. Owing to their physical properties, compact-sized optical circuits and photonic components can be constructed [8–10]. Moreover, due to their high-porosity structure, they can transport ions rapidly and are capable of aiding long-lasting electrolyte storage [11], resulting in a significant improvement in the efficiency of the energy-related nanofiber-based devices. Electrospun nanofibers have a refractive index of 1.5 to 2.5 at visible wavelengths, which is comparable with that of a normal optical fiber. When surrounded by air or a medium with a lower refractive index, they can direct light through total internal reflection [12].

Furthermore, nanofiber can be simply fabricated using electrospinning method as it is a straightforward technique that can control the nanofiber size and morphology to attain the preferred properties. The electrospinning method is the most favorable, because it is a cost-effective technique that can produce high surface-to-volume ratio, has tunable porosity and adaptability. The sol-gel method is among the common methods used in producing nanofiber solutions due to its versatile route for synthesizing inorganic and organic–inorganic structures, including glasses, ceramics and films [13].

Rare earth (RE) ions have attracted much attention in the development of optical amplifiers and solid-state lasers because of their sharp emission lines due to an electronic transition in the 4f band and their capacity to amplify weak signals [14–16]. Prior research has demonstrated that materials that include quantum dots and nanostructures can emit narrower photoluminescence (PL) peaks with a high quantum yield, larger absorption bands and significant effective Stokes shifts [17]. Qin et al. [18] demonstrated that the presence of an active RE ion in the 1D structure of Eu^{3+}/Tb^{3+} co-doped $LabBO_3$ nanofibers results in a luminescence intensity that is more intense and has a longer decay time compared with the powder sample of the same composites.

In the present work, we fabricated a multilayer nanofiber/thin-film structure composed of Tm^{3+}-doped silica-hafnia. The idea of stacking both nanofiber and thin-film structures is to enhance the emission intensity without causing an unwanted effect, such as concentration quenching. Thus, the optical properties of the fabricated hybrid structures were analyzed, and a comparison with theoretical results, which was obtained through IMD software, is provided. Additionally, the morphological, structural and optical properties of thin films and nanofibers are also compared.

2. Materials and Methods

2.1. Solution Preparation

In this study, the amounts of SiO_2 and HfO_2 used were 90 mol% and 10 mol%, respectively. A sol-gel solution was firstly made by combining two solutions, which are referred to as solution A (sol A) and solution B (sol B). Sol A represents 20 mL of a silica solution that was made by mixing tetraethylorthosilicate (TEOS), ethanol (EtOH), ultra-pure water (H_2O) and hydrochloric acid (HCl) with a molar ratio of TEOS:HCl:EtOH:H_2O of 1:0.01:37.9:2. The role of HCl in sol A was to act as a catalyst, whereas EtOH acted as a solvent. The solution was then hydrolyzed for 1 h with a stirring rate of 400 rpm at 65 °C. At the same time, 20 mL of HfO_2 solution (sol B) was prepared by mixing hafnium (IV) chloride ($HfCl_4$) powder with ethanol (EtOH). The solution was stirred at the rate of 400 rpm until $HfCl_4$ powder dissolved. Once sol A and sol B were done, aqueous thulium (III) chloride hexahydrate ($TmCl_3 \cdot 6H_2O$) was added into sol A. Sol B was then transferred into the sol A using a syringe. The resultant mixture was stirred for 16 h at room temperature with a stirring speed rate of 400 rpm. An additional step was required before depositing nanofibers, in which poly (vinyl) alcohol (PVA) solution was added into the sol-gel solution to produce continuous nanofiber strands.

2.2. Deposition of Nanofiber/Thin-Film Multilayer Structure

Nanofibers were deposited using a standalone NE-1000 Programmable Single Syringe Pump. The deposited samples were then annealed using a CARBOLITE Shimaden CWF 11/5 chamber furnace, in which (i) temperature was ramped up slowly at the rate of

5 °C/min to 950 °C, (ii) temperature was maintained at 950 °C for 1 h and (iii) temperature was then ramped down at the rate of 10 °C/min. Meanwhile, a single layer of the thin film comprising stacks of 20 nanoscale layers was deposited using a computerized KSV dip coater system. The dipping/withdrawal speed was kept constant at 40 mm/min. After the final nanoscale layer was deposited, the resulting film underwent heat treatment using CARBOLITE three-zone furnace for 30 min at 950 °C. The purpose of the annealing process is to eliminate the presence of the OH^- group, to improve the thin-film surface properties, and to strengthen the network structure through densification [19].

A few layers of nanofibers and thin films were deposited on a fused silica substrate and were arranged alternately. The deposited samples were identified as single-layered, nanofiber (NF), thin-film (TF), dual-layered NF–TF*, TF–NF* and three-layered NF–TF–NF* and TF–NF–TF*. The asterisk (*) symbol indicates the final top layer deposited on the substrate. Figure 1 illustrates the sequence of deposition of nanofibers and thin films on a fused silica glass substrate, as follows: (a) NF, (b) TF, (c) NF–TF*, (d) TF–NF*, (e) NF–TF–NF* and (f) TF–NF–NF*. Both structures were composed of 0.8 mol% of Tm^{3+}-doped $90SiO_2$–$10HfO_2$.

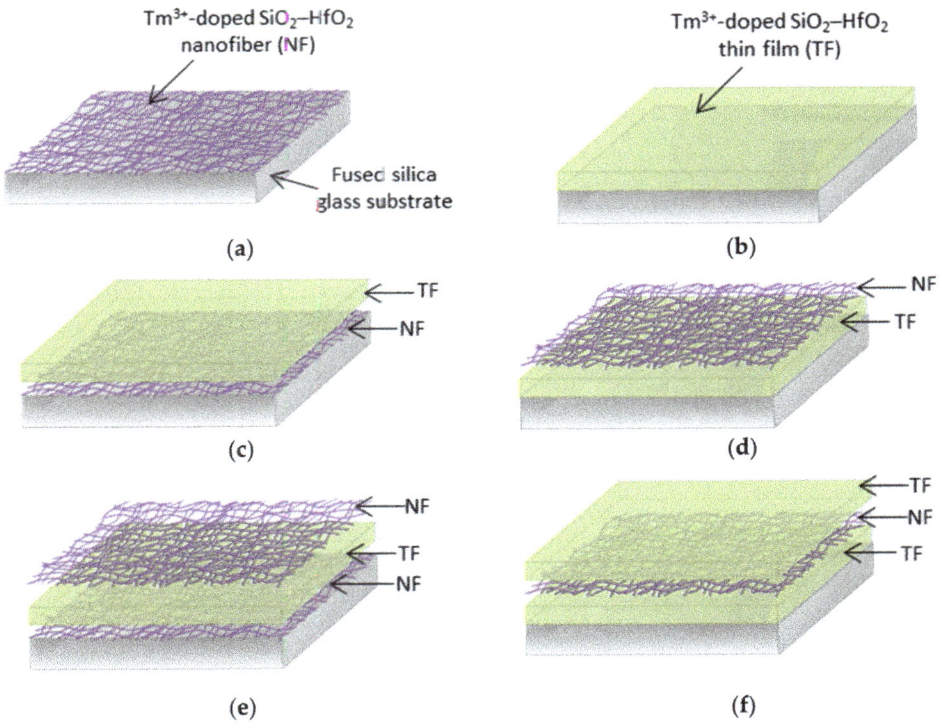

Figure 1. The sequence of deposition of nanofibers and thin films on fused SiO_2 glass substrate: (**a**) NF, (**b**) TF, (**c**) NF–TF*, (**d**) TF–NF*, (**e**) NF–TF–NF* and (**f**) TF–NF–NF*.

2.3. Sample Characterization

X-ray diffraction (PANalytical X'pert PRO, Almelo, The Netherlands) was used to examine and compare the crystallinity of the single-layered nanofibers and thin films, and Fourier transform infrared spectroscopy (FTIR) (PerkinElmer Spectrum One, Wellesley, MA, USA) was employed to obtain the infrared absorption spectra of the produced materials. Field-emission scanning electron microscopy (FESEM, Joel JSM-7600F, Tokyo, Japan) was utilized to examine the morphology of the electrospun nanofiber samples, and Image J software (Version 1.53k, U.S. National Institutes of Health, Bethesda, MD, USA) was used to determine the diameter of the deposited nanofibers. The elemental composition of the samples was determined using energy-dispersive X-ray spectroscopy (EDX). The optical transparency of the samples was evaluated using a UV–vis NIR spectrophotometer (Varian Cary 5000, Palo Alto, CA, US) at a range of 200–2000 nm. By using a xenon lamp as an excitation source with a wavelength of 350 nm, PL spectroscopy (FLS920 Edinburgh Instrument, Livingston, UK) was employed to determine the luminescence spectra.

2.4. Theoretical Aspects

The optical characteristics of the deposited samples were compared with theoretical results. The experimental results of transmission and absorbance of the samples were compared with the theoretical results obtained using IMD software, (Version 5.04, Bell Laboratories, Murray Hill, NJ, USA) whereas the PL emission intensity data were fitted to the Voigt function. The IMD software calculated the reflection, transmission and absorption of an electromagnetic plane wave at the interface of a multilayer structure based on the transfer matrix method (TMM) [20]. Fresnel's equations for *s*-polarization, where electric-field amplitude, *E*, is perpendicular to the plane of incidence, were as follows:

$$r_s = \frac{|E_r|}{|E_i|} = \frac{n_i \cos\theta_i - n_t \cos\theta_t}{n_i \cos\theta_i + n_t \cos\theta_t}, \tag{1}$$

$$t_s = \frac{|E_t|}{|E_i|} = \frac{2n_i \cos\theta_i}{n_i \cos\theta_i + n_t \cos\theta_t}, \tag{2}$$

Fresnel equations of *p*-polarization, where *E* is parallel to the plain of incidence, were as follows:

$$r_p = \frac{|E_r|}{|E_i|} = \frac{n_i \cos\theta_t - n_t \cos\theta_i}{n_i \cos\theta_t + n_t \cos\theta_i} \tag{3}$$

$$t_p = \frac{|E_t|}{|E_i|} = \frac{2n_i \cos\theta_i}{n_i \cos\theta_t + n_t \cos\theta_i} \tag{4}$$

where r represent the reflection coefficient and t is the coefficient of transmission. The refractive indices of a medium of incident light and transmitted light are denoted as n_i and n_t, respectively. The optical function of the multilayer stack was obtained by the derivation of Fresnel's equation by considering the thickness, interfacial roughness/diffuseness and optical constant of each layer.

3. Results and Discussion

3.1. Comparing Thin-Film and Nanofiber Results

The X-ray interference patterns of single-layered nanofiber and thin-film structures were compared, as shown in Figure 2a. The patterns of the two structures were identical, and neither showed a prominent high peak of crystallization, suggesting that the produced samples were in the amorphous phase. Both NF and TF exhibited the same two noticeable humps, which were situated at 2θ = 10° and 22°. However, NF samples had a substantially higher intensity than TF samples, which could be due to the present of PVA in the nanofiber samples. The appearance of such diffraction patterns in PVA, according to Gupta et al. [21], can be attributed to the presence of a very minor degree of crystallinity, which suggested the coexistence of a small PVA nanocrystalline phase with the amorphous phase material.

The OH^- group was not present in the single-layered TF and NF structures based on the FTIR analysis of the Tm^{3+}-doped SiO_2–HfO_2 samples shown in Figure 2b, as no apparent peaks were seen in the range of 3700–3200 cm^{-1}. Peaks in the TF sample were identical to those in the NF sample, suggesting that the bonds that existed in both samples were the same. Both samples had peaks located at 740, 614 and 450 cm^{-1} assigned to the presence of Hf–O bond, which was around 700–400 cm^{-1} [22,23]. Based on Figure 2b, both TF and NF samples had prominent peaks at around 810 and 1200 cm^{-1}, corresponding to the longitudinal optical (LO), whereas peaks located at 1100 and 1180 cm^{-1} were assigned to transverse optical (TO) components of asymmetric stretch of the SiO_4 unit [22–24].

Furthermore, both TF and NF samples also had a shoulder at around ~970 cm^{-1}, which corresponded to Si–O–Hf bond [23,25]. An additional peak was observed at ~2260 cm^{-1}, which was assigned to Si–H stretching [26], and this may be related to TEOS, as silicon alkoxide precursor may result in SiH after hydrolysis and condensation of the sol-gel process [27,28]. During the sol-gel process, the partial oxidation of Si–H bonds might result in the formation of SiO_4 units through the formation of new Si–O bonds [29,30].

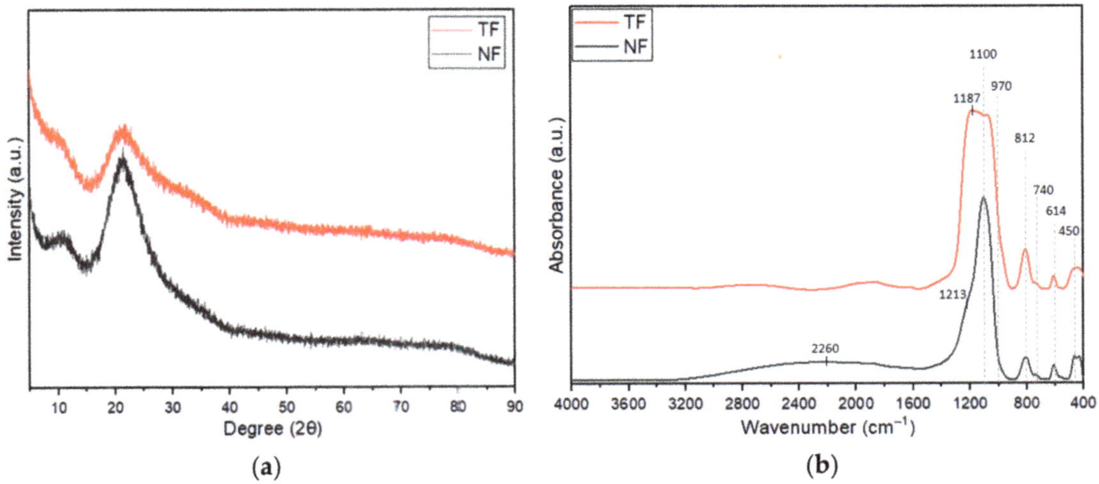

Figure 2. Both NFs and TFs were analyzed using (**a**) XRD and (**b**) FTIR.

The emission spectra of a single-layered Tm^{3+}-doped SiO_2–HfO_2 thin-film and nanofiber structure under excitation of 350 nm are shown in Figure 3. The full width at half maximum (FWHM) was labeled with a blue arrow. Both nanofiber and thin film structures were able to emit distinctive emission around 457, 512 and 634 nm, which corresponded to $^1D_2 \rightarrow {}^3F_4$, $^1D_2 \rightarrow {}^3H_5$ and $^1G_4 \rightarrow {}^3F_4$ transition, respectively. A few weak emission peaks were observed at around 578 nm, corresponding to $^1D_2 \rightarrow {}^3H_5$ transition. Despite having the same amount of Tm^{3+} ion, the emission produced by nanofibers was at a substantially higher intensity than that of the thin-film structure.

Based on Figure 3, the emission peaks produced by the thin-film structure demonstrated broader spectral bandwidth compared with nanofiber. As previously stated, the high-emission intensity of the NF sample was mostly attributable to its physical features, i.e., its high surface-to-volume ratio. The excited spontaneous emission in the light-emitting nanofibers can partly be explained as waveguides moving along the fibers and being transmitted into neighboring fibers, thus causing the emission produced to increase. This finding was also proven by previous studies, in which the absorption rate of the incident photons became greater when compared with thin films, because the excitation light was propagated through nanofibers [14,31]. Moreover, Sun [32] stated that when particle size decreased, the band gap enlarged, resulting in an increment in the photoluminescence (PL) and photoabsorption (PA) spectra.

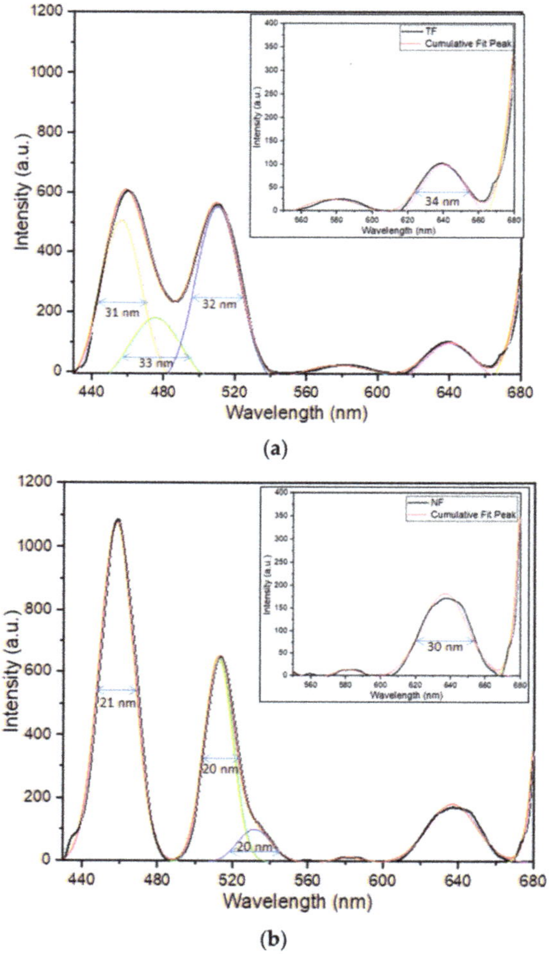

Figure 3. The emission spectra and full width at half maximum (FWHM) of Tm^{3+}-doped SiO_2–HfO_2 single-layered (**a**) thin-film (TF) and (**b**) nanofiber (NF) structures under excitation of 350 nm. Inset shows the emission around 578 and 634 nm. Coloured lines (yellow, green, blue and pink) represents the convoluted peak that were present in emission spectra (black line), whereas red line is the cumulative fit peak.

3.2. Morphological and Structural Properties of Multilayer Structures

A smooth and crack-free thin film of Tm^{3+}-doped SiO_2–HfO_2 was produced in all multilayer structures even after being annealed at 950 °C, as shown in Figure 4. All multilayer structure samples of the deposited electrospun nanofibers had uniform, smooth fibers with random orientation and ultrafine strands with a range diameter of 10–60 nm. Figure 5 shows the cross-section of the TF-NF-TF* multilayer structure. The thickness of the deposited thin film was around 1.85 μm, whereas the thickness of the nanofiber layer was approximately 100–120 nm. The thin film was expected to be much thicker compared with the nanofiber layer, as it contained a 20-nano layer of thin film.

Figure 6 shows the elemental analysis of the NF-TF-NF* multilayer structure comprised of Tm^{3+}-doped SiO_2–HfO_2, which was performed using EDS. The desired elements, Si, Hf, O and Tm, were present in the sample, whereas the appearance of element C in the EDS result was due to the carbon tape needed for FESEM-EDS analysis. Table 1 summarizes the elemental composition of samples of NF and NF-TF-NF*. The NF-TF-NF* sample contained more Tm than the NF sample, showing that the Tm^{3+} ion was effectively doped into each deposited layer of the NF-TF-NF* sample. Figure 7 depicts the EDS mapping of the individual element present in the scanned area, which confirmed that the ions were equally distributed throughout the host matrix.

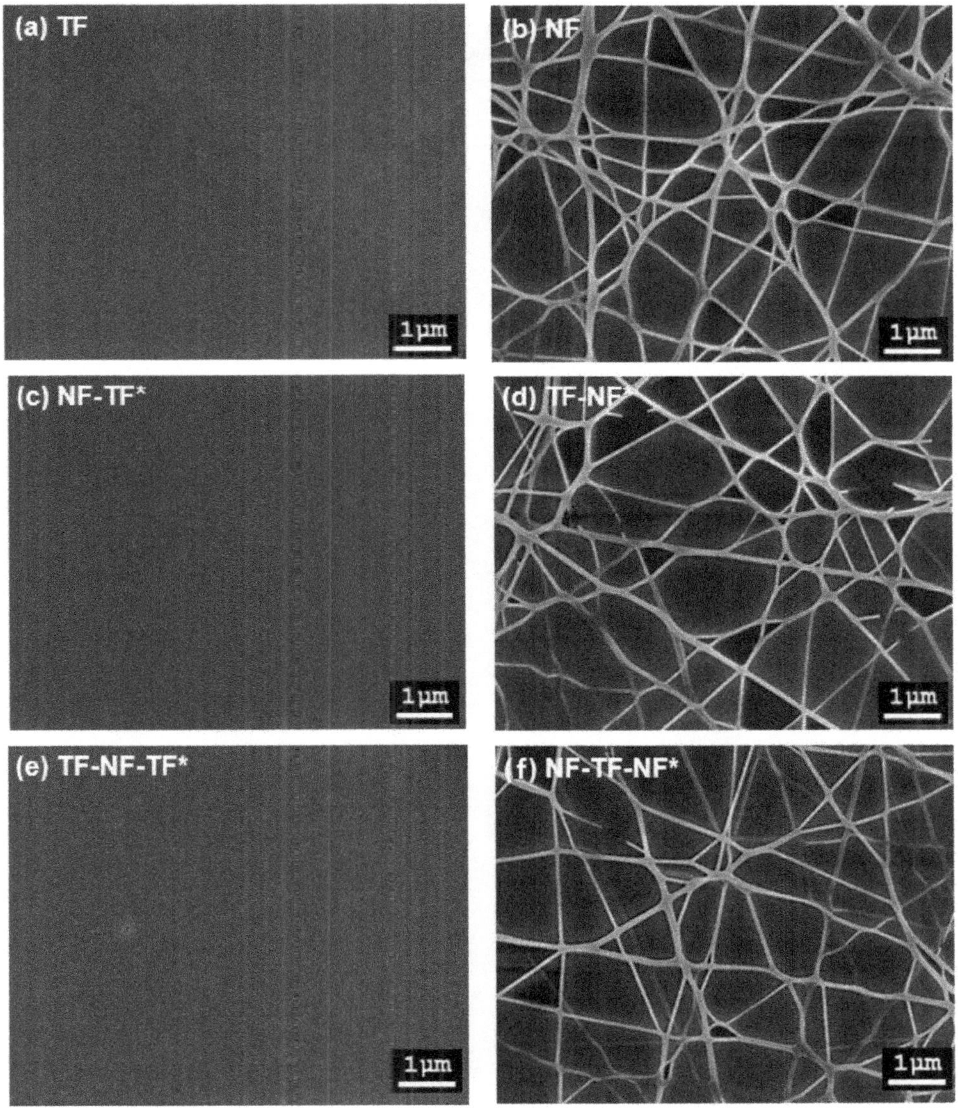

Figure 4. The field emission field scanning electron microscopy (FESEM) images of the top view of a multilayer structure sample at magnification of 15 kX: (**a**) TF, (**b**) NF, (**c**) NF-TF*, (**d**) TF-NF*, (**e**) TF-NF-TF* and (**f**) NF-TF-NF*.

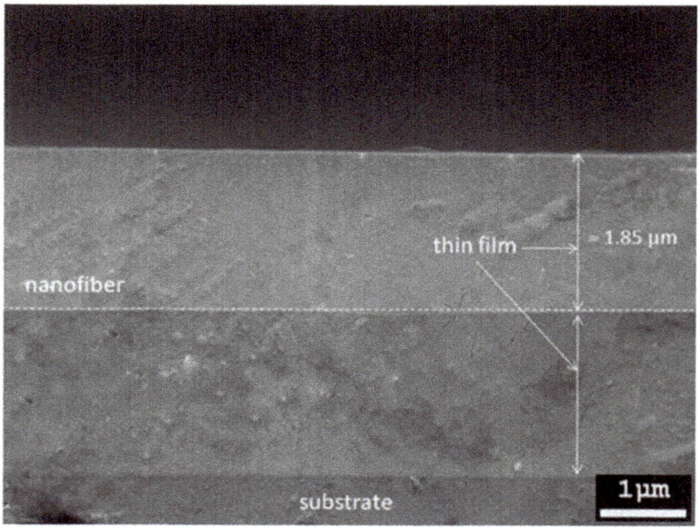

Figure 5. Cross-section of the TF-NF-TF* multilayer structure.

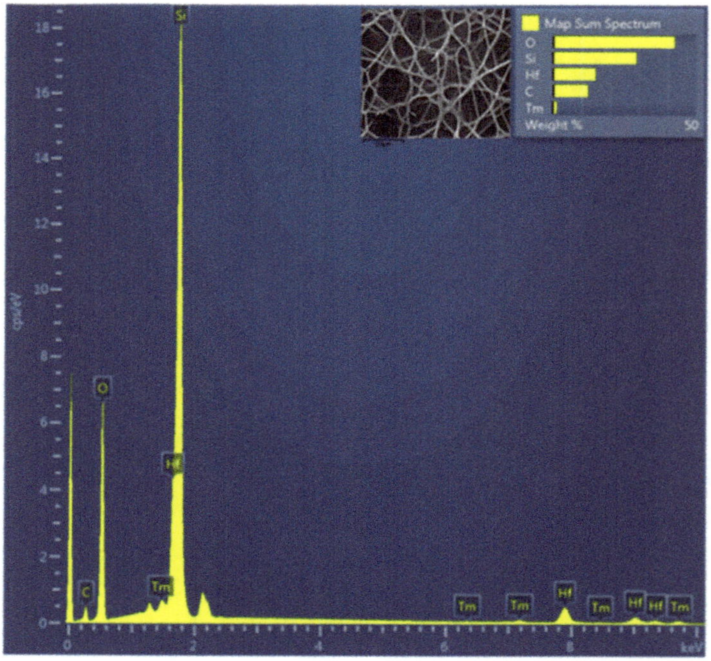

Figure 6. The energy dispersed spectroscopy (EDS) result of NF-TF-NF* multilayer structure Tm^{3+}-doped SiO_2-HfO_2. The inset image is the area scanned.

Table 1. The elemental composition of NF and NF-TF-NF* samples.

Element (Line Type)	Atomic%	
	NF	NF-TF-NF*
C K	8.10	9.04
O K	47.13	40.18
Si K	27.71	28.63
Tm L	0.99	1.67
Hf M	16.07	20.48

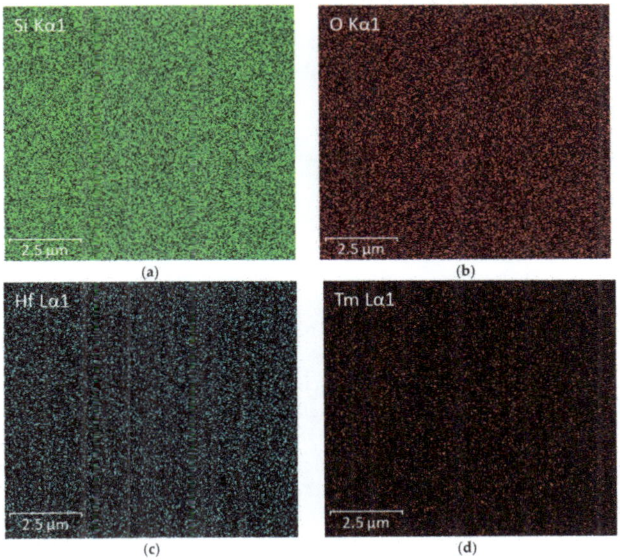

Figure 7. The element distribution of (**a**) Si, (**b**) O, (**c**) Hf and (**d**) Tm ions in the NF-TF-NF* sample.

3.3. Optical Properties of Multilayer Structures

The obtained photoluminescence emission peak of single-layered and multilayer structure samples is depicted in Figure 8. All samples were able to produce similar major peaks, which were around 457, 512 and 634 nm. The photoluminescence intensity of nanofibers significantly increased when paired with thin-film structures. From the obtained results, NF-TF-NF* had the most significant PL emission, followed by TF-NF-TF*, TF-NF*, NF-TF*, NF and finally TF. This indicated that the emission peak increased with an increasing number of layers of nanofiber/thin film. This also suggested that the increase in emission intensities was caused by the emergence of nanoscale nanofiber structures, which amplified emission intensities, whereas the increase in spectral bandwidth was caused by the presence of 2D thin-film structures. In earlier studies, Abd-Rahman and Razaki [14] discovered that the embodiment of nanostructured layers, which changed the energy levels of Tm^{3+} due to the confinement of the ions in a low-dimensional structure, combined with the effect from thin film, leading to the spectral expansion and enhancement of emission intensity.

The results obtained showed that the emission intensity of TF-NF* was higher compared with that of NF-TF*. This occurrence was probably due to the light trapped in the active nanofiber layer on the thin-film layer. Prior research by Chang et al. [7] revealed that the major enhancement of the emission intensity was related to the prolonged light path of the absorbed incoming light in the nanofiber layer [14], which caused multiple scattering events to occur inside the nanofiber layer. However, TF-NF-TF* still had a lower emission intensity when compared with NF-TF-NF* despite having nanofiber trapped between active thin films, indicating that the involvement of a nanofiber layer played a huge role in producing significantly high-emission intensity.

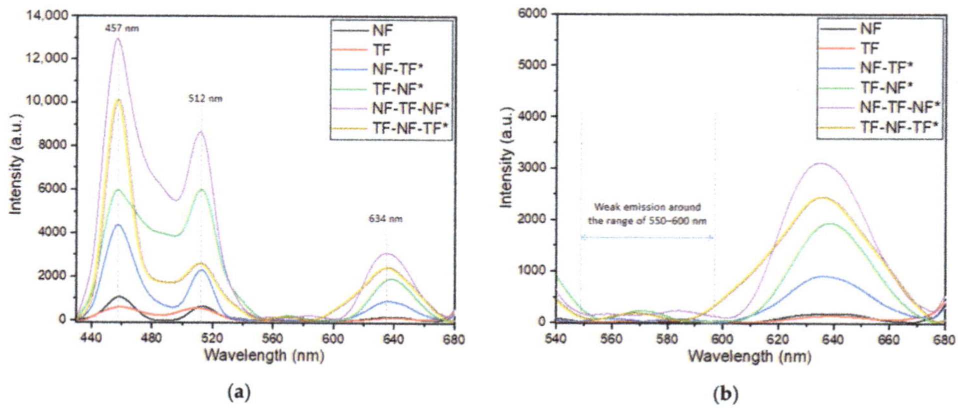

Figure 8. Emission spectra of Tm^{3+}-doped SiO_2–HfO_2 multilayer structure at range of (**a**) 430–680 nm and (**b**) 540–680 nm under an excitation of 350 nm.

Furthermore, NF-TF-NF* multilayer structure was discovered to have the largest intensity peak, which was about 10 times more than that of the single-layered structure. This was believed to be the outcome of an increase in active ion interactions, which increased the chance of net photon capture during the excitation process [14,31]. Moreover, this response was mediated by the surface contacts between fibers, thereby indicating that the two nanofiber layers in the sample were the main contributors to the enhancement of the emission intensity of the NF-TF-NF* sample. A few weak peaks were observed at around the 550–600 region (Figure 8b), which might correspond to $^1D_2 \rightarrow {}^3H_{4,5}$ and $^1G_4 \rightarrow {}^3F_4$ transition [33]. Figure 9 illustrates the mechanism of light excitation that propagated through active fiber strands in the nanofiber layers, thereby showing that the absorption points in nanofibers were much higher compared with that in the thin-film structures due to the high surface-to-volume of the nanofiber structures. This finding proved that the higher the absorption was, the higher the number of molecules promoted to the excited state was, thereby causing higher emission production.

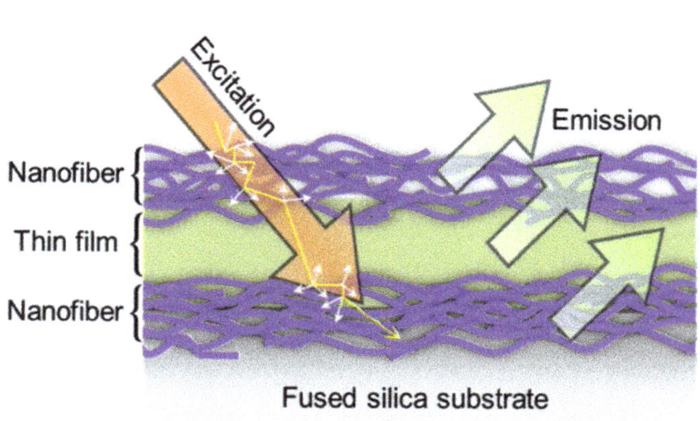

Figure 9. Mechanism of light excitation propagates through active fiber strands in nanofiber layers.

The transmittance of single-layered and multilayer structures at a region of 200–2000 nm is demonstrated in Figure 10. All fabricated samples exhibited a drastic increment of optical transparency in the UV–visible wavelength, whereas in the infrared region (700–2000 nm) the transparency remained relatively constant. The sample that had the highest transmittance percentage was NF, followed by TF, NF-TF*, TF-NF*, TF-NF-TF* and NF-TF-NF*. This demonstrated that the percentage of transmittance decreased as the number of layers increased, which was expected since the resulting sample structure becomes thicker. However, NF-TF-NF* was discovered to have a lower transmittance than TF-NF-TF*, despite its smaller resultant thickness. This was probably due to light scattering that contributed to the decline of transmittance percentage of the NF-TF-NF* structure. The low transmittance of the NF-TF-NF* sample was mainly due to the large surface-to-volume ratio of the nanofiber structure, which caused light to deflect in all directions, thus leading to the increment of the absorption rate on a single substrate. Considering that the NF-TF-NF* was composed of two nanofiber layers that "sandwiched" the thin-film layer, a higher absorption rate was obtained. Nevertheless, single-layered NF was still the most transparent, because it had a lower density and was far thinner than a single-layered thin film.

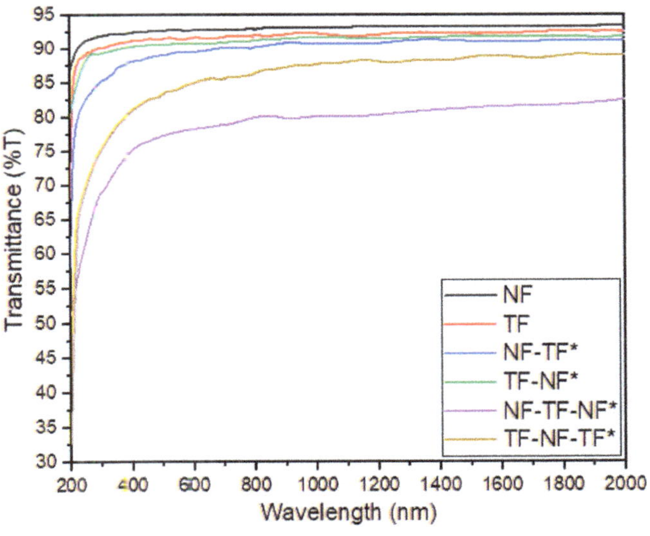

Figure 10. Transmittance of single-layered and multilayer structures at 200–2000 nm region.

3.4. *Comparison of Experimental and Theoretical Results*

The experimental data were compared with the multilayer stack computation produced by the IMD software. The thickness of thin films was fixed at 1.80 µm, whereas the thickness of nanofibers varied from 100–120 nm. Other parameters involved, such as optical constant, atomic weight, composition and density of each material, were taken from data stored in the IMD software and were kept constant. The transmittance results, both experimental and theoretical, of (a) single-layered, (b) dual-layered and (c) three-layered structures in the range of 200–2000 nm are shown in Figure 11. The pattern of the calculated transmittance shows a similar result to the experimental data of Tm^{3+}-doped SiO_2-HfO_2. The only difference was that the slope of the calculated transmittance gradually increased in the UV region towards the IR region, whereas the experimental result showed that the transmittance increased drastically towards the IR region before remaining constant throughout the IR region. This finding signified that the fabricated multilayer structure obeyed the modified Fresnel's equation where the interface imperfections were considered.

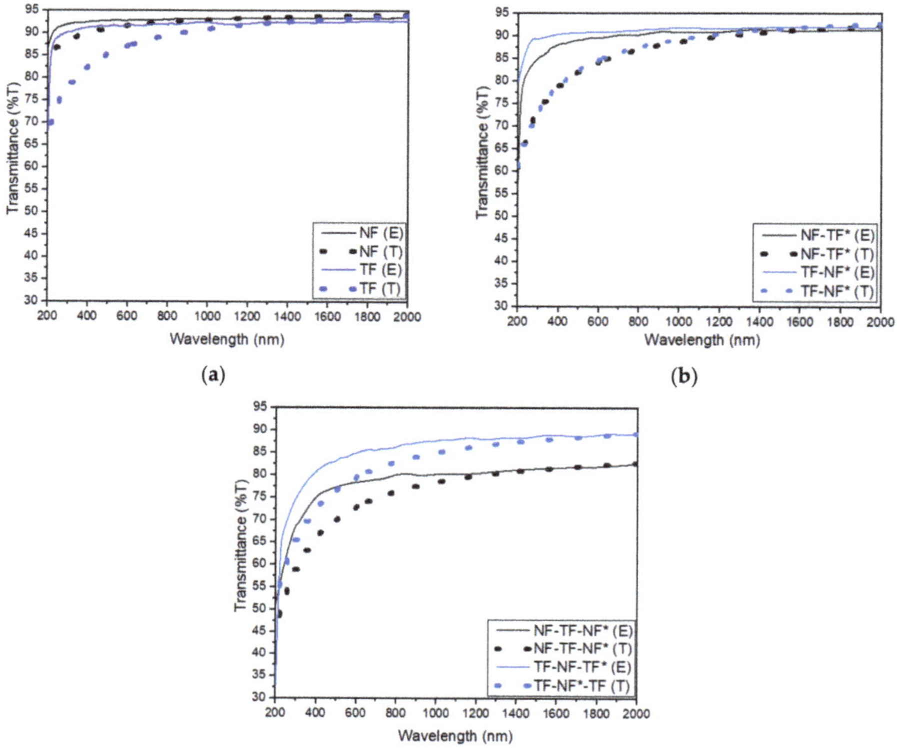

Figure 11. Transmittance of both experimental (E) and theoretical (T) of (**a**) single-layered structure, (**b**) dual-layered structure and (**c**) three-layered structure at the range of 200–2000 nm.

Similar to the experimental result, the transmittance decreased with the number of active layers due to light scattering. Killada [34] mentioned that the amount of light that is transmitted is influenced by the material's thickness and the way that photons interact with the structure of the material. Additionally, HfO_2 leads to the high refractive index of the layers. An increase in the refractive index concurrent with an increase in thickness demonstrates the dense uniformity and adhesion of both deposited structure layers. As a result, the absorbance and refractive index of the sample increases along with the number of layers, causing the percentage of light transmitted through the sample to be lowered, except for in the NF-TF-NF* structure, as discussed earlier.

The absorbance was also estimated using IMD software, and the results are as shown in Figure 12. In this study, the roughness of nanofibers was varied until they achieved the optimum degree, which was around 14.0–15.0 nm. Surface roughness significantly influences the rate of light absorption, as high roughness causes less light to reflect and more light to penetrate into the active layers. According to Liaparinos [35], the degree of roughness influences surface behavior, which affects the overall optical performance of the system. Based on Figure 12, the results obtained were similar to the experimental results. For single-layered samples, the absorbance of TF was higher compared with that of NF; however, it was notable that as the number of layers increased, the absorbance of the sample that consisted of nanofibers on the topmost layer had higher absorbance.

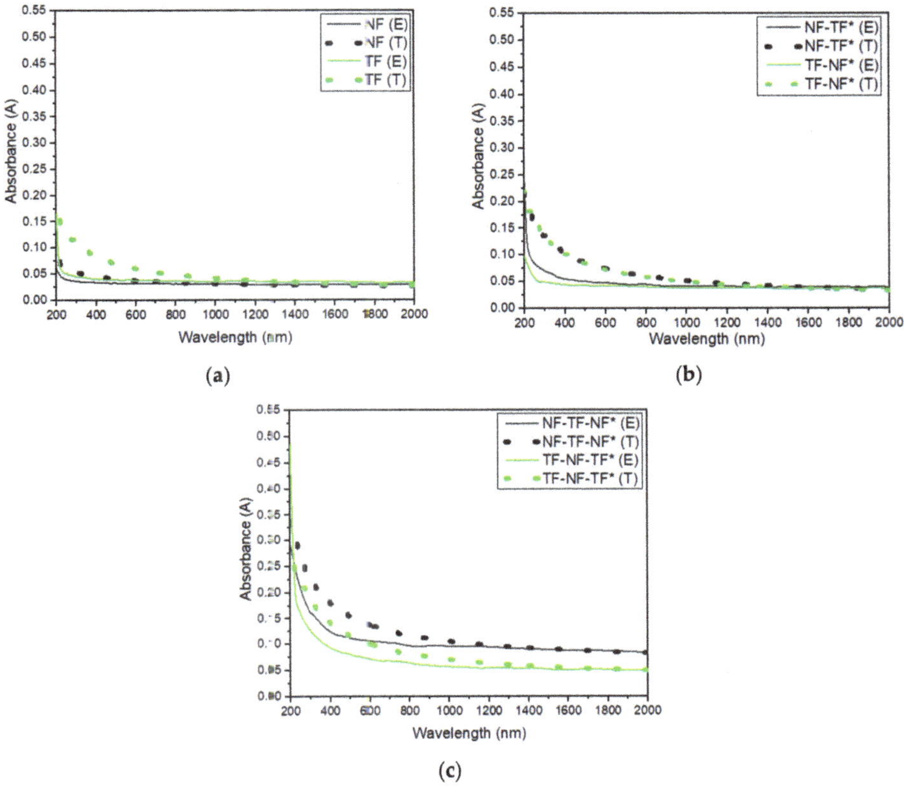

Figure 12. Absorbance of both experimental (E) and theoretical (T) of (**a**) single-layered structure, (**b**) dual-layered structure and (**c**) three-layered structure at the range of 200–2000 nm.

Therefore, a great variation of performance was present, and this depended on the resultant thickness of the sample and the surface roughness. The theoretical prediction and the experimental findings had a reasonable degree of coincidence. Photoluminescence intensity was fitted using the Voigt function. Figure 13 shows PL emission intensity data (dotted line) and the Voigt function profile (solid colored line). The parameters of the calculated data are tabulated in Table 2.

Based on Figure 13, the Voigt function provided the best match for the PL emission peak produced from the experiment, with just a small difference. Hence, the fitted graphs were acceptable. Based on tabulated data in Table 2, the peak that was present at 600–680 nm in the NF samples had a bigger area under the graph (peak 5 = 10,401 a.u.) when compared with the emissions of TF samples (peak 5 = 4854 a.u.). This finding showed that the low dimensional size indeed increased the emission intensity. However, the emission peaks emitted by the TF samples were broader compared with that of the NF samples. For instance, peak 5 of the NF sample had a FWHM of 29.9 nm, whereas that of the TF sample was 31.3 nm, which was 1.4 nm higher. The same held for peak 1 of the NF sample with a FWHM of 21.4 nm, which was 4.6 nm narrower compared with peak 1 of the TF sample (26.0 nm).

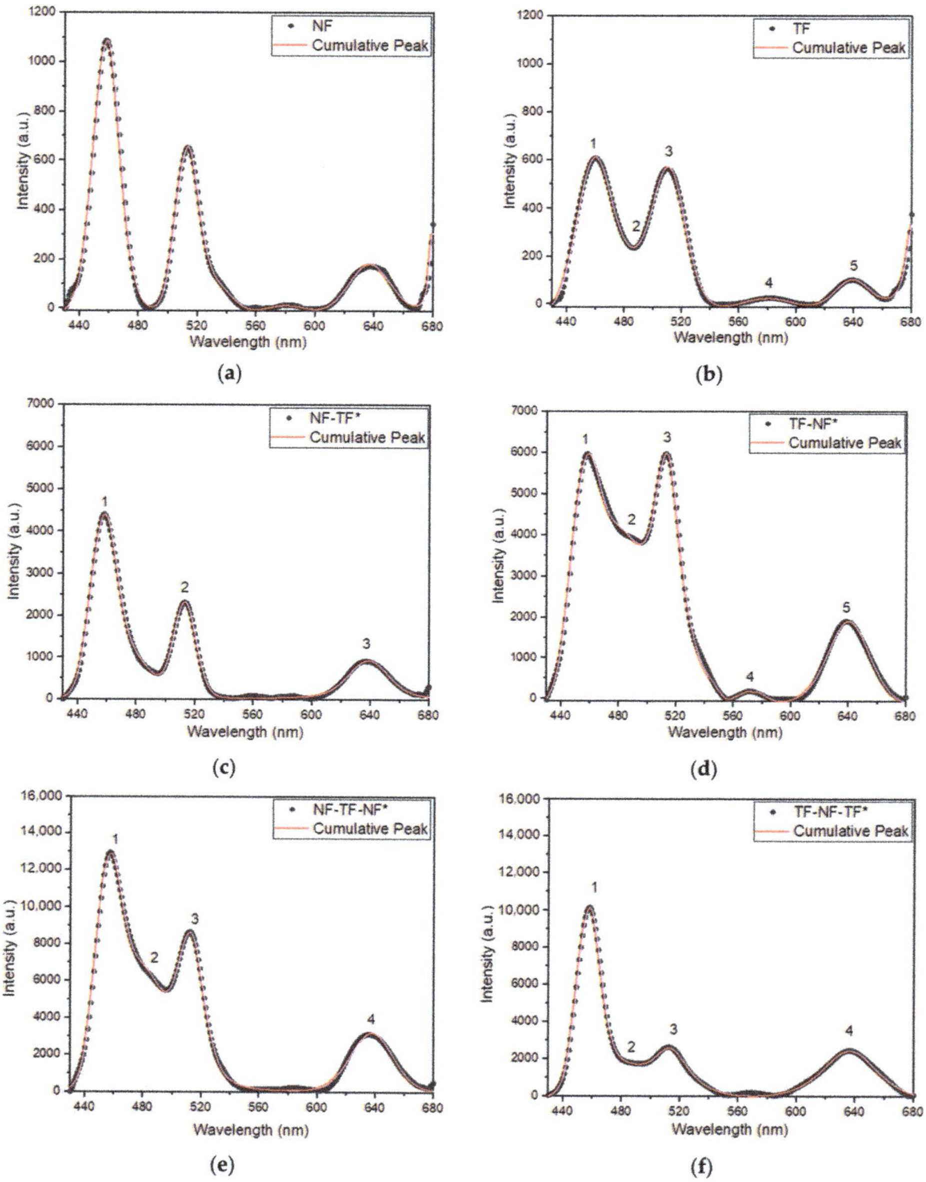

Figure 13. PL emission intensity data (dotted line) and the Voigt function profile (solid colored line) of (**a**) NF, (**b**) TF, (**c**) NF-TF*, (**d**) TF-NF*, (**e**) NF-TF-NF* and (**f**) TF-NF-TF* samples.

Table 2. Parameters obtained from Voigt function profile.

Sample	No. of Peak	Peak Centered (nm)	Area (a.u.)	Gaussian FWHM, wG (nm)	Lorentzian FWHM, wL (nm)	FWHM (nm)
NF	1	458.51 ± 0.07	27,398 ± 1698	19.13 ± 1.16	4.12 ± 2.13	21.4
	2	513.08 ± 0.43	13,052 ± 1140	18.56 ± 0.75	0	18.6
	3	532.95 ± 2.64	2528 ± 2835	20.68 ± 18.82	0	20.7
	4	580.41 ± 1.87	1132 ± 377	30.13 ± 5.49	0	30.1
	5	637.36 ± 0.46	10,401 ± 2629	9.74 ± 13.38	26.54 ± 10.68	29.9
TF	1	458.74 ± 1.66	17,969 ± 7419	25.99 ± 2.14	0	26.0
	2	485.11 ± 3.07	5888 ± 2341	30.91 ± 41.94	2.77 ± 1.68	32.4
	3	511.22 ± 1.46	18,428 ± 3535	25.02 ± 2.97	3.25 ± 3.29	26.8
	4	580.05 ± 2.05	3139 ± 1140	48.32 ± 5.21	0	48.3
	5	637.53 ± 0.46	4854 ± 4224	27.25 ± 14.24	7.23 ± 33.02	31.3
NF-TF*	1	458.87 ± 0.20	104,418 ± 3352	21.91 ± 0.97	0.68 ± 1.39	22.3
	2	511.82 ± 0.12	53,223 ± 2962	16.31 ± 1.42	5.67 ± 2.15	19.5
	3	480.16 ± 1.13	13,518 ± 2126	18.26 ± 1.98	25.69 ± 5.07	35.6
	4	637.91 ± 0.36	38,169 ± 4925	0	12.85 ± 7.54	12.9
TF-NF*	1	455.65 ± 0.10	91,357 ± 7848	21.69 ± 0.57	0	21.7
	2	477.17 ± 1.06	176,679 ± 11,901	43.43 ± 2.04	0	43.4
	3	513.86 ± 0.08	221,347 ± 5633	6.44 ± 1.72	23.24 ± 1.07	25.0
	4	575.29 ± 0.81	4302 ± 2304	17.17 ± 8.34	0.13 ± 15.30	17.2
	5	640.07 ± 0.13	101,948 ± 8665	25.73 ± 2.47	16.86 ± 4.15	35.9
NF-TF-NF*	1	455.88 ± 0.15	234,669 ± 15,351	20.95 ± 0.38	0	20.9
	2	478.79 ± 0.92	241,217 ± 19,747	36.35 ± 2.52	0	36.3
	3	512.73 ± 0.15	229,082 ± 6521	17.04 ± 0.78	9.52 ± 1.02	22.7
	4	637.27 ± 0.15	118,194 ± 6626	33.35 ± 1.92	2.02 ± 3.44	34.4
TF-NF-TF*	1	457.34 ± - -	209,130 ± - -	17.843 ± - -	2.29 ± - -	19.1
	2	475.96 ± 3.50	41,046 ± - -	28.86 ± 20.02	0	28.9
	3	510.91 ± 1.36	157,012 ± - -	5.0485 ± - -	34.67 ± - -	35.4
	4	637.43 ± 2.59	164,533 ± 35,765	31.57 ± 63.36	24.25 ± 122.79	46.5

The emission peak of the TF-NF* sample had a larger emission intensity when compared with that of the NF-TF* sample. Based on the computed data, the total area under the graph of the TF-NF* sample (peak 1 + peak 2 + peak 3 = 489,383 a.u.) at 430–550 nm region was 318,224 a.u. larger compared with that of the NF-TF* sample (peak 1 + peak 2 + peak 3 = 171,159 a.u.). In fact, the emission emitted at 600–680 nm in the TF-NF* sample (peak 5 = 101,948 a.u.) also had a bigger area under the graph, with a difference of 63,779 a.u. when compared with that of the NF-TF* sample (peak 4 = 38,169 a.u.). The FWHM of peak 5 of the TF-NF* sample (35.9 nm) was almost three times bigger when compared with the FWHM of peak 4 of NF-TF* (12.9 nm) samples. This finding showed that having nanofibers on the topmost layer prolonged the light path of the absorbed incoming light in the nanofiber layer compared with a single layer of nanofibers (NFs). However, when nanofibers were added onto NF-TF*, which produced NF-TF-NF*, the emission intensity was two times bigger compared with that of TF-NF-TF*.

The NF-TF-NF* multilayer structure had a larger emission intensity compared with other samples, as the areas under the graph for all peaks were larger compared with other peaks present in other samples. This finding was proven when compared with TF-NF-TF*, where both emissions had broad emission peaks at around 430–550 nm. NF-TF-NF* had three peaks (peaks 1, 2 and 3) at the mentioned emission region. The total areas under the graph of each peak were added, thereby producing an area of 704,968 a.u. Similar to NF-TF-NF*, TF-NF-TF* also had three peaks (peaks 1, 2 and 3) and had a total area under the graph of 407,188 a.u. When single-layered samples were compared with three-layered samples, the area under the graph of the single-layered samples increased by a factor of about 10, indicating that a significant increase in emission intensity was possible as the amount of multilayer increased. Hence, it was theoretically shown that the intensity of the PL emission generated increased with the number of layers in the multilayer structure. However, the FWHM of TF-NF-TF* was the widest among the samples. This was observed at the 600–680 nm region, where the FWHM of peak 4 of the TF-NF-TF* sample was 46.5 nm, which was 12.1 nm bigger when compared with the FWHM of peak 4 of the NF-TF-NF* sample (34.4 nm). This occurrence was likely due to the involvement of two thin-film layers, which caused the emissions to have a wide FWHM. The broad emission intensity could also be due to the presence of an overlapping emission peak at that particular region.

4. Conclusions

Significantly enhance PL emission intensity was successfully achieved owing to the structural properties of both nanofiber and thin-film structures and also a high amount of active ions doped in a single substrate. A comparison between nanofibers and thin films shows that nanofibers are able to produce higher emission intensity compared with thin films due to the following: (i) nanofibers had higher absorption points due to their high surface-to-volume ratio and (ii) the prolonged light path of the absorbed incoming light in the nanofiber layer. Nanofibers produced a narrow emission intensity compared with thin films. Therefore, specific applications were possible when constructing a multilayer structure and choosing the composite material. Moreover, the experimental and computed results reached a satisfactory agreement with accurate values of parameters such as the emission peak, FWHM and area under the graph.

Author Contributions: Conceptualization, N.I.Z.Z., S.A.K. and A.-B.M.A.I.; methodology, N.I.Z.Z., S.A.K. and A.-B.M.A.I.; software, N.I.Z.Z. and A.-B.M.A.I.; validation, N.I.Z.Z., S.A.K. and A.-B.M.A.I.; formal analysis, N.I.Z.Z., S.A.K., N.A.M., R.M.Y. and A.-B.M.A.I.; investigation, N.I.Z.Z.; resources, N.I.Z.Z., S.A.K. and A.-B.M.A.I.; data curation, N.I.Z.Z. and S.A.K.; writing—original draft preparation, N.I.Z.Z.; writing—review and editing, N.I.Z.Z., S.A.K., N.A.M., R.M.Y. and A.-B.M.A.I.; visualization, N.I.Z.Z.; supervision, S.A.K., N.A.M. and A.-B.M.A.I.; project administration, N.I.Z.Z., S.A.K. and A.-B.M.A.I.; funding acquisition, S.A.K. All authors have read and agreed to the published version of the manuscript.

Funding: This project was financially supported by the Ministry of Higher Education Malaysia (MOHE) under the Fundamental Research Grant Scheme (FRGS, Sponsorship File No. FRGS/1/2019/STG02/UITM/02/11; 600-IRMI/FRGS 5/3 (431/2019)).

Data Availability Statement: Not applicable.

Acknowledgments: The authors express gratitude to the Research Management Centre (RMC), Universiti Teknologi MARA, Malaysia for assistance throughout the research. We would like to acknowledge the Faculty of Applied Sciences, UiTM and i-CRIM, Universiti Kebangsaan Malaysia (UKM) for providing great service and facilities.

Conflicts of Interest: The authors declare no conflict of interest.

References

1. Rao, M.; Shekhawat, M. A brief Survey on Basic Properties of Thin Films for Device Application. *Int. J. Mod. Phys. Conf. Ser.* **2013**, *22*, 576–582. [CrossRef]
2. Richter, F.; Kupfer, H.; Schlott, P.; Gessner, T.; Kaufmann, C. Optical properties and mechanical stress in SiO_2/Nb_2O_5 multilayers. *Thin Solid Film.* **2001**, *389*, 278–283. [CrossRef]
3. Sullivan, B.T.; Dobrowolski, J. Deposition error compensation for optical multilayer coatings. I. Theoretical description. *Appl. Opt.* **1992**, *31*, 3821–3835. [CrossRef]
4. Sullivan, B.T.; Dobrowolski, J. Deposition error compensation for optical multilayer coatings. II. Experimental results—Sputtering system. *Appl. Opt.* **1993**, *32*, 2351–2360. [CrossRef]
5. Rahmani, F.; Ardyanian, M. Fabrication and characterization of ZnO/TiO_2 multilayers, deposited via spin coating method. *J. Mater. Sci. Mater. Electron.* **2018**, *29*, 4285–4293. [CrossRef]
6. Camposeo, A.; Di Benedetto, F.; Stabile, R.; Neves, A.A.; Cingolani, R.; Pisignano, D. Laser emission from electrospun polymer nanofibers. *Small* **2009**, *5*, 562–566. [CrossRef]
7. Chang, C.-C.; Huang, C.-M.; Chang, Y.-H.; Kuo, C. Enhancement of light scattering and photoluminescence in electrospun polymer nanofibers. *Opt. Express* **2010**, *18*, A174–A184. [CrossRef]
8. Xia, H.; Chen, T.; Hu, C.; Xie, K. Recent advances of the polymer micro/nanofiber fluorescence waveguide. *Polymers* **2018**, *10*, 1086. [CrossRef]
9. Kirchain, R.; Kimerling, L. A roadmap for nanophotonics. *Nat. Photonics* **2007**, *1*, 303–305. [CrossRef]
10. Tong, L.; Gattass, R.R.; Ashcom, J.B.; He, S.; Lou, J.; Shen, M.; Maxwell, I.; Mazur, E. Subwavelength-diameter silica wires for low-loss optical wave guiding. *Nature* **2003**, *426*, 816–819. [CrossRef] [PubMed]
11. Lim, C.T. Nanofiber technology: Current status and emerging developments. *Prog. Polym. Sci.* **2017**, *70*, 1–17.
12. Camposeo, A.; Persano, L.; Pisignano, D. Light-Emitting Electrospun Nanofibers for Nanophotonics and Optoelectronics. *Macromol. Mater. Eng.* **2013**, *298*, 487–503. [CrossRef]
13. Milea, C.; Bogatu, C.; Duta, A. The influence of parameters in silica sol-gel process. *Bull. Transilv. Univ. Brasov. Eng. Sciences. Ser. I* **2011**, *4*, 59.
14. Abd-Rahman, M.K.; Razaki, N.I. Effect of nanofiber/thin-film multilayers on the optical properties of thulium-doped silica-alumina. *J. Lumin.* **2018**, *196*, 442–448. [CrossRef]
15. Zhao, J.; Jia, C.; Duan, H.; Sun, Z.; Wang, X.; Xie, E. Structural and photoluminescence properties of europium-doped titania nanofibers prepared by electrospinning method. *J. Alloy Compd.* **2008**, *455*, 497–500. [CrossRef]
16. Lucas, J.; Lucas, P.; Mercier, T.; Rollat, A.; Davenport, W. *Rare Earth Doped Lasers and Optical Amplifiers*; Marcel Dekker Inc.: New York, NY, USA, 2015; pp. 319–332.
17. Farias, P.M.; Santos, B.S.; Chaves, C.R.; Figueiredo, R.C.; Ferreira, R.C.; Fontes, A. Hybrid organic/II–VI quantum dots: Highly luminescent nanostructures for bioimaging. In *Molecular Probes for Biomedical Applications II*; International Society for Optics and Photonics: San Jose, CA, USA, 2008; Volume 6867, p. 68670G.
18. Qin, C.; Qin, L.; Chen, G.; Lin, T. One-dimensional Eu^{3+} and Tb^{3+} doped $LaBO_3$ nanofibers: Fabrication and improved luminescence performances. *Mater. Lett.* **2013**, *106*, 436–438. [CrossRef]
19. Righini, G.C.; Chiappini, A. Glass optical waveguides: A review of fabrication techniques. *Opt. Eng.* **2014**, *53*, 071819. [CrossRef]
20. Windt, D.L. IMD—Software for modeling the optical properties of multilayer films. *Comput. Phys.* **1998**, *12*, 360–370. [CrossRef]
21. Gupta, S.; Pramanick, A.; Kailath, A.; Mishra, T.; Guha, A.; Nayar, S.; Sinha, A. Composition dependent structural modulations in transparent poly(vinyl alcohol) hydrogels. *Colloids Surf. B Biointerfaces* **2009**, *74*, 186–190. [CrossRef]
22. Khomenkova, L.; An, Y.-T.; Labbé, C.; Portier, X.; Gourbilleau, F. Hafnia-based luminescent insulator for phosphor applications. *ECS Trans.* **2012**, *45*, 119. [CrossRef]
23. Neumayer, D.; Cartier, E. Materials characterization of ZrO_2–SiO_2 and HfO_2–SiO_2 binary oxides deposited by chemical solution deposition. *J. Appl. Phys.* **2001**, *90*, 1801–1808. [CrossRef]
24. Gonçalves, R.; Carturan, G.; Zampedri, L.; Ferrari, M.; Montagna, M.; Chiasera, A.; Righini, G.; Pelli, S.; Ribeiro, S.; Messaddeq, Y. Sol-gel Er-doped SiO_2–HfO_2 planar waveguides: A viable system for 1.5 μm application. *Appl. Phys. Lett.* **2002**, *81*, 28–30. [CrossRef]

25. Khomenkov, D.; An, Y.-T.; Portier, X.; Labbe, C.; Gourbilleau, F.; Khomenkova, L. Light-emitting and structural properties of Si-rich HfO_2 thin films fabricated by RF magnetron sputtering. *MRS Online Proc. Libr.* **2013**, *1617*, 85–91. [CrossRef]
26. Dirè, S.; Ceccato, R.; Facchin, G.; Carturan, G. Synthesis of Ni metal particles by reaction between bis (cyclooctadiene) nickel (0) and sol–gel SiO_2 modified with Si–H groups. *J. Mater. Chem.* **2001**, *11*, 678–683. [CrossRef]
27. Soraru, G.D.; D'Andrea, G.; Campostrini, R.; Babonneau, F. Characterization of methyl-substituted silica gels with Si–H functionalities. *J. Mater. Chem.* **1995**, *5*, 1363–1374. [CrossRef]
28. Pauthe, M.; Phalippou, J.; Corriu, R.; Leclercq, D.; Vioux, A. Silica xerogels containing a functional group at silicon. *J. Non-Cryst. Solids* **1989**, *113*, 21–30. [CrossRef]
29. Thami, T.; Tauk, L.; Flaud, V. Controlled structure and hydrophilic property of polymethylhydrosiloxane thin films attached on silicon support and modified with phosphorylcholine group. *Thin Solid Film* **2020**, *709*, 138196. [CrossRef]
30. Karakuscu, A. Synthesis and Characterization of Luminescent Nanostructured SiOC Thin Films. Ph.D. Thesis, University of Trento, Trento, Italy, 2010.
31. Teki, R.; Koratkar, N.; Karabacak, T.; Lu, T.-M. Enhanced photoemission from nanostructured surface topologies. *Appl. Phys. Lett.* **2006**, *89*, 193116. [CrossRef]
32. Sun, C.Q. Size dependence of nanostructures: Impact of bond order deficiency. *Prog. Solid State Chem.* **2007**, *35*, 1–159. [CrossRef]
33. De Azevedo Marques, A.P.; Künzel, R.; Umisedo, N.K.; Latini, R.M.; Yoshimura, E.M.; Okuno, E. Tm3+ doped barium molybdate: A potential long-lasting blue phosphor. *J. Alloy Compd.* **2018**, *735*, 707–717. [CrossRef]
34. Killada, S. *Fabrication of Polymer Surfaces with Varying Roughness and Its Effect on Anti-Reflective Properties*; Indian Institute of Technology: Hyderabad, India, 2014.
35. Liaparinos, P. Influence of surface roughness on the light transmission through the boundaries of luminescent materials in radiation detectors. *Opt. Spectrosc.* **2021**, *129*, 1257–1265. [CrossRef]

nanomaterials

Review

Thermochromic Smart Windows Assisted by Photothermal Nanomaterials

Yong Zhao, Haining Ji *, Mingying Lu, Jundong Tao, Yangyong Ou, Yi Wang, Yongxing Chen, Yan Huang, Junlong Wang and Yuliang Mao *

School of Physics and Optoelectronics, Xiangtan University, Xiangtan 411105, China
* Correspondence: sdytjhn@126.com (H.J.); ylmao@xtu.edu.cn (Y.M.)

Abstract: Thermochromic smart windows are optical devices that can regulate their optical properties actively in response to external temperature changes. Due to their simple structures and as they do not require other additional energy supply devices, they have great potential in building energy-saving. However, conventional thermochromic smart windows generally have problems with high response temperatures and low response rates. Owing to their great effect in photothermal conversion, photothermal materials are often used in smart windows to assist phase transition so that they can quickly achieve the dual regulation of light and heat at room temperature. Based on this, research progress on the phase transition of photothermal material-assisted thermochromic smart windows is summarized. In this paper, the phase transition mechanisms of several thermochromic materials (VO_2, liquid crystals, and hydrogels) commonly used in the field of smart windows are introduced. Additionally, the applications of carbon-based nanomaterials, noble metal nanoparticles, and semiconductor (metal oxygen/sulfide) nanomaterials in thermochromic smart windows are summarized. The current challenges and solutions are further indicated and future research directions are also proposed.

Keywords: thermochromic; smart window; photothermal materials; phase change

Citation: Zhao, Y.; Ji, H.; Lu, M.; Tao, J.; Ou, Y.; Wang, Y.; Chen, Y.; Huang, Y.; Wang, J.; Mao, Y. Thermochromic Smart Windows Assisted by Photothermal Nanomaterials. *Nanomaterials* **2022**, *12*, 3865. https://doi.org/10.3390/nano12213865

Academic Editor: Enyi Ye

Received: 9 October 2022
Accepted: 31 October 2022
Published: 2 November 2022

1. Introduction

According to the report published by the China Association of Building Energy Efficiency in 2021, building energy consumption accounted for 46.5% of the country's total energy in 2018 of China. Among them, 21.7% is used for energy consumption in the building operation stage [1], that is, the energy consumed by terminal equipment, such as heating, refrigeration, ventilation, and lighting in the building. Additionally, the proportion is still increasing. In the environment of a global energy shortage, the increase in building energy consumption has become a weakness that has restricted the development of the Chinese economy. The percentage of energy lost from windows directly or indirectly is 60%. Therefore, scholars have invented a variety of smart windows [2–7] to improve the energy-saving effect of building windows.

In recent years, with energy-saving and green development being vigorously promoted, smart windows have attracted much attention due to their outstanding energy-saving effect. The concept of "smart windows" was introduced by Granqvist in the 1980s [8]. They are energy-saving windows that regulate solar radiation by combining a dimming material [9–11] with a base material, such as glass. As optical devices, their core is the sensitive material attached to the glass. Under external excitation by light, electromagnetic radiation, and temperature, the sensitive materials will color or fade, thereby changing the windows' colors and other optical properties so that they can automatically adjust the indoor temperature and light intensity according to the surrounding environment and ultimately achieve the purpose of saving energy consumption.

At present, most typical smart windows require an external power supply or heating device to dynamically tune the optical properties of materials in response to external stimuli. However, these additional devices have significant disadvantages, such as a complex structure, low lifespan, and high energy consumption, which also greatly hinder their commercial utilization. In contrast, solar-driven thermochromic smart windows can adjust the optical properties actively through the strong absorption of solar energy by photothermal materials, achieving a dual response to light and temperature. The features of low cost, simple structure, and low energy consumption are more conducive to large-scale applications in the future.

Smart windows assisted by photothermal materials are extremely important for the development of smart windows. However, systematic summaries of this aspect are relatively rare. Therefore, in this review, several thermochromic smart windows with different substrates are presented, the auxiliary effects of different kinds of photothermal materials on the phase change of smart windows are reported, and, finally, future research trends are proposed. This will provide references and suggestions for overcoming the problems of a high response temperature and low response rate of traditional thermochromic smart windows.

2. Thermochromic Materials for Smart Windows

There are many kinds of thermochromic materials that play an important role as substrates for smart windows. However, only those materials with phase change temperatures adjustable to room temperature (about 28 °C) can be used for building energy efficiency [12]. Therefore, the most widely used photothermal materials for smart windows are vanadium dioxide (VO_2), thermochromic hydrogels, and liquid crystals, as shown in Figure 1.

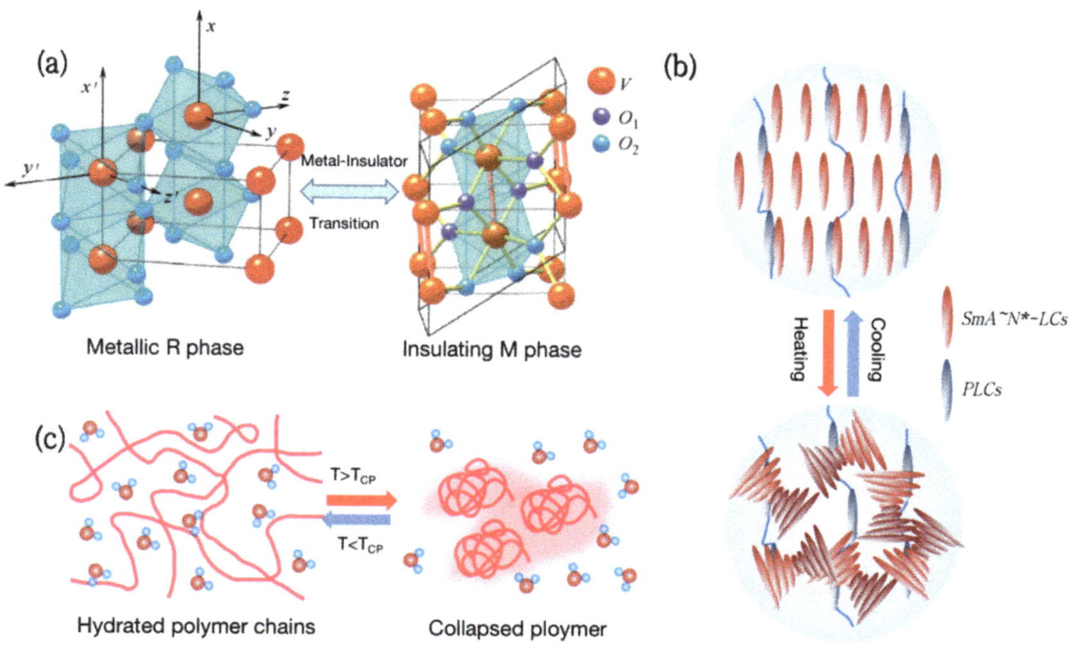

Figure 1. Types of thermochromic smart windows. (**a**) Metal-insulator phase transition (MIT) of vanadium dioxide (VO_2). (**b**) Liquid crystal changes transmittance by adjusting molecular orientation in response to temperature. (**c**) Thermochromic hydrogels change transparency reversibly with temperature.

Vanadium dioxide (VO_2) is a kind of metal-insulator phase transition (MIT) material, which is extensively utilized in thermochromic smart windows [13–16]. Its phase transition temperature is close to room temperature at 68 °C [17]. As the temperature rises, VO_2 will undergo a phase transition from a semiconductor to a metal (Figure 1a), with a sudden change in infrared transmittance while keeping the visible transmittance unchanged [18]. Although VO_2 films have great practical value, which are used in smart windows [19], photoelectric switches [20,21], and infrared stealth [22–24], the high phase transition temperature and poor optical properties are still problems to be solved.

Liquid Crystal (LC) is a special material between traditional liquids and crystalline solids. Its optical, electrical, and mechanical characteristics are anisotropic, and it can also be employed in smart windows [25–28]. From the molecular point of view, the thermochromism of Liquid Crystals can be achieved by anisotropic molecules responding to temperature and adjusting their orientation (Figure 1b). For example, the molecules inside the polymer network Liquid Crystals (PNLC) are arranged in order at low temperatures, and the refractive index of the liquid crystal and the polymer are the same, showing a highly transparent state. As the temperature rises, the liquid crystals transform into a blurred state due to the disorder of molecular orientation, and the transmittance, including near-infrared light, can change by more than 20%.

Hydrogels are both hydrophilic and hydrophobic and have been commonly employed in the field of smart windows [29–32]. Thermochromic hydrogels, such as poly(N-isopropylacrylamide) (PNIPAm) and hydroxypropyl methyl cellulose (HPMC), have a lower critical solution temperature (LCST) of around 32 °C to 40 °C. When the temperature is below the LCST, hydrophilic groups in the hydrogel form hydrogen bonds with water molecules, presenting a highly transparent state. Above the LCST, the hydrogen bonds are broken and polymers aggregate, resulting in a significant decrease in transmittance in the entire spectral range [33] (Figure 1c). Thermochromic hydrogels can reversibly change transparency with an increase in temperature, making them ideal materials for smart windows.

3. Common Photothermal Conversion Materials

Photothermal and thermochromic materials are often combined for sunlight-driven thermochromic smart windows. The strong light-absorption characteristics or localized surface plasmon resonance effects [34,35] of photothermal materials can convert solar energy into heat energy, increase the surrounding temperature, and achieve the purpose of assisting the phase transition.

3.1. Carbon-Based Nanomaterials

Carbon-based nanomaterials are a new type of photothermal material developed in recent years, such as carbon dots [36,37], graphene, and graphene oxide [38–40]. They have strong absorption in visible and near-infrared light and can transform solar energy into heat energy rapidly owing to their unique structures. Therefore, they show a wide range of applications in smart windows [41–46], photocatalysis [47], and tumor-targeted therapy [48].

Graphene oxide (GO) is a two-dimensional nanomaterial with a single atomic thickness obtained by the oxidation and dispersion of natural graphite. Under sunlight, most of the energy photons in visible light can be absorbed by electrons, so they are excited. When the excited-state electrons fall back to the ground state, they release heat and raise the local temperature, producing a significant photothermal effect.

The excellent photothermal performance of GO was verified by Kim [49] through the control experiment. In the experiment, the aqueous graphene oxide (GO) could convert solar energy into heat energy under sunlight, raising the solution temperature by 5 °C in 8 min, while the temperature of pure water had little change (Figure 2a).

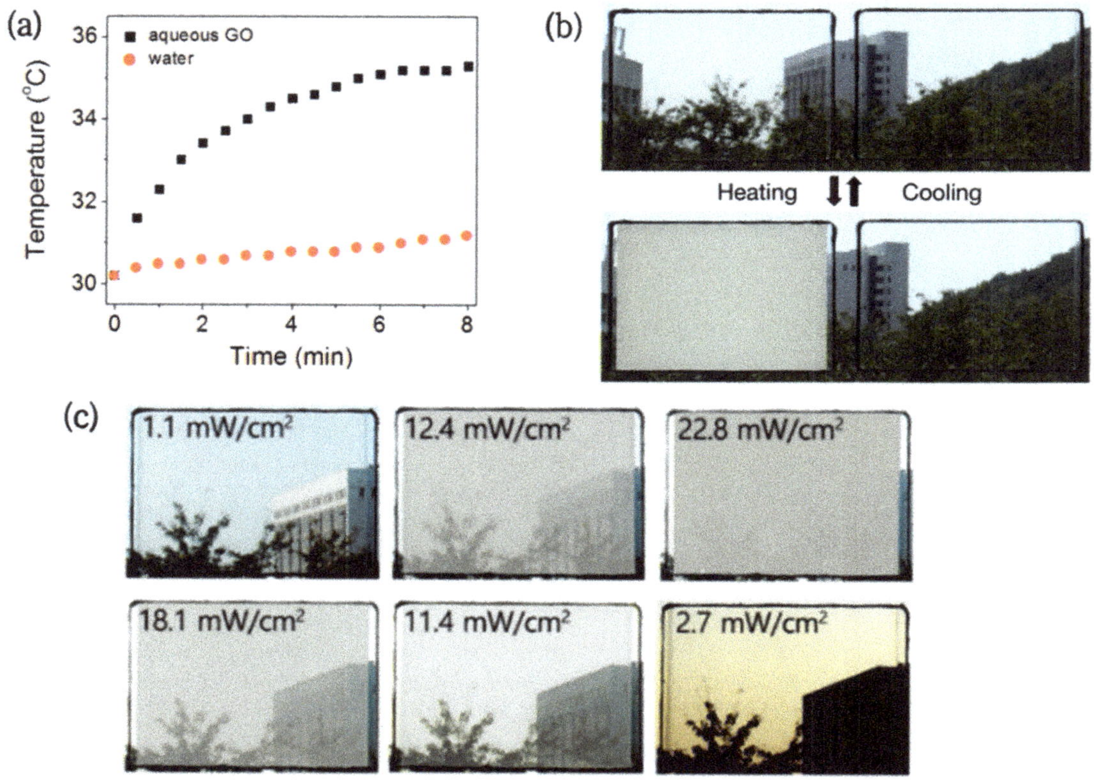

Figure 2. (**a**) Temperature change of pure water and 2.5 mg/mL GO aqueous solution under visible light irradiation [49]. (**b**) Photo of the transparency transition of PNIPAMm/GO hydrogel window (left) and PNIPAm hydrogel window (right) under sunlight [41]. (**c**) Photo of the PNDV/GO hydrogel window transmittance gradient change with light intensity [43].

Based on this, GO/PNIPAm hydrogels were successfully prepared [41]. The GO dispersed in the PNIPAm hydrogels allowed the originally transparent films to reach the LCST at room temperature (30 °C) and complete the phase change in 2 min (Figure 2b). It is worth noting that similar research was conducted by Zhu [42]. The results showed that the heating effect and rate were positively correlated with the GO concentration, but had no significant effect on the conversion temperature of the composite hydrogel.

In order to obtain a wider phase transition temperature range, Lee [43] et al. prepared multicomponent copolymers with a composition gradient. Then, NIPAm-co-NDEAm-co-VP(PNDV) hydrogel and graphene oxide (GO) composites were sealed in double-layer glass to assemble a new type of embedded glass. Compared with ordinary hydrogel glass, this novel glass achieved step-by-step solar control of the smart window in response to sunlight intensity (Figure 2c), providing users with more choices.

Although hydrogels have been widely studied in the field of smart windows, the poor mechanical properties of traditional hydrogels greatly limit their applications. Near-infrared light (NIR)-responsive (L-PNIPAm/GO) hydrogels were prepared [44] by the in situ polymerization of N-isopropylacrylamide (NIPAm), in which laponite (L) was the crosslinking agent and graphene oxide (GO) was the photothermal conversion agent. The addition of GO not only improved the hydrogels' photothermal sensitivity, but also raised the system's cross-linking point, resulting in a more flawless structure. Through the characterization of multiple samples, it was found that the prepared L-PNIPAm/GO-12.0-

0.21 hydrogels had high strength and good toughness when the contents of laponite and GO were 12.0% and 0.21% of the mass of NIPAm. Under NIR irradiation, it could be heated from 20.3 °C to 48.5 °C within 300 s, with good temperature sensitivity and photothermal effect. Thus, it has great potential in the fields of smart windows and light-controlled switches.

Graphene oxide can also be an effective auxiliary in the phase transition of VO_2. A series of tungsten-doped vanadium dioxide/graphene composites was prepared by a hydrothermal method using graphene oxide powders (GO), vanadium pentoxide, oxalic acid, and ammonium tungstate as raw materials [45]. Due to the doping of tungsten and the addition of GO, VO_2 is uniformly loaded on the surface of graphene as spherical particles, the particle size is significantly reduced, and the agglomeration is improved. When the atomic percentage of tungsten was 2.5% and the content of GO was 4%, the phase transition temperature of VO_2 was reduced from 66.0 °C to 32.2 °C. In addition, the thermal conductivity reached the maximum value of 16.341 W/(m·K), which effectively improved the heat dissipation effect of the coating. Such an improvement in properties allows it to better meet the required functionality of an intelligent thermal insulation coating.

There are many varieties of carbon-based nanomaterials, and different structures afford them different characteristics. Among them, carbon dots (CDS) are identical to GO in many ways, including the chemical structure and physical qualities. As a new zero-dimensional carbon nanomaterial with a size of less than 10 nm, it has good photothermal properties. A light-responsive material was reported by Shen [46], which was obtained by incorporating carbon dots (CDs) into hydroxypropylmethyl cellulose (HPMC). Through a series of studies, it was found that the ultrasmall size of CDs guaranteed the considerable transparency of CDs/HPMC and enhanced light absorption. Moreover, chloride-modified CDs (Cl-CDs) can greatly shorten the response time and lower the LCST, which also provides a new idea to improve the material's function by regulating the surface characteristics of CDs.

3.2. Noble Metal Nanoparticles

Noble metal nanoparticles are the most researched near-infrared photothermal materials, such as Au [50,51], Ag [52], Pd [53,54], etc. Their photothermal effect is mainly derived from the strong localized surface plasmon resonance effect of nanoparticles, which is a unique phenomenon occurring in metal structures. When the frequency of the incident light matches the eigenfrequency of the free electrons in the metal, the electrons will be collectively excited and resonated. Vibrating electrons will convert kinetic energy into heat energy due to the damping effect, thus increasing the local temperature [55].

The most common noble metal nanomaterials exhibiting the plasmon resonance effect are gold nanocrystals (AuNCs). Sunlight-responsive thin HPMC/AuNCs films, combining the thermochromic properties of HPMC and the photothermal effect of AuNCs, have been prepared successfully via a simple and low-cost method [56]. Under laser irradiation, the temperature of HPMC films containing AuNRs was observed to increase rapidly and reach a steady state within 80 s. Their steady-state temperature was close to 69 °C, however, the temperature of the film without AuNRs only increased by 1 °C under the same conditions (Figure 3a). Additionally, the switching temperature and colors of the film could be flexibly controlled by adjusting the concentration of AuNCs and the type of thermochromic material (Figure 3b).

Compared with gold nanospheres (AuNPs), aspheric AuNRs are anisotropic and exhibit different LSPR properties. Based on this, novel, random-filled composite nanohydrogels were obtained by an in situ reduction method. The nanohydrogels were mainly composed of AuNRs and PNIPAm [59]. It was found that there were two completely separated plasmon resonance absorption peaks on the absorption spectra of AuNRs. With an increase in temperature, the transverse LSPR peaks altered slightly, while the longitudinal LSPR peak had a strong red shift. Additionally, the amplitude of the resonance enhancement increased with an increase in the length-to-diameter ratio of AuNRs.

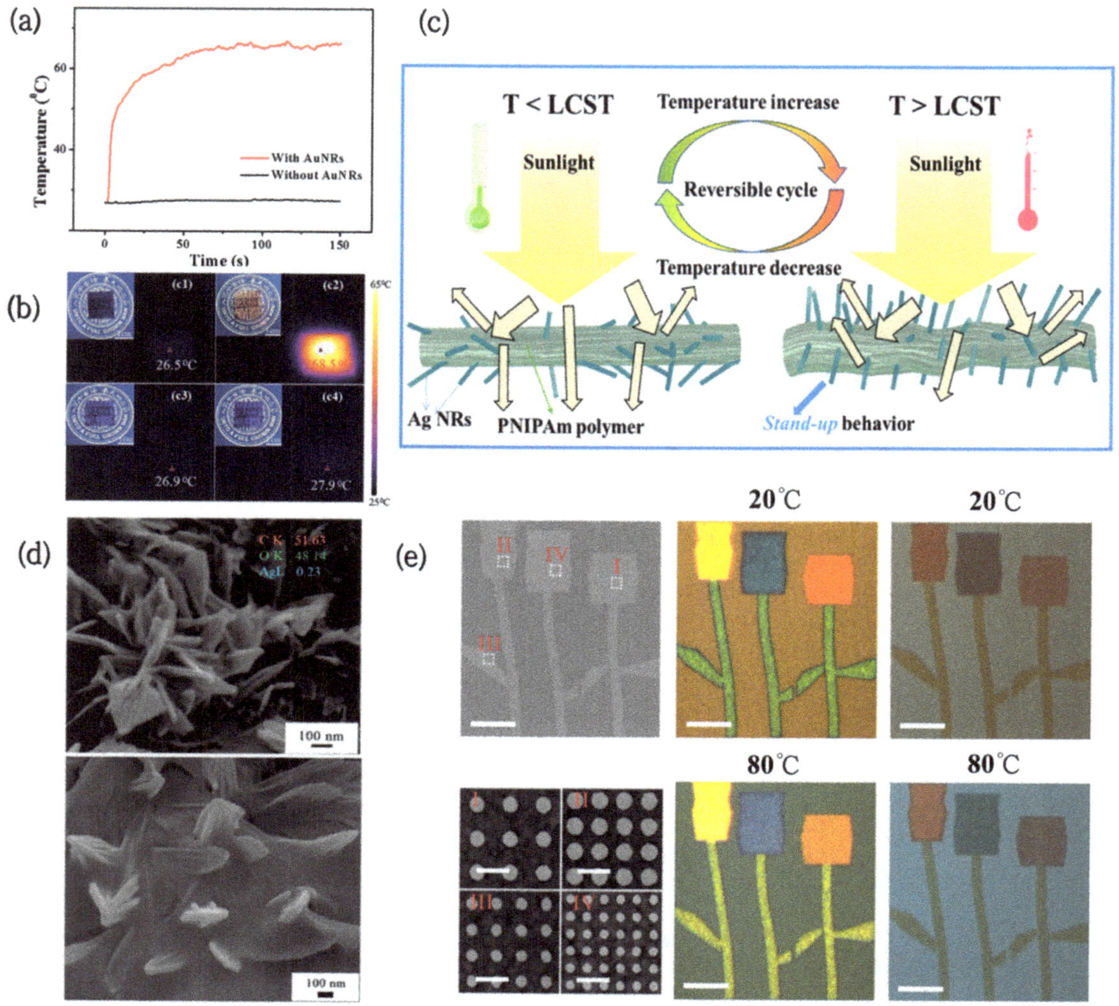

Figure 3. (**a**) Temperature rise curves of PVA/dye films with and without AuNRs [56]. (**b**) Digital photographs and infrared photographs of PVA/dye/AuNRs films and PVA/dye films before and after laser irradiation [56]. (**c**) Schematic representation of the structural shrinkage and deformation of PNIPAm acrylic/AgNRs when the temperature is higher than LCST [57]. (**d**) SEM images of PNIPAm-acrylic acid/AgNRs hydrogels at different temperatures [57]. (**e**) Color patterns constructed from arrays of Ag nanodiscs on silica films at different temperatures [58].

Furthermore, noble metal nanoparticles can be improved to exert their photothermal effects more efficiently by building particular structures or coatings. Wei et al. developed a new strategy for thermal-responsive PNIPAm–acrylic/Ag NR hybrid hydrogels [57]. The hydrogels had full-wavelength thermal management functions, which can be used for smart windows. Initially, AgNRs are coated with polymers and point in different directions. When the temperature rises above the LCST, the hydrogel structure shrinks and deforms, causing the AgNRs to stand up like flowers (Figure 3d,e), and their IR emissivity slightly increases from 0.947 to 0.958. In addition to this, the hydrogels can be fully transparent at room temperature, with a high solar modulation capability (ΔT_{sol} = 59.24%) and ultrahigh luminous transmission (T_{lum} = 61.36%) after phase transition. Therefore, they can be used for solar radiation modulation between the visible, near-infrared (NIR), and mid-infrared (MIR) regions.

The combination of noble metal nanoparticles and VO_2 also has significant application potential in the field of smart windows. Periodic arrays of Ag nanodiscs on VO_2 films were fabricated by Shu [58], and their LSPR effect brought afforded absorption to the films in the visible regions. Moreover, by tuning the diameter of the Ag nanodiscs and spatial periodicity of the arrays, various colors could be actively adjusted in the whole visible spectrum (Figure 3c), greatly increasing the color diversity of the VO_2 films.

Moreover, a vanadium dioxide (VO_2) and gold nanotriangle composite sandwich structure (VO_2/Au/VO_2) was prepared via nanosphere lithography [60]. The gold nanotriangles in the films exhibited periodic arrangement, and their sharp edges and tips promoted the occurrence of LSPR. This sandwich architecture may provide enlightenment for the design of temperature-sensitive smart windows.

With the in-depth study of nanoscience, it has been found that composites prepared by mixing two or more noble metal nanoparticles often combine their respective advantages and have superior physicochemical properties than single nanoparticles. Au@AgNR@PNIPAM microgels, with Au@AgNR as the core and cross-linked PNIPAM as the shell, were prepared via seeded precipitation polymerization [61]. The combination of the two metals not only improves the antioxidant capacity of AgNRs, but also overcomes the problem of the weak enhancement of the electric field on the surface of AuNRs.

Although noble metal materials, such as Au, Ag, and Pd, have an excellent theoretical basis in the fields of catalysis [62,63], photothermal devices [64], biomedicine [65], etc., their scarce reserves and high prices seriously limit their large-scale promotion. This is also the greatest problem in the practical application of noble metal nanomaterials.

3.3. *Semiconductor Nanomaterials*

Due to the wide energy gap, conventional semiconductor materials need to absorb UV light with higher energy to excite the electrons and release heat in the process of falling back to the ground state. However, with the study of more kinds of semiconductor materials, it has been found that localized surface plasmon resonance (LSPR) exists not only in noble metals, but also in semiconductor materials with appreciable free carrier density, such as tin-doped indium oxide (ITO) [66,67], copper sulfide (Cu_xS) [68,69], titanium nitride (TiN) [70,71], etc. Compared with conventional noble metals, semiconductors can exhibit LSPR in both the ultraviolet-visible (UV-vis) and near–mid-infrared (IR) spectral regions, which significantly extends the light absorption range.

3.3.1. Metal Oxide

Visible light transmittance (T_{lum}) is an important indicator for smart windows. Therefore, when selecting photothermal materials, their impact on transparency should also be given priority. Antimony-doped tin dioxide (ATO) is a transparent N-type semiconductor with special photoelectric effects and excellent light absorption properties. Its UV absorption is caused by the material's inherent wide bandgap, and the NIR absorption is attributed to the localized surface plasmon resonance (LSPR) induced by N-type doping [72].

A novel, fully autonomous light-driven thermochromic material was proposed, consisting of hybridizing poly-N-isopropylacrylamide (PNIPAM) hydrogel and antimony–tin oxide (ATO) [72]. Even when the outdoor temperature is far lower than the transition temperature, PNIPAM/ATO (PATO) composite hydrogels can also use ATO as a nanoheater to induce the optical switching of the hydrogel under sunlight. Furthermore, by increasing the Sb dopant content, the free electron concentration can be increased and the NIR absorption efficiency can be enhanced, which means that the ATO can convert more solar energy into heat. It is worth noting that the 10 atom % Sb-doped PATO not only showed the best response speed and solar modulation ability, but also achieved a subtle balance between thermal comfort and visuality. On this basis, Xu [73] prepared supramolecular nanocomposite hydrogel films by integrating ethylene glycol-modified pillar arene (EGP5) and antimony–tin oxide (ATO) nanoparticles (Figure 4a). The obtained films displayed good initial luminous transmission (77.2%) and excellent solar modulation ability (56.1%) under 100 mW/cm^2 xenon lamp irradiation owing to the thermal responsiveness of EGP5 and plasma heating induced by NIR absorption of ATO (Figure 4b). At the same time, the host–guest interaction between EGP5 and pyridinium units efficiently avoids the collapse of and damage to the hydrogel structure, resulting in a film that possessed high repeatability and durability.

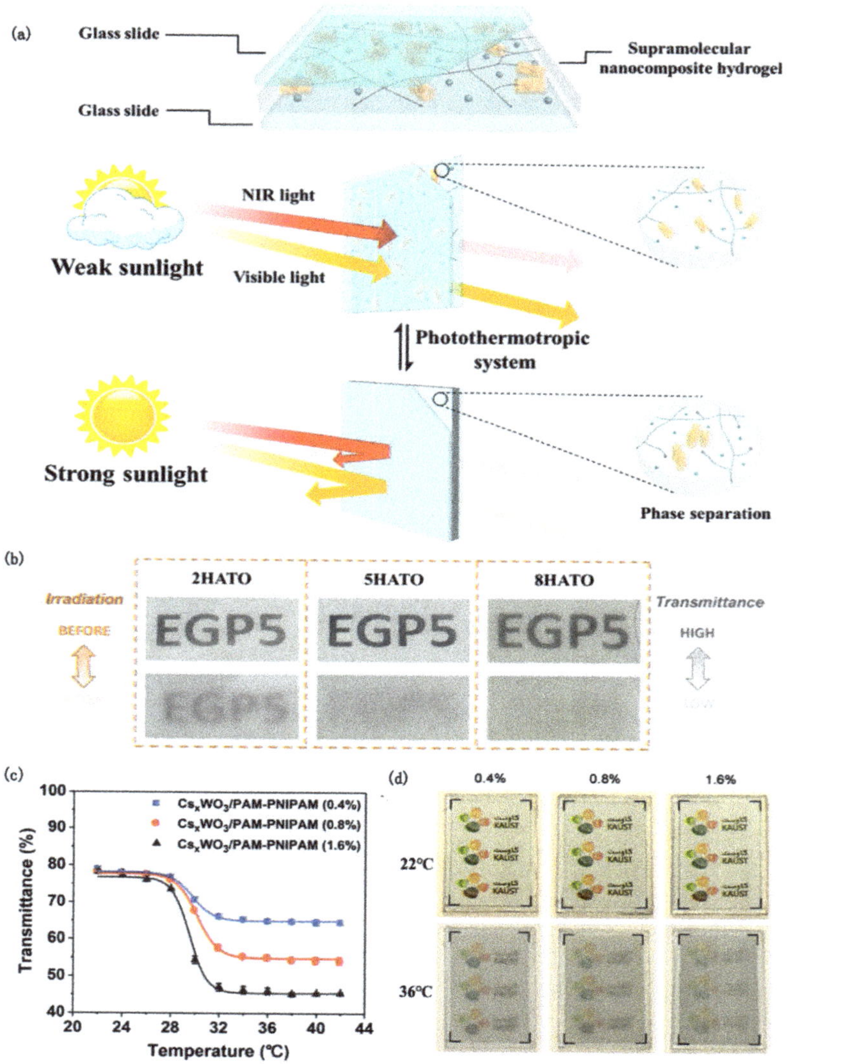

Figure 4. (**a**) Schematic diagram of the chemical structure and working principle of the photochromic supramolecular hydrogel smart window [73]. (**b**) Photographs of ATO composite hydrogel films with EGP5 contents of 2, 5, and 8 mol%, respectively, before and after irradiation at 100 mW/cm^2 for 10 min [73]. (**c**) Transmittance variation of Cs_xWO_3/PAM-PNIPAM film with temperature at 550 nm wavelength [74]. (**d**) Transparency changes of Cs_xWO_3/PAM-PNIPAM films with different PNIPAM concentrations at 22 °C (above) and 36 °C (below) [74].

ATO can also be used in other types of smart windows. Novel smart window films were successfully prepared by doping ATO nanoparticles into polymer-stabilized Liquid Crystal (PSLC) [75]. These films had a wider waveband modulation function, covering the visible and infrared regions (380–5500 nm), and the transmittance could be altered in the visible region from a highly transparent (78.5%) state to a strong light-scattering (10%) state. In addition, owing to the LSPR of ATO nanoparticles, up to 80.7% of the IR light could be effectively shielded.

Tin-doped indium oxide (ITO) is a promising NIR shielding material similar to ATO. A flexible, multi-responsive material was invented by creating a compatible interface between ITO nanocrystals and polar syrup. Polar syrup is a Liquid Crystal material that initially combines the good mechanical strength of PDLC with the high transparency of PSLC [76]. The Liquid Crystal molecules in it can change from homeotropically oriented to a focal–conic texture in response to the electric field and temperature, achieving a rapid transition from the transparent state (78%) to the opaque state (1.5%).

Cesium tungsten bronze (Cs_xWO_3) is an excellent photothermal material with high absorption in both the UV and NIR regions. A spectrally selective smart window using Cs_xWO_3 as the photothermal component and PAM hydrogel matrix embedded with thermally responsive PNIPAM microgels as the photocontrol switch was designed [74] (Figure 4c,d). Under solar irradiation, the steady-state surface temperature of the Cs_xWO_3/PAM-PNIPAM window can remain at 47 °C. It can shield about 96.2% of the near-infrared radiation while ensuring good visible light transmission, so that the room temperature is maintained at about 25°C.

Moreover, a novel roll-to-roll fabrication of multi-responsive flexible Liquid Crystal material was presented [77]. It can not only control the transmission of visible light by the stimulus of voltage, heat, or NIR light, but also shield nearly 95% of the NIR radiation in the 800–2500 nm band. Excellent properties of the films were achieved by forming a compatible poly(vinylpyrrolidone) (PVP) tuning layer between Cs_xWO_3 nanorods (NRs) and polymeric syrup containing liquid crystals. It also offers a wide working temperature range, improved flexibility, long-term stability, high mechanical strength, and large-area processability. These studies on the combination of Cs_xWO_3 and smart windows point toward a new route for the development of multifunctional and energy-efficient smart windows.

3.3.2. Metal Sulfide

Metal sulfides, such as PbS [78], MoS_2 [79], and Cu_7S_4 [80,81], have the advantages of low cost, good photothermal stability, and shape-controlled particle sizes. These features are beneficial to reduce the cost and the difficulty of the large-scale preparation of smart window films.

PbS is a narrow band gap semiconductor with a bandwidth of 0.4 eV. Its absorption band covers the main area of solar energy, making it an excellent light-absorbing coating. The PVA-PbS-VO_2 (M) composite film was fabricated by Qi [82] using the casting film method. It can effectively absorb sunlight from 0.3 to 2.5 μm through the interaction of solar photons with PbS nanostructured phonons and convert it into heat energy so that VO_2 undergoes a phase transition at room temperature.

Molybdenum disulfide (MoS_2) is a semiconductor material with a two-dimensional layered structure similar to graphene, which has strong absorption of near-infrared light. The MoS_2/PNIPAM composite hydrogels were prepared by incorporating chemically exfoliated MoS_2 nanosheets into poly(N-isopropylacrylamide) (PNIPAM) hydrogels [83]. It not only inherits the good thermal-responsive properties from the PNIPAM hydrogel, but also significantly shortens the response time. However, due to the black color of MoS_2, the coordination of the MoS_2 content and the transparency of smart windows will be the focus of future research.

Cu_7S_4 nanocrystals are a p-type semiconductor nanomaterial with an indirect bandgap. Because of their localized surface plasmon resonance (LSPR) properties in the NIR region, they have broad application prospects in solar cells, environmental pollution treatment, and photothermal treatment. Cu_7S_4/poly (N-isopropylacrylamide) (PNIPAM) composite films with fast optical/thermal response were prepared by Zhu [84]. When the temperature exceeds 32 °C, the smart window can quickly change from transparent to opaque within 3 min based on the wide absorption of sunlight and the high photothermal conversion efficiency of Cu_7S_4 nanoparticles. As long as the sunlight is strong enough, this smart window can even raise the indoor temperature by absorbing and converting sunlight in cold weather.

Through research on the shape-controlled synthesis of Cu_7S_4, it was found that there was an obvious size dependence on its localized surface plasmon resonance [85]. That is, by increasing the crystal size, the maximum resonance absorption peak shifted to red and the absorption intensity also increased. This research is beneficial to further improve the light absorption properties of Cu_7S_4 nanomaterials.

3.4. Others

Transition metal nitrides, such as titanium nitride, are versatile metal–ceramic materials with high-temperature durability, chemical stability, corrosion resistance, electrical conductivity, and also localized surface plasmon resonance properties [86–88]. It can replace noble metal plasmas to generate LSPR and promote material phase changes under light conditions, resulting in lower-cost thermochromic materials that are available at room temperature.

Vanadium dioxide/titanium nitride (VO_2/TiN) smart coatings were produced by Hao [89] via coating TiN nanoparticles with pure monoclinic-phase (M-phase) VO_2 films. These coatings can control the infrared (IR) radiation dynamically, depending on the ambient temperature and light intensity. The photothermal conversion effect of TiN nanomaterials will induce VO_2 to change from a monoclinic phase to a tetragonal phase under strong light or higher temperatures (28 °C), which can effectively block IR radiation by nearly 70%; conversely, these coatings show good IR transparency at 20 °C (Figure 5a). In addition, the visible light transmittance of these coatings at 2000 nm is 51%, and the infrared conversion efficiency is 48%. These unique features result in VO_2-TiN having great potential in the field of energy-saving windows.

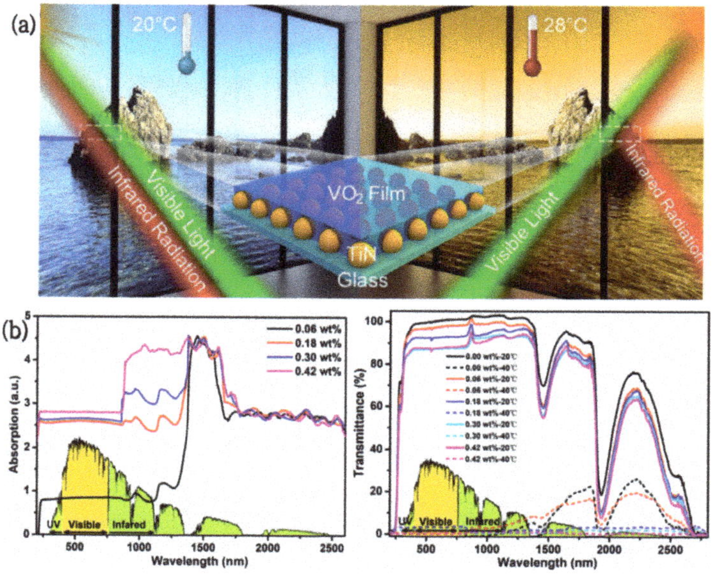

Figure 5. (**a**) Schematic diagram of a VO_2/TiN smart window [86]. (**b**) UV-Vis-NIR absorption and transmission spectra of PNIPAm hydrogels containing different levels of PDAPs at 20 °C and 40 °C [87].

Not only can inorganic materials be used as photothermal materials, but also organic polymers, such as polydopamine nanoparticles (PDAPs) [90,91]. Compared with conventional materials, PDAPs have excellent near-infrared responsivity and high thermal conductivity, as well as good biocompatibility and oxidation resistance. They also have the characteristics of easy preparation, low cost, and high yields.

Based on this, a new, fully autonomous thermochromic smart window containing PDAPs was constructed via a one-step in situ polymerization method (Figure 5b) [92]. It was shown that the addition of PDAPs not only ameliorated the microstructure and mechanical properties of the hydrogel, but also improved its photothermal conversion ability. It could also trigger the smart window to quickly adjust the indoor light and temperature, even if the ambient temperature was below the LCST of the hydrogel. This bold attempt provides a new idea for the research into multifunctional thermochromic smart windows.

4. Conclusions and Perspectives

In recent years, great progress has been made in the research into the phase transition of photothermal material-assisted thermochromic smart windows. In this review, an overview of several thermochromic materials from VO_2, Liquid Crystal, and hydrogel is provided. The photothermal conversion efficiency of carbon-based nanomaterials, noble metal nanoparticles, and semiconductor (metal-oxygen/sulfide) nanomaterials (Table 1), as well as their applications in thermochromic smart windows (Table 2), were summarized. The current challenges and solutions are further pointed out and future research directions are also proposed.

Table 1. Conversion efficiency of photothermal materials.

Type	Power of Excitation Light Source	Efficiency	Reference
GO NSs	Simulated sunlight (0.1 W/cm^2)	63.2%	[93]
CDs	AM1.5 Standard sunlight (0.1 W/cm^2)	23.46%	[46]
Au NRs	808 nm laser (1.27 W/cm^2)	39.2%	[94]
Au STs	808 nm laser (2 W/cm^2)	42%	[95]
ATO Hollow ball	1064 nm laser (-)	30.69%	[96]
$Cs_{0.33}WO_3$	980 nm laser (1.66 W/cm^2)	23.1%	[97]
Cu_9S_5 NCs	980 nm laser (0.51 W/cm^2)	25.7%	[98]
Cu_7S_4 NFs	808 nm NIR laser (1 W/cm^2)	56.32%	[99]
Cu_7S_4 Nano Superlattice	808 nm NIR laser (1 W/cm^2)	65.7%	[99]
Au@Cu_7S_4 Yolk–shell NPs	980 nm laser (0.51 W/cm^2)	63%	[100]
TiN	808 nm NIR laser (1.5 W/cm^2)	38.6%	[101]
PDAPs	808 nm NIR laser (2 W/cm^2)	40%	[102]

Table 2. Performance comparison of thermochromic smart window.

Thermochromic Material	Photothermal Material	$T_{lum,l}$	$T_{lum,h}$	Δ_{Tsol}	T_C	Reference
HPMC	-	98.94%	89.32%	6.52%	40°C	[46]
PNIPAM	-	73.2%	48.58%	26.88%	31.4°C	[58]
VO_2	-	47%	52%	ΔT_{NIR} = 51%	68°C	[86]
PNIPAM	GO	97.8%	3.0%	-	32.5°C	[41]
HPMC	CDs	82.11%	3.50%	65.55%	45°C	[46]
HPMC	Au NCs	85.4%	3.5%	62.6%	40°C	[56]
VO_2	Au	74.1%	66.6%	13.2%	50°C	[60]
PNIPAM	Ag NRs	61.36%	3.43%	59.24%	32.9°C	[58]

Table 2. *Cont.*

Thermochromic Material	Photothermal Material	$T_{lum,l}$	$T_{lum,h}$	Δ_{Tsol}	T_C	Reference
PNIPAM (10%Sb)	ATO NPs	83.0%	55.3%	35.7%	33.6°C	[72]
PNIPAM	ATO NPs	77.2%	18.6%	56.1%	32.1°C	[73]
PSLC	ATO NPs	78.5%	10%	-	-	[74]
PDLC-PSLC	ITO NCs	78%	1.5%	-	28.3°C	[75]
PAM-PNIPAM	Cs_xWO_3	78.2%	45.3%	-	30°C	[76]
PT-LCs	Cs_xWO_3 NRs	67%	1.5%	-	30.5°C	[77]
PNIPAM	Cu_7S_4 NPs	79%	41.1%	31.2%	32°C	[84]
VO_2	TiN NPs	49%	51%	ΔT_{NIR} = 48%	28°C	[85]
PNIPAM	PDAPs	92.58%	3.41%	88.9%	33°C	[92]

The use of thermochromic smart windows has been discussed for many decades. Typically, the high response temperature, poor solar modulation, and slow optical switching are considered as the main drawbacks of traditional thermochromic smart windows. Therefore, researchers are trying to introduce photothermal materials into smart windows and use the photothermal conversion effect to assist the phase transition of smart windows. This can not only achieve the dual regulation of light and heat at room temperature, but also greatly shorten the response time and drive the development of smart windows toward low energy consumption and high universality. The outlooks and prospects for future research on the phase transition of photothermal material-assisted thermochromic smart windows are suggested as follows:

(1) Improvement of morphology (size and shape) of photothermal materials.

Owing to the size and shape dependence of photothermal nanomaterials, the optical absorption characteristics can be effectively adjusted by changing the particle size or adopting different structures, such as cages, core–shells, and so on. For example, in Table 1, the photothermal efficiency of a Cu_7S_4 Nano Superlattice is significantly higher than that of Cu_7S_4 NPs [99]. In addition, photothermal nanoparticles of different sizes can also be mixed together so that their extinction spectrum can be better matched with the solar radiation spectrum, and as much solar energy can be harvested as possible.

(2) Composites of different kinds of photothermal materials.

Based on the absorption spectrum characteristics of various photothermal materials, composite photothermal materials could be incorporated into the smart window system. Au@Cu_7S_4 yolk–shell NPs [100] not only retain the respective absorption peaks of Au and Cu_7S_4 in UV-Vis-NIR spectra, but also show significantly enhanced absorption in the near-infrared region. Additionally, the photothermal conversion efficiency is also higher, which shows that the total is greater than the sum of parts.

(3) Construction of bionic anti-reflection structure.

In addition to improving the conversion efficiency of photothermal materials, the solar absorption efficiency is also an important parameter. In the field of solar cells, bionic anti-reflection periodic structures, such as butterfly wings and moth eyes, are often used on the surface of solar panels [103]. These rough nanostructures can effectively eliminate reflection and help them absorb more sunlight than smooth surfaces. Similarly, this structure is also applicable to smart windows, which will significantly shorten the response time.

Author Contributions: Investigation, Y.Z., H.J., M.L. and J.T.; Writing, Y.Z. and H.J.; Review and editing, Y.Z., H.J., M.L., J.T., Y.O., Y.W., Y.C., Y.H., J.W. and Y.M. All authors have read and agreed to the published version of the manuscript.

Funding: This work was financially supported by the National Natural Science Foundation of China (No. 51902276), the Natural Science Foundation of Hunan Province (No. 2019JJ50583), Scientific Research Fund of Hunan Provincial Education Department (No. 21B0111), and Hunan Provincial Innovation Foundation for Postgraduate (No. QL20220158).

Institutional Review Board Statement: Not applicable.

Informed Consent Statement: Not applicable.

Data Availability Statement: Not applicable.

Conflicts of Interest: The authors declare no conflict of interest.

References

1. China Building Energy Efficiency. China building energy consumption annual report 2020. *Build. Energy Effic.* **2021**, *49*, 1–6.
2. Xu, F.; Cao, X.; Luo, H.; Jin, P. Recent advances in VO_2-based thermochromic composites for smart windows. *J. Mater. Chem. C* **2018**, *6*, 1903–1919. [CrossRef]
3. Cao, X.; Chang, T.; Shao, Z.; Xu, F.; Luo, H.; Jin, P. Challenges and opportunities toward real application of VO_2-based smart glazing. *Matter* **2020**, *2*, 862–881. [CrossRef]
4. Zou, X.; Ji, H.; Zhao, Y.; Lu, M.; Tao, J.; Tang, P.; Liu, B.; Yu, X.; Mao, Y. Research Progress of Photo-/Electro-Driven Thermochromic Smart Windows. *Nanomaterials* **2021**, *11*, 3335. [CrossRef]
5. Tao, X.; Liu, D.; Yu, J.; Cheng, H. Reversible metal electrodeposition devices: An emerging approach to effective light modulation and thermal management. *Adv. Opt. Mater.* **2021**, *9*, 2001847. [CrossRef]
6. Luo, L.; Liang, Y.; Feng, Y. Recent Progress on Preparation Strategies of Liquid Crystal Smart Windows. *Crystals* **2022**, *12*, 1426. [CrossRef]
7. Zhou, Y.; Dong, X.; Mi, Y.; Fan, F.; Xu, Q.; Zhao, H.; Wang, S.; Long, Y. Hydrogel smart windows. *J. Mater. Chem. A* **2020**, *8*, 10007–10025. [CrossRef]
8. Hamberg, I.; Granqvist, C. Evaporated Sn-doped In_2O_3 films: Basic optical properties and applications to energy-efficient windows. *J. Appl. Phys.* **1986**, *60*, 123–160. [CrossRef]
9. Jing, T. Selective Reflection of Cholesteric Liquid Crystals. *Sci. Insights* **2022**, *40*, 515–517. [CrossRef]
10. Ming, Y.; Sun, Y.; Liu, X. Optical evaluation of a Smart Transparent Insulation Material for window application. *Energy Convers. Manag. X* **2022**, 100315. [CrossRef]
11. Huang, Z.; Chen, S.; Lv, C.; Huang, Y.; Lai, J. Infrared characteristics of VO_2 thin films for smart window and laser protection applications. *Appl. Phys. Lett.* **2012**, *101*, 191905. [CrossRef]
12. Rui, S.; Lin, Y.; Junhui, H.; Jie, L. Thermochromic smart coatings. *Prog. Chem.* **2019**, *31*, 1712.
13. Cui, Y.; Ke, Y.; Liu, C. Thermochromic VO_2 for energy-efficient smart windows. *Joule* **2018**, *2*, 1707–1746. [CrossRef]
14. Barimah, E.; Boontan, A.; Steenson, D. Infrared optical properties modulation of VO_2 thin film fabricated by ultrafast pulsed laser deposition for thermochromic smart window applications. *Sci. Rep.* **2022**, *12*, 1–10. [CrossRef]
15. Ren, H.; Hassna, O.; Li, J.; Arigong, B. A patterned phase-changing vanadium dioxide film stacking with VO_2 nanoparticle matrix for high performance energy-efficient smart window applications. *Appl. Phys. Lett.* **2021**, *118*, 051901. [CrossRef]
16. Riapanitra, A.; Asakura, Y.; Yin, S. One-step hydrothermal synthesis and thermochromic properties of chlorine-doped VO_2(M) for smart window application. *Funct. Mater. Lett.* **2020**, *13*, 1951008. [CrossRef]
17. Ji, H.; Liu, D.; Cheng, H.; Zhang, C.; Yang, L.; Ren, D. Infrared thermochromic properties of monoclinic VO_2 nanopowders using a malic acid-assisted hydrothermal method for adaptive camouflage. *RSC Adv.* **2017**, *7*, 5189–5194. [CrossRef]
18. Ji, H.; Liu, D.; Cheng, H.; Zhang, C. Inkjet printing of vanadium dioxide nanoparticles for smart windows. *J. Mater. Chem. C* **2018**, *6*, 2424–2429. [CrossRef]
19. Ke, Y.; Zhou, C.; Zhou, Y.; Wang, S.; Chan, S.H.; Long, Y. Emerging thermal-responsive materials and integrated techniques targeting the energy-efficient smart window application. *Adv. Funct. Mater.* **2018**, *28*, 1800113. [CrossRef]
20. Markov, P.; Marvel, R.E.; Conley, H.J.; Miller, K.J.; Haglund, R.F., Jr.; Weiss, S.M. Optically monitored electrical switching in VO_2. *ACS Photonics* **2015**, *2*, 1175–1182. [CrossRef]
21. Ji, H.; Liu, D.; Cheng, H.; Tao, Y. Large area infrared thermochromic VO_2 nanoparticle films prepared by inkjet printing technology. *Sol. Energy Mater. Sol. Cells* **2019**, *194*, 235–243. [CrossRef]
22. Ji, H.; Liu, D.; Cheng, H. Infrared optical modulation characteristics of W-doped VO_2(M) nanoparticles in the MWIR and LWIR regions. *Mater. Sci. Semicond. Process.* **2020**, *119*, 105141. [CrossRef]
23. Ji, H.; Liu, D.; Cheng, H.; Zhang, C.; Yang, L. Vanadium dioxide nanopowders with tunable emissivity for adaptive infrared camouflage in both thermal atmospheric windows. *Sol. Energy Mater. Sol. Cells* **2018**, *175*, 96–101. [CrossRef]
24. Ji, H.; Liu, D.; Zhang, C.; Cheng, H. VO_2/ZnS core-shell nanoparticle for the adaptive infrared camouflage application with modified color and enhanced oxidation resistance. *Sol. Energy Mater. Sol. Cells* **2018**, *176*, 1–8. [CrossRef]
25. Meng, W.; Gao, Y.; Hu, X.; Tan, L.; Li, L.; Zhou, G.; Jiang, L. Photothermal Dual Passively Driven Liquid Crystal Smart Window. *ACS Appl. Mater. Interfaces* **2022**, *14*, 28301–28309. [CrossRef] [PubMed]

26. Meng, C.; Chen, E.; Wang, L.; Tang, S.; Tseng, M.; Guo, J.; Kwok, H. Color-switchable liquid crystal smart window with multi-layered light guiding structures. *Opt. Express* **2019**, *27*, 13098–13107. [CrossRef]
27. Kragt, A.; Loonen, R.; Broer, D.; Debije, M.; Schenning, A. 'Smart' light-reflective windows based on temperature responsive twisted nematic liquid crystal polymers. *J. Polym. Sci.* **2021**, *59*, 1278–1284. [CrossRef]
28. Oh, S.; Nam, S.; Kim, S.; Yoon, T.; Kim, W. Self-regulation of infrared using a liquid crystal mixture doped with push–pull azobenzene for energy-saving smart windows. *ACS Appl. Mater. Interfaces* **2021**, *13*, 5028–5033. [CrossRef]
29. Li, K.; Meng, S.; Xia, S.; Ren, X.; Gao, G. Durable and controllable smart windows based on thermochromic hydrogels. *ACS Appl. Mater. Interfaces* **2020**, *12*, 42193–42201. [CrossRef]
30. Xiao, X.; Shi, D.; Yang, Z.; Yu, Q.; Kaneko, D.; Chen, M. Near infrared-sensitive smart windows from Au nanorod–polymer hybrid photonic hydrogels. *New J. Chem.* **2021**, *45*, 4016–4023. [CrossRef]
31. Zhang, L.; Xia, H.; Xia, F.; Du, Y.; Wu, Y.; Gao, Y. Energy-saving smart windows with HPC/PAA hybrid hydrogels as thermochromic materials. *ACS Appl. Energy Mater.* **2021**, *4*, 9783–9791. [CrossRef]
32. Tian, J.; Peng, H.; Du, X.; Wang, H.; Cheng, X.; Du, Z. Hybrid thermochromic microgels based on UCNPs/PNIPAm hydrogel for smart window with enhanced solar modulation. *J. Alloy. Compd.* **2021**, *858*, 157725. [CrossRef]
33. Pelton, R. Poly (N-isopropylacrylamide)(PNIPAM) is never hydrophobic. *J. Colloid Interface Sci.* **2010**, *348*, 673–674. [CrossRef] [PubMed]
34. Sherry, L.; Jin, R.; Mirkin, C.; Schatz, G.; Van Duyne, R. Localized surface plasmon resonance spectroscopy of single silver triangular nanoprisms. *Nano Lett.* **2006**, *6*, 2060–2065. [CrossRef]
35. Agrawal, A.; Cho, S.; Zandi, O.; Ghosh, S.; Johns, R.; Milliron, D. Localized surface plasmon resonance in semiconductor nanocrystals. *Chem. Rev.* **2018**, *118*, 3121–3207. [CrossRef]
36. Kang, Z.; Lee, S. Carbon dots: Advances in nanocarbon applications. *Nanoscale* **2019**, *11*, 19214–19224. [CrossRef]
37. Mintz, K.; Bartoli, M.; Rovere, M.; Zhou, Y.; Hettiarachchi, S.; Paudyal, S.; Leblanc, R. A deep investigation into the structure of carbon dots. *Carbon* **2021**, *173*, 433–447. [CrossRef]
38. Dideikin, A.; Vul', A. Graphene oxide and derivatives: The place in graphene family. *Front. Phys.* **2019**, *6*, 149. [CrossRef]
39. Wang, Y.; Li, S.; Yang, H.; Luo, J. Progress in the functional modification of graphene/graphene oxide: A review. *RSC Adv.* **2020**, *10*, 15328–15345.
40. Farjadian, F.; Abbaspour, S.; Sadatlu, M.; Mirkiani, S.; Ghasemi, A.; Hoseini-Ghahfarokhi, M.; Hamblin, M. Recent developments in graphene and graphene oxide: Properties, synthesis, and modifications: A review. *ChemistrySelect* **2020**, *5*, 10200–10219.
41. Kim, D.; Lee, E.; Lee, H.S.; Yoon, J. Energy efficient glazing for adaptive solar control fabricated with photothermotropic hydrogels containing graphene oxide. *Sci. Rep.* **2015**, *5*, 1–6. [CrossRef] [PubMed]
42. Zhu, C.; Lu, Y.; Peng, J.; Chen, J.; Yu, S. Photothermally sensitive poly (N-isopropylacrylamide)/graphene oxide nanocomposite hydrogels as remote light-controlled liquid microvalves. *Adv. Funct. Mater.* **2012**, *22*, 4017–4022. [CrossRef]
43. Lee, E.; Kim, D.; Yoon, J. Stepwise activation of switchable glazing by compositional gradient of copolymers. *ACS Appl. Mater. Interfaces* **2016**, *8*, 26359–26364. [CrossRef] [PubMed]
44. NIWAER, A.E.A.; Wang, H.; Meng, X.; He, X.; Yang, S.; Yu, H.; Zuo, F. Preparation and preliminary application of PNIPAM/laponite/graphene oxide near-infrared light responsive hydrogels. *Fine Chem.* **2021**, *3*, 6.
45. Zhou, T.; Zhou, Q.; Hua, Y.; Huang, C.; Li, L.; LÜ, W. Preparation and properties of tungsten-doped vanadium dioxide/graphene composites. *Fine Chem.* **2020**, *37*, 1574–1579.
46. Chang, Q.; Shen, Z.; Guo, Z.; Xue, C.; Li, N.; Yang, J.; Hu, S. Hydroxypropylmethyl cellulose modified with carbon dots exhibits light-responsive and reversible optical switching. *ACS Appl. Mater. Interfaces* **2021**, *13*, 12375–12382. [CrossRef]
47. Qin, J.; Li, R.; Lu, C.; Jiang, Y.; Tang, H.; Yang, X. Ag/ZnO/graphene oxide heterostructure for the removal of rhodamine B by the synergistic adsorption–degradation effects. *Ceram. Int.* **2015**, *41*, 4231–4237. [CrossRef]
48. Zhang, M.; Zhang, X.; Zhao, K.; Dong, Y.; Yang, W.; Liu, J.; Li, D. Assembly of gold nanorods with L-cysteine reduced graphene oxide for highly efficient NIR-triggered photothermal therapy. *Spectrochim. Acta Part A Mol. Biomol. Spectrosc.* **2022**, *266*, 120458. [CrossRef]
49. Kim, D.; Lee, H.S.; Yoon, J. Remote control of volume phase transition of hydrogels containing graphene oxide by visible light irradiation. *RSC Adv.* **2014**, *4*, 25379–25383. [CrossRef]
50. Chen, H.; Shao, L.; Ming, T.; Sun, Z.; Zhao, C.; Yang, B.; Wang, J. Understanding the photothermal conversion efficiency of gold nanocrystals. *Small* **2010**, *6*, 2272–2280. [CrossRef]
51. Kim, H.; Lee, D. Near-infrared-responsive cancer photothermal and photodynamic therapy using gold nanoparticles. *Polymers* **2018**, *10*, 961. [CrossRef]
52. Chen, M.; He, Y.; Huang, J.; Zhu, J. Synthesis and solar photo-thermal conversion of Au, Ag, and Au-Ag blended plasmonic nanoparticles. *Energy Convers. Manag.* **2016**, *127*, 293–300. [CrossRef]
53. Xiao, J.; Fan, S.; Wang, F.; Sun, L.; Zheng, X.; Yan, C. Porous Pd nanoparticles with high photothermal conversion efficiency for efficient ablation of cancer cells. *Nanoscale* **2014**, *6*, 4345–4351. [CrossRef]
54. Tang, S.; Chen, M.; Zheng, N. Multifunctional ultrasmall Pd nanosheets for enhanced near-infrared photothermal therapy and chemotherapy of cancer. *Nano Res.* **2015**, *8*, 165–174. [CrossRef]
55. Jain, P.K.; Huang, X.; El-Sayed, I.H.; El-Sayed, M.A. Noble metals on the nanoscale: Optical and photothermal properties and some applications in imaging, sensing, biology, and medicine. *Acc. Chem. Res.* **2008**, *41*, 1578–1586. [CrossRef]

56. Cao, D.; Xu, C.; Lu, W.; Qin, C.; Cheng, S. Sunlight-Driven Photo-Thermochromic Smart Windows. *Sol. RRL* **2018**, *2*, 1700219. [CrossRef]
57. Wei, G.; Yang, D.; Zhang, T.; Yue, X.; Qiu, F. Thermal-responsive PNIPAm-acrylic/Ag NRs hybrid hydrogel with atmospheric window full-wavelength thermal management for smart windows. *Sol. Energy Mater. Sol. Cells* **2020**, *206*, 110336. [CrossRef]
58. Shu, F.Z.; Yu, F.F.; Peng, R.W.; Zhu, Y.Y.; Xiong, B.; Fan, R.H.; Wang, Z.H.; Liu, Y.; Wang, M. Dynamic plasmonic color generation based on phase transition of vanadium dioxide. *Adv. Opt. Mater.* **2018**, *6*, 1700939. [CrossRef]
59. Zhou, J.; Dong, X.; Zha, L. Research Progress of Gold Nanorods Based Intelligent Composite Nanogels. *Polym. Bull.* **2015**, *6*, 1–9.
60. Liang, J.; Song, X.; Li, P.; Zhou, L.; Guo, J. Localized surface plasmon resonance tunability of VO_2/Au/VO_2 composite nanotriangle sandwich array. *Mater. Res. Express* **2019**, *6*, 0850f0856. [CrossRef]
61. Dong, X.; Liu, X.; Zhou, J.; Zhang, L.; Lin, D.; Zha, L. Fabrication and characterization of thermo-responsive composite microgels based on Au@Ag nanorod as core. *Chem. Res. Appl.* **2015**, *27*, 836–841.
62. Li, M.; Liu, D.; Cheng, H.; Peng, L.; Zu, M. Manipulating metals for adaptive thermal camouflage. *Sci. Adv.* **2020**, *6*, eaba3494. [CrossRef] [PubMed]
63. Haruta, M.; Daté, M. Advances in the catalysis of Au nanoparticles. *Appl. Catal. A Gen.* **2001**, *222*, 427–437. [CrossRef]
64. Irshad, M.; Arshad, N.; Wang, X. Nanoenabled photothermal materials for clean water production. *Glob. Chall.* **2021**, *5*, 2000055. [CrossRef]
65. Chang, M.; Hou, Z.; Wang, M.; Yang, C.; Wang, R.; Li, F.; Jun, L. Single-Atom Pd Nanozyme for Ferroptosis-Boosted Mild-Temperature Photothermal Therapy. *Angew. Chem.* **2021**, *133*, 202101924. [CrossRef]
66. Blemker, M.; Gibbs, S.; Raulerson, E.; Milliron, D.; Roberts, S. Modulation of the visible absorption and reflection profiles of ITO nanocrystal thin films by plasmon excitation. *ACS Photonics* **2020**, *7*, 1188–1196. [CrossRef]
67. Wei, W.; Hong, R.; Jing, M.; Shao, W.; Tao, C.; Zhang, D. Thickness-dependent surface plasmon resonance of ITO nanoparticles for ITO/In-Sn bilayer structure. *Nanotechnology* **2017**, *29*, 015705. [CrossRef]
68. Kalanur, S.; Seo, H. Synthesis of Cu_xS thin films with tunable localized surface plasmon resonances. *ChemistrySelect* **2018**, *3*, 5920–5926. [CrossRef]
69. Song, G.; Han, L.; Zou, W.; Xiao, Z.; Huang, X.; Qin, Z.; Hu, J. A novel photothermal nanocrystals of Cu_7S_4 hollow structure for efficient ablation of cancer cells. *Nano-Micro Lett.* **2014**, *6*, 169–177. [CrossRef]
70. Zhang, J.; Chen, T.; Li, X.; Liu, Y.; Liu, Y.; Yang, H. Investigation of localized surface plasmon resonance of TiN nanoparticles in TiN_xO_y thin films. *Opt. Mater. Express* **2016**, *6*, 2422–2433. [CrossRef]
71. Wang, L.; Zhu, G.; Wang, M.; Yu, W.; Zeng, J.; Yu, X.; Li, Q. Dual plasmonic Au/TiN nanofluids for efficient solar photothermal conversion. *Sol. Energy* **2019**, *184*, 240–248. [CrossRef]
72. Lee, H.Y.; Cai, Y.; Bi, S.; Liang, Y.N.; Song, Y.; Hu, X.M. A Dual-Responsive Nanocomposite toward Climate-Adaptable Solar Modulation for Energy-Saving Smart Windows. *Acs Appl. Mater. Interfaces* **2017**, *9*, 6054. [CrossRef] [PubMed]
73. Xu, Z.; Wang, S.; Hu, X.; Jiang, J.; Wang, L. Sunlight-Induced Photo-Thermochromic Supramolecular Nanocomposite Hydrogel Film for Energylogaving Smart Window. *Sol. RRL* **2018**, *2*, 1800204. [CrossRef]
74. Wu, M.; Shi, Y.; Li, R.; Wang, P. Spectrally selective smart window with high near-infrared light shielding and controllable visible light transmittance. *ACS Appl. Mater. Interfaces* **2018**, *10*, 39819–39827. [CrossRef] [PubMed]
75. Zhang, Z.; Zhang, R.; Xu, L.; Li, J.; Yang, L.; Yang, Y.; Bolshakov, A.; Zhu, J. Visible and infrared optical modulation of PSLC smart films doped with ATO nanoparticles. *Dalton Trans.* **2021**, *50*, 10033–10040. [CrossRef]
76. Liang, X.; Guo, S.; Chen, M.; Li, C.; Wang, Q.; Zou, C.; Zhang, C.; Zhang, L.; Guo, S.; Yang, H. A temperature and electric field-responsive flexible smart film with full broadband optical modulation. *Mater. Horiz.* **2017**, *4*, 878–884. [CrossRef]
77. Liang, X.; Guo, C.; Chen, M.; Guo, S.; Zhang, L.; Li, F.; Guo, S.; Yang, H. A roll-to-roll process for multi-responsive soft-matter composite films containing Cs_xWO_3 nanorods for energy-efficient smart window applications. *Nanoscale Horiz.* **2017**, *2*, 319–325. [CrossRef]
78. Ding, B.; Shi, M.; Chen, F.; Zhou, R.; Deng, M.; Wang, M.; Chen, H. Shape-controlled syntheses of PbS submicro-/nano-crystals via hydrothermal method. *J. Cryst. Growth* **2009**, *311*, 1533–1538. [CrossRef]
79. Chou, S.; Kaehr, B.; Kim, J.; Foley, B.; De, M.; Hopkins, P.; Dravid, V. Chemically exfoliated MoS_2 as near-infrared photothermal agents. *Angew. Chem. Int. Ed.* **2013**, *52*, 4160–4164. [CrossRef]
80. Wang, X.; He, Y.; Hu, Y.; Jin, G.; Jiang, B.; Huang, Y. Photothermal-conversion-enhanced photocatalytic activity of flower-like CuS superparticles under solar light irradiation. *Sol. Energy* **2018**, *170*, 586–593. [CrossRef]
81. Xi, M.; Xu, L.; Li, N.; Zhang, S.; Wang, Z. Plasmonic $Cu_{27}S_{24}$ nanocages for novel solar photothermal nanoink and nanofilm. *Nano Res.* **2022**, *15*, 3161–3169. [CrossRef]
82. Ji, Q. Design of Nanomaterials System for Smart Thermal Controlling. Master's Thesis, Zhengzhou University, Zhengzhou, China, 2014.
83. Zhang, J.; Du, P.; Xu, D.; Li, Y.; Peng, W.; Zhang, G.; Zhang, F.; Fan, X. Near-infrared responsive MoS_2/poly (N-isopropylacrylamide) hydrogels for remote light-controlled microvalves. *Ind. Eng. Chem. Res.* **2016**, *55*, 4526–4531. [CrossRef]
84. Zhu, H.; Wang, L. Smart window based on Cu_7S_4/hydrogel composites with fast photothermal response. *Sol. Energy Mater. Sol. Cells* **2019**, *202*, 110109. [CrossRef]
85. Zhang, C.; Yan, C.; Xue, Z.; Yu, W.; Wang, T. Shape-Controlled Synthesis of High-Quality Cu_7S_4 Nanocrystals for Efficient Light-Induced Water Evaporation. *Small* **2016**, *12*, 5320–5328. [CrossRef]

86. Patsalas, P.; Kalfagiannis, N.; Kassavetis, S. Optical properties and plasmonic performance of titanium nitride. *Materials* **2015**, *8*, 3128–3154. [CrossRef]
87. Gschwend, P.; Conti, S.; Kaech, A.; Maake, C.; Pratsinis, S. Silica-coated TiN particles for killing cancer cells. *ACS Appl. Mater. Interfaces* **2019**, *11*, 22550–22560. [CrossRef] [PubMed]
88. Jiang, W.; Fu, Q.; Wei, H.; Yao, A. TiN nanoparticles: Synthesis and application as near-infrared photothermal agents for cancer therapy. *J. Mater. Sci.* **2019**, *54*, 5743–5756. [CrossRef]
89. Hao, Q.; Li, W.; Xu, H.; Wang, J.; Yin, Y.; Wang, H.; Ma, L.; Ma, F.; Jiang, X.; Schmidt, O. VO_2/TiN plasmonic thermochromic smart coatings for room-temperature applications. *Adv. Mater.* **2018**, *30*, 1705421. [CrossRef]
90. Yang, L.; Tong, R.; Wang, Z.; Xia, H. Polydopamine Particle-Filled Shape-Memory Polyurethane Composites with Fast Near-Infrared Light Responsibility. *ChemPhysChem* **2018**, *19*, 2052–2057. [CrossRef]
91. Song, M.; Wang, Y.; Liang, X.; Zhang, X.; Zhang, S.; Li, B. Functional materials with self-healing properties: A review. *Soft Matter.* **2019**, *15*, 6615–6625. [CrossRef]
92. Tian, J.; Gu, J.; Peng, H.; Wang, H.; Du, Z.; Cheng, X.; Du, X. Sunlight-driven photo-thermochromic hybrid hydrogel with fast responsiveness and durability for energy efficient smart windows. *Compos. Part A Appl. Sci. Manuf.* **2021**, *149*, 106538. [CrossRef]
93. Deng, C. Graphene oxide Aggregate Modification and Photothermal Complex Properties of Phase Change Materials. Master's Thesis, East China University of Technology, Fuzhou, China, 2021.
94. Lebepe, T.; Oluwafemi, O. Thermal and Medium Stability Study of Polyvidone-Modified Graphene Oxide-Coated Gold Nanorods with High Photothermal Efficiency. *Nanomaterials* **2022**, *12*, 3382. [CrossRef]
95. Shan, G.; Xu, S.; Shi, W. Measuring the thermal conversion efficiency of gold nanoparticle solution. *Phys. Exp.* **2018**, *38*, 10–15.
96. Balitskii, O.; Mashkov, O.; Barabash, A. Ligand Tuning of Localized Surface Plasmon Resonances in Antimony-Doped Tin Oxide Nanocrystals. *Nanomaterials* **2022**, *12*, 3469. [CrossRef]
97. Huang, L.; Tang, H.; Bai, Y. Preparation of monodispersed $Cs_{0.33}WO_3$ nanocrystals by mist chemical vapor deposition for near-infrared shielding application. *Nanomaterials* **2020**, *10*, 2295. [CrossRef]
98. Tian, Q.; Jiang, F.; Zou, R.; Liu, Q.; Hu, J. Hydrophilic Cu_9S_5 nanocrystals: A photothermal agent with a 25.7% heat conversion efficiency for photothermal ablation of cancer cells in vivo. *Acs Nano* **2011**, *5*, 9761. [CrossRef]
99. Cui, J.; Jiang, R.; Xu, S.; Hu, G.; Wang, L. Cu_7S_4 nanosuperlattices with greatly enhanced photothermal efficiency. *Small* **2015**, *11*, 4183–4190. [CrossRef]
100. Zhang, J.; Liu, G.; He, F.; Chen, L.; Huang, Y. Au@Cu_7S_4 yolk–shell nanoparticles as a 980 nm laser-driven photothermal agent with a heat conversion efficiency of 63%. *RSC Adv.* **2015**, *5*, 87903–87907. [CrossRef]
101. Kaur, M.; Ishii, S.; Shinde, S.L.; Nagao, T. All-ceramic microfibrous solar steam generator: TiN plasmonic nanoparticle-loaded transparent microfibers. *ACS Sustain. Chem. Eng.* **2017**, *5*, 8523–8528. [CrossRef]
102. Li, Y.; Wang, D. Application of polydopamine in tumor nano-drug delivery. *Chin. J. Pharm.* **2021**, *19*, 81–86.
103. Boden, S.; Bagnall, D. Optimization of moth-eye antireflection schemes for silicon solar cells. *Prog. Photovolt. Res. Appl.* **2010**, *18*, 195–203. [CrossRef]

nanomaterials

MDPI

Review

Research Progress of Photo-/Electro-Driven Thermochromic Smart Windows

Xiaotong Zou, Haining Ji *, Yong Zhao, Mingying Lu, Jundong Tao, Pinghua Tang, Bin Liu, Xitao Yu and Yuliang Mao *

School of Physics and Optoelectronics, Xiangtan University, Xiangtan 411105, China;
xiaotongzou_0705@163.com (X.Z.); yongzhao_0912@163.com (Y.Z.); 202021001488@smail.xtu.edu.cn (M.L.);
JundongTao_314@163.com (J.T.); pinghuatang@xtu.edu.cn (P.T.); bl987@uowmail.edu.au (B.L.);
yxt1240710760@163.com (X.Y.)
* Correspondence: sdytjhn@126.com (H.J.); ylmao@xtu.edu.cn (Y.M.)

Abstract: Thermochromic smart windows can automatically control solar radiation according to the ambient temperature. Compared with photochromic and electrochromic smart windows, they have a stronger applicability and lower energy consumption, and have a wide range of application prospects in the field of building energy efficiency. At present, aiming at the challenge of the high transition temperature of thermochromic smart windows, a large amount of innovative research has been carried out via the principle that thermochromic materials can be driven to change their optical performance by photothermal or electrothermal effects at room temperature. Based on this, the research progress of photo- and electro-driven thermochromic smart windows is summarized from VO_2-based composites, hydrogels and liquid crystals, and it is pointed out that there are two main development trends of photo-/electro-driven thermochromic smart windows. One is exploring the diversified combination methods of photothermal materials and thermochromic materials, and the other is developing low-cost large-area heating electrodes.

Keywords: photo-/electro-driven; thermochromic; smart windows; research progress

Citation: Zou, X.; Ji, H.; Zhao, Y.; Lu, M.; Tao, J.; Tang, P.; Liu, B.; Yu, X.; Mao, Y. Research Progress of Photo-/Electro-Driven Thermochromic Smart Windows. *Nanomaterials* **2021**, *11*, 3335. https://doi.org/10.3390/nano11123335

Academic Editor: Alessia Irrera

Received: 10 November 2021
Accepted: 6 December 2021
Published: 8 December 2021

Publisher's Note: MDPI stays neutral with regard to jurisdictional claims in published maps and institutional affiliations.

1. Introduction

Rapid increasing energy consumption leads to an energy shortage, accompanied by environmental pollution. In developed countries, building energy consumption accounts for 20–40% of the total energy consumption [1]. Nowadays, some new technologies have been developed to adjust the indoor temperature and reduce building energy consumption, such as smart windows. In 1985, M. Lampert and C. G. Granqvist et al. [2] first proposed electrochromic materials for smart windows. Smart windows are composed of glass or transparent materials as substrates and dimming materials. Under certain conditions, the transparency can be adjusted to regulate the amount of sunlight, which can effectively save energy. According to different excitation means, smart windows can be divided into electrochromic, photochromic and thermochromic smart windows. Thermochromic smart windows have been widely investigated by researchers due to their simple structure, low preparation cost and active response to external temperature stimulation. However, when the external temperature is lower than the switching threshold, the optical performance of the thermochromic smart window is difficult to change. Therefore, many experts and scholars have combined photothermal or electric heating control to assist in driving the thermochromic smart window to achieve the transformation of optical performance at room temperature, further enhancing the applicability of the thermochromic smart windows.

At present, photo-/electro-driven thermochromic smart windows are mainly divided into VO_2-based composites, hydrogels and liquid crystals.

VO_2 is the most widely used inorganic thermochromic material, and is used for its metal-to-insulator transition in smart windows, which is accompanied by significant changes in its electrical, optical and magnetic properties. This transition is due to a structural change from a monoclinic semiconductor phase to a metallic tetragonal rutile structure when the sample temperature is above 68 °C. Based on its excellent metal–insulator transition (MIT) properties, VO_2 has attracted widespread attention from researchers in the field of smart windows [3–6] and infrared camouflage stealth [7–12]. However, in practical applications, VO_2 still has challenges, such as its high phase transition temperature and poor optical performances.

Hydrogels with a low critical temperature have a wide range of tunable chemical and physical properties, making them ideal for use in smart windows. Hydrogels can reversibly shift between transparent/opaque states through hydrophilic/hydrophobic phase transitions, responding to changes in the ambient temperature. At low temperatures, the hydrogels show a transparent state due to the formation of hydrogen bonds between hydrophilic groups and water molecules. When the temperature rises to a critical temperature, the hydrophobic group leads to an opaque state due to hydrogen bond breakage. Compared with VO_2, hydrogels are easy to prepare and have a low critical temperature. The lower critical solution temperature (LCST) of PNIPAm is only 32 °C. Therefore, they are also widely used in the field of smart windows [13,14].

However, the visible light transmittance of hydrogels at a high temperature is low, so it is necessary to adjust its thickness and design a suitable glass panel to increase the transparency of hydrogels before it can be used in the field of smart windows [15].

Similar to hydrogels, liquid crystals are mainly controlled by visible light. In thermodynamics, liquid crystals are between the crystalline solid state and the isotropic liquid state, and can simultaneously show the anisotropy of the crystals and the flow characteristics of the liquids. Thermochromic liquid crystals can change the arrangement and orientation of anisotropic molecules in response to the stimulation of temperature and voltage at the same time. Liquid crystal/polymer materials have multiple responsiveness and good mechanical properties, which make them widely used in smart windows [16].

Based on this, the paper mainly reviews photo- and electro-driven thermochromic smart windows, from the three aspects of VO_2, hydrogels and liquid crystals, as shown in Figure 1.

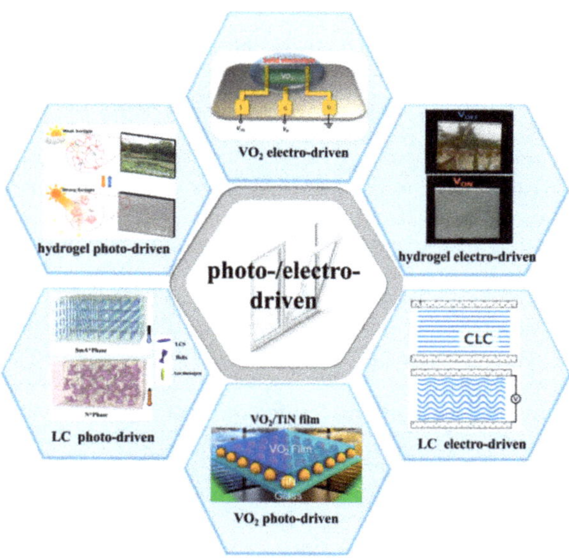

Figure 1. Types and working principle of photo- and electro-driven thermochromic smart windows.

2. Photo-Driven Thermochromic Smart Windows

The photo-driven thermochromic smart windows can actively adjust its optical properties in response to the changes in light radiation intensity and temperature, which is a completely passive way of light modulation. Compared with the single response of the traditional thermochromic smart windows, the photo-driven thermochromic smart windows realize the dual response of light and heat. It can respond to stimulation in areas with strong light radiation and a low temperature, and can regulate the sunlight, which further expands the application range of thermochromic smart windows.

2.1. VO_2-Based Smart Windows

In order to be close to practical applications, researchers usually reduce the phase transition temperature of VO_2 via doping elements, but this will also lead to a decrease in its optical properties [17–19]. However, by combining VO_2 with photothermal materials and using photothermal conversion, the sunlight is converted into heat energy, and VO_2 can be driven to change the optical performance at room temperature. Ji et al. [20] designed and assembled a composite film of PbS and VO_2. The UV-Vis-NIR light can be absorbed through the interaction of solar photons and PbS phonons and converted to heat energy, thus allowing for VO_2 to undergo a phase change at room temperature. Subsequently, in order to further improve the optical performance of the photo-driven VO_2 film, Hao et al. [21] prepared VO_2/TiN smart coatings for room temperature applications by hybridizing thermochromic VO_2 with plasmonic TiN nanoparticles (Figure 2a). The VO_2 phase transition was accelerated by the strong plasma absorption in the near-infrared region of the TiN plasmonic nanoarray (Figure 2b). In addition, the VO_2/TiN coating had a visible light transmittance of approximately 50% and an infrared conversion efficiency of 48% at 2000 nm, which effectively improved the optical performance of the light-driven VO_2 composite films (Figure 2c).

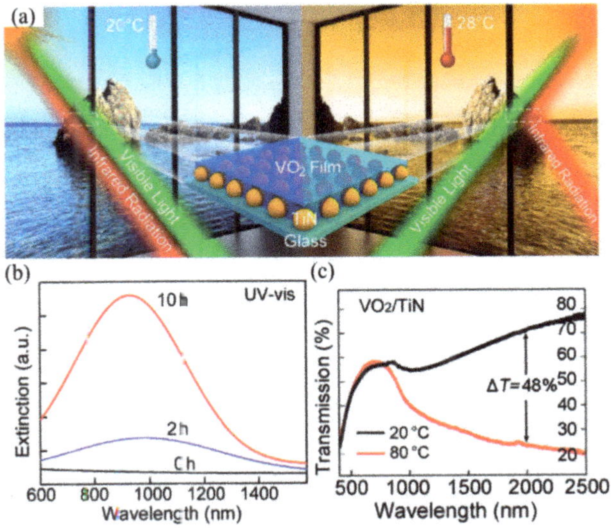

Figure 2. (**a**) Schematic diagram of VO_2/TiN smart windows. (**b**) UV-Vis spectra of TiO_x (0 h), TiO_xN_y (2 h), TiN (10 h) nanoarrays on quartz substrate. (**c**) Transmission spectra of VO_2/TiN coatings at 20 °C and 30 °C. Reproduced with permission from [21]. Copyright 2018, Wiley.

2.2. Hydrogel Smart Windows

Hydrogels are popular materials for photo-driven thermochromic smart windows. The combination of hydrogels with graphene oxide (GO) [22–25], antimony-doped tin oxide (ATO) [26–28], Cs_xWO_3 [29] and other high-absorption materials can stimulate the transformation of the optical properties of the hydrogels, thereby improving the conversion rate of the hydrogels.

GO is a common photothermal material. Under medium-intensity visible light irradiation, the fully reversible volume phase transition of the hydrogels can be triggered by the photothermal effect of GO [22]. Kim et al. [22] first studied the optical properties of hydrogels driven by the heat generated via GO. Thus, we can develop the switchable glazing of a novel photothermotropic mechanism that screens strong sunlight and heat radiation in response to the sunlight intensity, as well as the temperature. Subsequently, Kim et al. [23] prepared a photo-driven thermochromic smart window by using GO and poly(N-isopropylacrylamide) (PNIPAm). Lee et al. [24] prepared a gradient copolymer hydrogel containing GO by manipulating the monomer composition to control the hydrophilic–hydrophobic balance of the copolymer. Copolymer hydrogels exhibit different thermal behaviors according to the monomer composition. With changes in the temperature and light radiation, copolymer hydrogels can be gradually dehydrated, resulting in almost linear transmittance changes. Besides, in order to create a more comfortable and colorful life, the color change in thermochromic smart windows, such as the warm/cool-tone, is also the focus of research. As can be seen from Figure 3a,b, GO embedded within the thermotropic hydrogels can absorb the colored organic solvent, can prepare the smart glasses with arbitrary color and can further promote the development of color-tuning smart windows [25].

When choosing photothermal materials, it is necessary to consider its influence on the visible light transmittance of windows and to avoid materials that act on the visible light region. ATO mainly regulates near-infrared light. Compared with PNIPAm/GO film, PNIPAm/ATO film has better solar modulation in the near-infrared region [26]. Besides, the effect of Sb doping in ATO on its photothermal properties was also studied. The results showed that the PNIPAm/ATO film with 10% Sb doping had the best response speed and solar modulation ability [26,27] (Figure 3c). With the increase in ATO content and film thickness, the solar modulation ability and response speed will also increase [28]. Subsequently, a supramolecular nanocomposite hydrogel film was prepared by integrating ethylene glycol-modified pillar [5] arene (EGP5) and ATO [28]. Owing to the thermo-responsiveness of EGP5 and plasmonic heating induced by the near-infrared absorption of ATO, the film exhibited an outstanding photo-thermochromic effect, with an excellent solar modulation ability (56.1%) and initial luminous transmittance (77.2%) (Figure 3d). The dynamics and reversibility of the host–guest interaction between EGP5 and the pyridinium unit can avoid the collapse and damage of the polymeric hydrogel structure, which solved the problem of its poor durability and the repeatability of the hydrogels. Then, Wu et al. [29] prepared Cs_xWO_3/PAM-PNIPAm smart windows, in which Cs_xWO_3 was the photothermal component, PNIPAm was the optical control switch and polyacrylamide (PAM) hydrogel was the skeleton of the hydrogel matrix, so as to prevent any damage to the hydrogel structure. This smart window system was mainly heated by the near-infrared(NIR) light. While maintaining a good visible light transmittance, nearly 96% of the near-infrared light can be shielded by this window, and the indoor temperature can be maintained at approximately 25 °C. It can greatly reduce energy consumption from sources such as air conditioning and heating.

Photothermal materials, such as GO and Cs_xWO_3, mainly use high absorption characteristics to convert light energy into heat energy. However, AuNRs and Cu_7S_4 used a local surface plasmon resonance effect to convert absorbed light energy into the kinetic energy of electron resonance, and then converted it into the vibration energy of the lattice through the lattice scattering of electrons [30]. The heat energy of lattice vibration was transmitted to the surrounding environment, thereby increasing the temperature of the environment. Cao et al. (Figure 3e,f) [31] prepared polyvinylalcohol/thermochromic dyes/AuNRs sunlight-responsive smart window films through combining the thermal discoloration property of the thermochromic material and the photothermal effect of AuNRs.

Then, this composite material was combined with hydroxypropyl methylcellulose (HPMC) to prepare a smart window model with good optical properties. This model can achieve a stable optical performance transition under sunlight, and by controlling the content of AuNRs and the type of dye, a prototype of color smart windows with different switching temperatures can be achieved. (Figure 3g). Using a similar principle, Zhu et al. [32] prepared a smart window based on Cu_7S_4/PNIPAm hydrogel with a lower price and cost (Figure 3h). Based on the good light and heat effect of Cu_7S_4, this smart window can not only respond quickly at room temperature, but can also improve the indoor temperature in the cold.

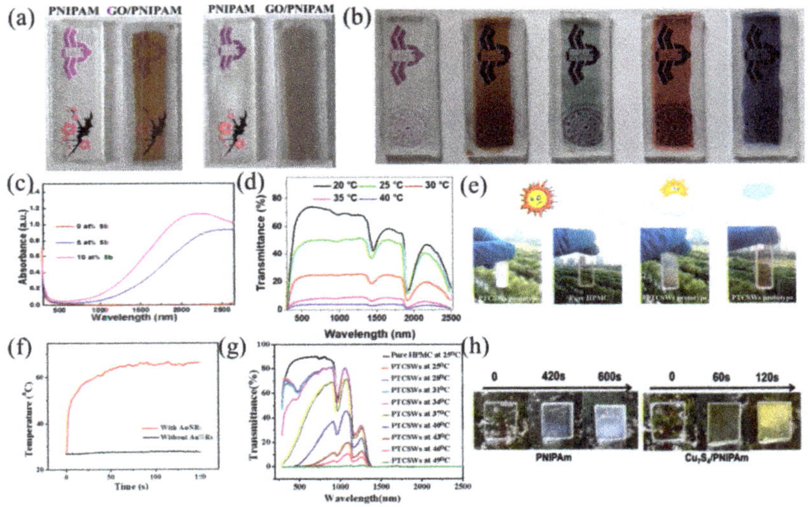

Figure 3. (**a**) Photos of the transparency/opacity transition of PNIPAm glass and GO/PNIPAm glass before and after 5 min of sunlight. Reproduced with permission from [25]. Copyright 2017, Elsevier. (**b**) Color GO/PNIPAm glass. Reproduced with permission from [25]. Copyright 2017, Elsevier. (**c**) Ultraviolet–visible–near infrared absorption spectra of 0.025 wt% SnO_2 with Sb doping content of 0, 5 and 10%. Reproduced with permission from [27]. Copyright 2017, American Chemical Society. (**d**) Transmission spectra of 5HATO films at different temperatures. Reproduced with permission from [28]. Copyright 2018, Elsevier. (**e**) PTCSWs prototype and pure HPMC in the hot summer (left), digital photos of soft sunny summer days (middle) and cloudy days (right). Reproduced with permission from [31]. Copyright 2018, Elsevier. (**f**) Temperature rise trajectory of PVA/dye film with or without AuNRs. Reproduced with permission from [31]. Copyright 2018, Elsevier. (**g**) UV-Vis-NIR spectra of PTCSWs prototype. Reproduced with permission from [31]. Copyright 2018, Elsevier. (**h**) Photos of Cu_7S_4/PNIPAm and PNIPAm hydrogels under real sunlight in summer (31 °C). Reproduced with permission from [32]. Copyright 2019, Elsevier.

2.3. *Liquid Crystal Smart Windows*

Liquid crystals respond to temperature stimulation by adjusting the orientation of anisotropic molecules. They show a high transparency at a low temperature and become blurred due to strong light scattering at a high temperature. They have broad application prospects in the field of smart windows. Through combining the liquid crystal polymer and the photothermal properties of Cs_xWO_3, a flexible multi-response smart film with excellent mechanical strength was obtained [33]. In actual application, the transmittance of visible light through temperature and infrared light can be controlled, and 95% of near-infrared radiation in the range of 800~2500 nm can be shielded by this film. This film can achieve a mass production by using a roll-to-roll process, which had great significance in the field of smart windows.

In addition, liquid crystals are often combined with azo dyes to prepare UV-driven liquid crystal devices. Azo dye is an ultraviolet (UV) photochromic material that can change its molecular shape when irradiated by light. The trans-formed form azo derivatives is rod-shaped, which can stabilize the liquid crystal phase, whereas the cis form is curved, and when it exists, will destroy the stability of the liquid crystal phase [34]. Under UV irradiation, the directional alignment of liquid crystal molecules can be induced by the photoisomerization of azobenzene molecules, thereby driving its phase transition. Oh et al. [35] proposed a photoelectrically adjustable cholesteric liquid crystal doped with push-pull azobenzene. These liquid crystals can be stimulated by light or heat to achieve a reversible conversion between transparency and opacity. Then, Oh et al. [36] further doped push-pull azobenzene and chiral liquid crystals to prepare liquid crystal smart windows controlled by ultraviolet light and temperature. Under UV irradiation, the transformation of the liquid crystals, from the transparent chiral smectic phase (SmA* phase) to the opaque chiral phase sequence (N* phase), was induced by the trans-cis photoisomerization of push-pull azobenzene (Figure 4a). In addition, Kuang et al. [37] prepared a stable azobenzene copolymer brush, which was used as a substrate to prepare UV-driven polymer-stabilized liquid crystal (PSLC) smart windows (Figure 4b).

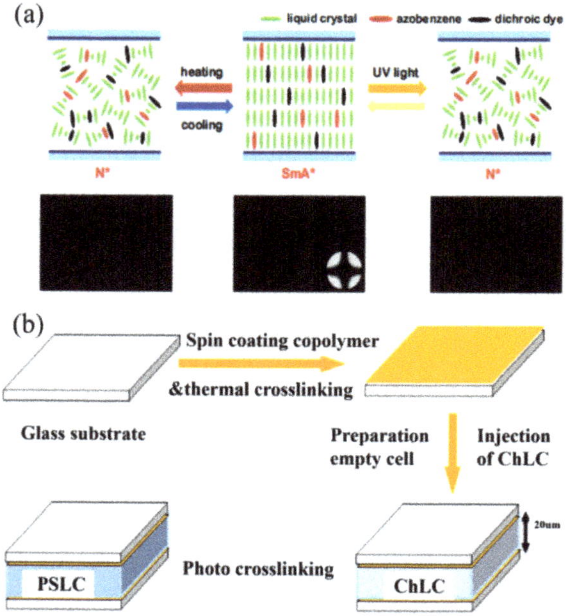

Figure 4. (**a**) Self-occlusion shutter schematic and POM images [36]. (**b**) Preparation process of polymer-stabilized liquid crystal (PSLC) smart window. Reproduced with permission from [37]. Copyright 2017, American Chemical Society.

2.4. Chapter Summary

A recent progression of photo-driven thermochromic smart windows is introduced in this section. In order to achieve the transformation of the optical properties of thermochromic materials at room temperature, through using the photothermal conversion performance of photothermal materials, the surface temperature of the composite device was increased, and the optical performance of the composite device was changed. Research from single-band light-absorbing materials to multi-band light-absorbing materials has improved the optical performance of smart windows to the greatest extent.

At present, the research on photo-driven thermochromic smart windows is still in the preliminary stage, and its research is mainly concentrated on three types of VO_2, hydrogels and liquid crystals. In the future, it is still necessary to explore the combination of different photothermal materials and thermotropic materials to further improve the optical performances and durability of smart windows, and to realize large-scale, industrialized production. So far, most of the research is still focused on the photothermal performance and optical performance of composite devices, whereas there is less research on the durability and mechanical performance of the composite device. This is an aspect that needs to be focused on.

3. Electro-Driven Thermochromic Smart Windows

Compared with the photo-driven thermochromic smart windows, the electro-driven thermochromic smart windows adds active regulation, which can independently adjust the transparency of the windows. The selection of electric heating materials and the structure of the electric heating layer were mainly researched in electro-driven thermochromic smart windows. The electrode materials with a low cost, mature preparation technology and high electrical and thermal conductivity tended to be selected. The structure of the electro-driven thermochromic smart windows was more complicated than those that were photo-driven, and its structure was also the focus of research. Further research is still needed in order to avoid the influence of the electric heating layer on the optical performance of the composite device and to increase the electric heating rate.

3.1. VO_2-Based Smart Windows

Due to the advantages of high transparency, high thermal conductivity and high electrical conductivity, electrothermal materials such as ITO and AgNWs have been integrated into different kinds of electro-driven thermochromic smart windows. Beydaghyan et al. [38] prepared VO_2 films with a high transmittance and excellent thermochromic switching performance on the ITO layer. A VO_2 phase transition was induced by Joule heating with ITO layer as the conductive layer. The transmittance of VO_2/ITO films at 2500 nm has a reversible conversion from as high as 65% to near to zero. The particle size of VO_2 deposited on ITO was smaller than that deposited on the glasses, which reduced the phase transition temperature of VO_2 to a certain extent. Compared with the heating plate, the phase transition temperature of VO_2 deposited on ITO is lowered by 4~7 °C. On this basis, Li et al. [39,40] studied the optical properties of VO_2-based electro-driven thermochromic smart windows with AgNWs and ITO as conductive materials. The decrease in the infrared spectrum can be observed in the AgNWs/VO_2 device at the voltage of 6.5–8 V (Figure 5a,b), and a further increasing of the voltage no longer leads to a decrease in the infrared spectrum. The same phenomenon was produced in ITO/VO_2 devices at approximately 12.5–15 V. However, when the voltage was applied, the ITO substrate had a strong absorption rate at 2500 nm, and its infrared switching performance could be extended to near infrared regions.

In practical applications, how to reduce production costs and realize large-area heating is a problem to be considered in electric-driven thermochromic smart windows. The preparation technology of ITO has become mature, but the price is expensive. Therefore, using ITO as an electric heating device will greatly increase the preparation cost. Generally, the use of cheap and stable electrodes, such as fluorine-doped SnO_2 (FTO) [41,42], Al: ZnO (AZO) [43,44], will be more conducive to the development of electrically driven thermochromic smart windows.

Both FTO and ATO were transparent conductive oxides that were easy to prepare and had a low cost. The optical constant of FTO is between VO_2 and glass, and the dielectric constant is similar to VO_2. Therefore, the optical performance of VO_2 can be effectively improved by introducing an FTO buffer layer between glass and VO_2. It has been proven that that VO_2 phase transition can be driven by applied voltage on FTO/VO_2/FTO multilayer films [38]. However, the threshold voltage also increased with the increase in the area of FTO and VO_2 [41,42]. Then, Xu et al. [42] directly deposited VO_2 on FTO glass. The experiment results showed that the rapid phase transition of VO_2 in a large area can be achieved by applying a voltage below 6 V.

Similarly, studies have shown that, as long as a weak electric field is applied to the AZO/VO_2 film, the fine control of the phase transition of the VO_2 film can also be realized [43]. Furthermore, the structure of the AZO/VO_2 film is also the focus of research. A suitable structure can reduce the voltage demand and further reduce energy consumption. The optical properties of AZO/VO_2 multilayer films with different structures were studied [44] (Figure 5c). The results showed that depositing AZO on the edge of VO_2 can maintain the Joule heating effect and the phase transition amplitude. The phase transition temperature of VO_2 was reduced by introducing strain at the same time. Compared with depositing VO_2 on AZO, this structure was more suitable for electric-driven thermochromic smart windows.

In addition to using the Joule heating effect to drive the phase transition of thermochromic materials, the phase transition can also be regulated by directly using bias voltage. In order to overcome the limitation of the high phase transition temperature of VO_2, Chen et al. [45] adopted the method of directly growing VO_2 on layered mica sheets and integrating it with highly transparent single wall carbon nanotube (SWNT) films (Figure 5d). By adjusting the bias current, it is possible to change the starting local temperature and shift the initial situation close to the "phase transition boundary", resulting in the decreased energy barrier in order to trigger the MIT behavior. This device has important application prospects in the future. Further, Chen et al. [46] modulated the reversible form of H from the VO_2 lattice with a solid electrolyte layer assisting gating treatment. The insulation–metal–insulation tristate phase transitions of VO_2 were realized (Figure 5e). The transition between H_xVO_2 and HVO_2 can produce an obvious electrochromic effect at room temperature. The dramatic increase in the visible/infrared transmittance due to the phase transition from the metallic (lightly H-doped) to the insulating (heavily H-doped) phase results in an increased solar energy regulation ability of up to 26.5%, while maintaining a 70.8% visible luminous transmittance (Figure 5f). This effectively overcame the defects of the traditional VO_2 intelligent windows. In addition, Lee et al. [47] integrated electrochromic materials and thermochromic materials into a single device to achieve a single or dual response, but this device failed to achieve effective control of the VO_2 phase transition. Therefore, Jia et al. [48] combined electrochromic and thermochromic materials to design an all-solid-state VO_2-based multilayer device with a $VO_2/LiTaO_3/WO_3$ sandwich structure. The reversible doping of Li in the VO_2 lattice was controlled by the bias voltage. This not only avoids the degradation of the optical performance of VO_2 caused by doping, but also effectively realizes the regulation of the phase transition temperature of VO_2.

3.2. *Hydrogel Smart Windows*

Traditional hydrogels, such as PNIPAm, are easily dehydrated in harsh environments. Therefore, they have a poor thermal stability. At 60 °C, the total weight of PNIPAm decreased by 90% after 30 min due to the poor water retention ability [49]. In the photo-driven part, some researchers have already proposed solutions for this [28,29], and in the electro-driven part, Gyenes et al. [50] prepared an electrically driven polymer gel smart window. The polymer gel was composed of two thermally induced gel layers. The active hydrogel layer was used to adjust the optical performance of the window, and the inactive gel layer prevented the active gel from generating spatial separation after phase separation, which was used to improve the stability of the device. The optical properties of the gel layer can be controlled by changing the audio frequency alternating current. In this way, the electrolysis and gas generation of hydrogels can be avoided, and the durability of hydrogel smart windows can be further improved. Besides, Chen et al. [51] designed a thermo- and electro- dual response smart window system based on P(NIPAM-Dav) (diallyl-viologen) ionic liquid (IL) gel. The transmittance change in the fabricated devices was observed to be greater than 50% at a wavelength of 580 nm. After 20 continuous cycles, the transmittance had no obvious decay, indicating that the device had good stability.

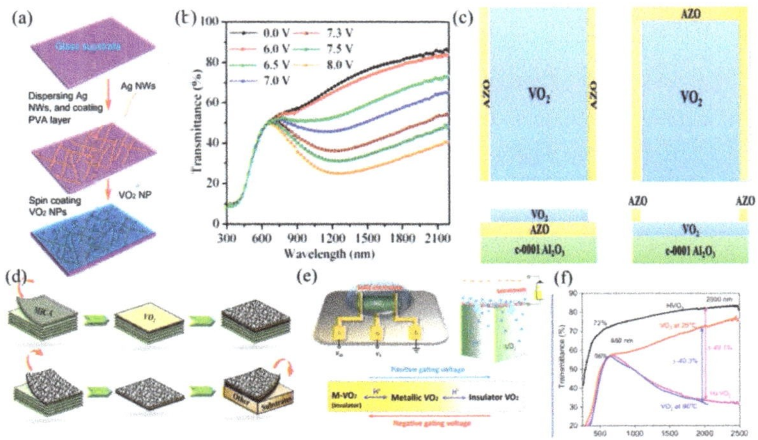

Figure 5. (**a**) The preparation process diagram of VO_2NF/AgNWS electrochromic film on glass substrate. Reproduced with permission from [39]. Copyright 2014, Royal Society of Chemistry. (**b**) Transmission spectra of VO_2NF/AgNWS electrochromic films under different applied voltages. Reproduced with permission from [39]. Copyright 2014, Royal Society of Chemistry. (**c**) VO_2/AZO multilayer structure, AZO at the bottom and edge diagram. Reproduced with permission from [44]. Copyright 2020, Springer Nature. (**d**) CNT-VO_2-MICA film preparation process schematic diagram. Reproduced with permission from [45]. Copyright 2017, Elsevier. (**e**) Gate control diagram of VO_2 device with source, drain and gate. Reproduced with permission from [46]. Copyright 2019, The American Association for the Advancement of Science. (**f**) Optical transmission spectra of VO_2, H_xVO_2 and HVO_2 films. Reproduced with permission from [46]. Copyright 2019, The American Association for the Advancement of Science.

Furthermore, in order to improve their optical performance and the response rate of electro-driven hydrogels smart windows, it is necessary to focus on the materials and structures of the electric heating layers [52–57]. Zhou et al. [52] prepared transparent conductive grids by using silver nanoparticles through a simple room temperature preparation process (Figure 6a) and studied the electrothermal effects of conventional grids and honeycomb structures. The results showed that the initial light transmittance of the honeycomb was 30% lower than that of the grid structure, but the heating efficiency was higher (Figure 6b). Then, in order to seek lower cost electrode materials, using Sn [53] and Cu [54] as conductive materials, a high connectivity metal wire mesh was prepared on the HPMC hydrogel layer by means of crack lithography. (Figure 6c,d). The Sn electrode had a light transmittance of approximately 80%. With the 8 V voltage applied, the transformation of the optical properties can be driven by the Joule heat generated by the Sn electrode. Similar electro-driven characteristics can be produced in the Cu electrode at a voltage of 4.5 V. Importantly, the variation of hydrogel optical properties can be driven by the two electrodes, with approximately only a 0.2 W/cm^2 power consumption. They are inexpensive, and can be used as a substitute for electrodes such as Ag and ITO.

3.3. Liquid Crystal Smart Windows

Cholesteric liquid crystals are also called spiral liquid crystals, and can selectively reflect the light incident along the spiral axis. Chen et al. [58] demonstrated a generally transparent smart window based on a cholesteric liquid crystal with negative dielectric anisotropy (Figure 7). The transparency of this window can be controlled through the field strength. When there was no voltage, it showed a well-arranged plane cholesteric texture, showing a transparent state. When applying voltage, it presented a diffuse state and became blurred.

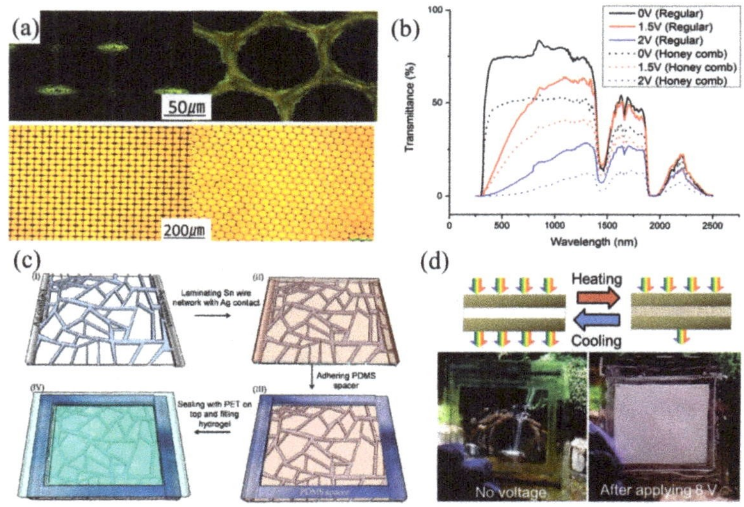

Figure 6. (**a**) Ag mesh electrode square and honeycomb arrays. Reproduced with permission from [52]. Copyright 2016, Wiley. (**b**) Sunlight transmittance of conventional and honeycomb arrays [52]. (**c**) Schematic diagram of the manufacturing steps of Sn/hydrogel devices. Reproduced with permission from [53]. Copyright 2017, Elsevier. (**d**) Cu/HPMC hydrogel before and after Joule heating [53]. Reproduced with permission from [53]. Copyright 2017, Royal Society of Chemistry.

In recent years, liquid crystal polymers have attracted extensive research interest due to their excellent physical and chemical properties. Liquid crystal polymers are mainly divided into polymer-dispersed liquid crystal (PDLC) and polymer stabilized liquid crystal (PSLC). PDLC has good mechanical properties, but, in the absence of electric field, its molecular arrangement is disordered and presents a fuzzy state. PSLC can stabilize the initial orientation of liquid crystal molecules through the network, showing a transparent state. Based on this, mesogenic functionalized graphene (MFG) is integrated with PSLC to prepare a film with a self-assembled chiral structure. The optical properties of this film can be controlled by electricity heat and near-infrared light [59]. Later, the good mechanical properties of PDLC were combined with the high initial transparency of PSLC to prepare a composite film of a phase-deposited polymer network [60], which had a good mechanical strength and processability. This film can regulate the visible light and near-infrared by responding to the stimulation of electricity, heat and ultraviolet light, has a region of 400–2500 nm and has excellent optical properties.

3.4. Chapter Summary

Compared with the photo-driven thermochromic smart window, the electro-driven thermochromic smart window added active regulation, which can independently adjust the transparency of the windows. The selection of electric heating materials and the structure of the electric heating layer were mainly researched in electro-driven thermochromic smart windows. Electrode materials with a low cost, mature preparation technology and high electrical and thermal conductivity tended to be selected. The structure of the electro-driven thermochromic smart windows was more complicated than that of the photo-driven; in order to avoid the influence of the electric heating layer on the optical performance of the composite device and to increase the electric heating rate, its structure was also the focus of research, and further research is still needed.

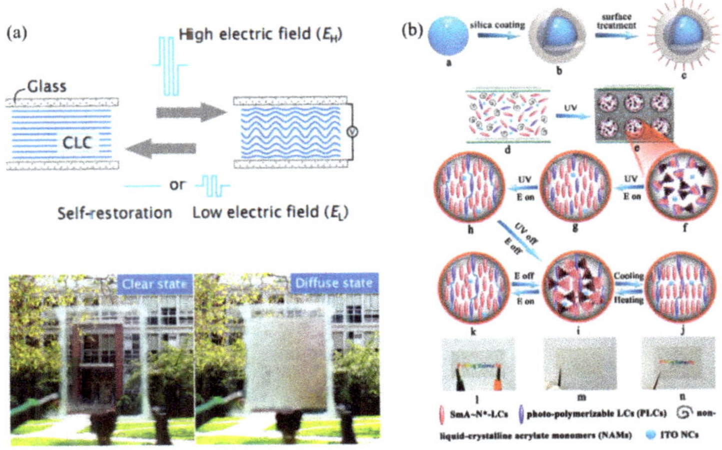

Figure 7. (**a**) Schematic diagram and photo of CLC smart window (d ≈ 45 μm). Reproduced with permission from [58]. Copyright 2018, Optica. (**b**) Schematic diagram of preparation and response of ITOncs/liquid crystal. (**a–c**): a single ITO NC is encapsulated with a hydrophilic silicon barrier to form a core/shell structure, and then subjected to methacryloylpropyl-trimethoxysilane (MPTMS) surface treatment; (**d–k**): Preparation procedures for the smart film: a homogenous polymeric syrup is sandwiched between two plastic transparent substrates (**d**), and the film is irradiated with ultraviolet light to form a porous polymer network and LCs droplets (**e**,**f**). Then, an electric field is applied to perpendicularly orient the LCs (**g**), meantime, a second step of UV polymerization is carry out to complete the cross-linking between PLCs in the LCs droplets to form orientated liquid-crystalline polymer networks in the porous structure (**h**). According to temperature or electric field, the original film (l) can be reversibly changed between transparent (**n**) and opaque (**m**). Reproduced with permission from [60]. Copyright 2017, Royal Society of Chemistry.

4. Conclusions

Aimed at the problem of thermochromic smart windows responding to high temperatures in practical applications, this paper summarized the research progress of photo-/electro-driven thermochromic smart windows. It is generally accepted that photothermal materials or electrothermal materials were used to cooperate with thermochromic materials to control the incidence of sunlight at room temperature. This method can effectively solve the problem of an excessively high response temperature of thermochromic smart windows, further increasing the application range of thermochromic smart windows. This can promote the development of smart windows in terms of no energy consumption and a high popularity. However, the current photo-/electric-driven thermochromic smart windows are still in the preliminary stage of development. To commercialize the photo-/electric-driven smart windows as soon as possible, future research should focus on the following three aspects.

(1) Photo-/electro-driven thermochromic smart windows were still limited by the characteristics of the materials, and each material used in reports now has certain shortcomings. In the future, it is still necessary to modify these materials and develop novel photothermal materials, and it is urgent to explore the combination of different photothermal materials and thermotropic materials in order to improve the optical performance of smart windows;

(2) To promote the commercialization of photo-/electro-driven thermochromic smart windows, smart windows that have good mechanical effects, a low cost, are easy to promote and have long-term stability need to be prepared. Therefore, reducing the cost of materials

and improving the stability and mechanical properties of devices are crucial in promoting commercial production;

(3) Photo-/electro-driven thermochromic smart windows have greatly promoted the development of smart windows in the direction of no energy consumption. The next development trend is to combine them with energy storage devices to collect energy and further promote energy conservation.

Author Contributions: Investigation, X.Z., H.J., Y.Z. and M.L.; Writing, X.Z. and H.J.; Review and editing, X.Z., H.J., J.T., P.T., B.L., X.Y. and Y.M. All authors have read and agreed to the published version of the manuscript.

Funding: This work was financially supported by the National Natural Science Foundation of China (Grant No. 51902276), the Natural Science Foundation of Hunan province (Grant No. 2019JJ50583).

Institutional Review Board Statement: Not applicable.

Informed Consent Statement: Not applicable.

Data Availability Statement: Not applicable.

Conflicts of Interest: The authors declare no conflict of interest.

References

1. Pérez-Lombard, L.; Ortiz, J.; Pout, C. A review on buildings energy consumption information. *Energy Build.* **2008**, *40*, 394–398. [CrossRef]
2. Svensson, J.; Granqvist, C.G. Electrochromic coatings for smart windows: Crystalline and amorphous WO_3 films. *Thin Solid Films* **1985**, *126*, 31–36. [CrossRef]
3. Ji, H.; Liu, D.; Cheng, H.; Zhang, C. Inkjet printing of vanadium dioxide nanoparticles for smart windows. *J. Mater. Chem. C.* **2018**, *6*, 2424–2429. [CrossRef]
4. Cui, Y.; Ke, Y.; Liu, C.; Chen, Z.; Wang, N.; Zhang, L.; Zhou, Y.; Wang, S.; Gao, Y.; Long, Y. Thermochromic VO_2 for energy-efficient smart windows. *Joule* **2018**, *6*, 1707–1746. [CrossRef]
5. Ji, H.; Liu, D.; Cheng, H.; Tao, Y. Large area infrared thermochromic VO_2 nanoparticle films prepared by inkjet printing technology. *Sol. Energy Mater. Sol. Cells* **2019**, *194*, 235–243. [CrossRef]
6. Kim, J.; Paik, T. Recent advances in fabrication of flexible, thermochromic vanadium dioxide films for smart windows. *Nanomaterials* **2021**, *11*, 23674. [CrossRef] [PubMed]
7. Ji, H.; Liu, D.; Zhang, C.; Yang, L.; Cheng, H.; Zheng, W. Snowflake-Like monoclinic VO_2 powders: Hydrothermal synthesis, characterization and in situ monitoring phase-transition behavior. *Sci. Adv. Mater.* **2017**, *9*, 861–867. [CrossRef]
8. Ji, H.; Liu, D.; Zhang, C.; Yang, L. Vanadium dioxide nanopowders with tunable emissivity for adaptive infrared camouflage in both thermal atmospheric windows. *Sol. Energy Mater. Sol. Cells* **2018**, *175*, 96–101. [CrossRef]
9. Ji, H.; Liu, D.; Cheng, H.; Zhang, C. VO_2/ZnS core-shell nanoparticle for the adaptive infrared camouflage application with modified color and enhanced oxidation resistance. *Sol. Energy Mater. Sol. Cells* **2018**, *176*, 1–8. [CrossRef]
10. Liu, D.; Ji, H.; Peng, R.; Cheng, H.; Zhang, C. Infrared chameleon-like behavior from VO_2(M) thin films prepared by transformation of metastable VO_2(B) for adaptive camouflage in both thermal atmospheric windows. *Sol. Energy Mater. Sol. Cells* **2018**, *185*, 210–217. [CrossRef]
11. Ji, H.; Liu, D.; Cheng, H.; Zhang, C.; Yang, L.; Ren, D. Infrared thermochromic properties of monoclinic VO_2 nanopowders using a malic acid-assisted hydrothermal method for adaptive camouflage. *RSC. Adv.* **2017**, *7*, 5189–5194. [CrossRef]
12. Li, M.; Liu, D.; Cheng, H.; Peng, L.; Zu, M. Manipulating metals for adaptive thermal camouflage. *Sci. Adv.* **2020**, *6*, eaba3494. [CrossRef] [PubMed]
13. Li, K.; Meng, S.; Xia, S.; Ren, X.; Gao, G. Durable and controllable smart windows based on thermochromic hydrogels. *ACS Appl. Mater. Interfaces* **2020**, *12*, 42193–42201. [CrossRef] [PubMed]
14. Zhou, Y.; Dong, X.; Mi, Y.; Fan, F.; Xu, Q.; Zhao, H.; Wang, S.; Long, Y. Hydrogel smart windows. *J. Mater. Chem. A* **2020**, *8*, 10007. [CrossRef]
15. Zhou, Y.; Cai, Y.; Hu, X.; Long, Y. Temperature-responsive hydrogel with ultra-large solar modulation and high luminous transmission for "smart window" applications. *J. Mater. Chem. A* **2014**, *2*, 13550. [CrossRef]
16. Keller, H.; Hatz, R. *Handbook of Liquid Crystals*; Verlag Chemie: Hongkong, China, 1980.
17. Ji, H.; Liu, D.; Cheng, H. Infrared optical modulation characteristics of W-doped VO_2(M) nanoparticles in the MWIR and LWIR regions. *Mater. Sci Semicond. Process.* **2020**, *119*, 105141. [CrossRef]
18. Kurajica, S.; Mandić, V.; Panžić, I.; Gaboardi, M.; Mužina, K.; Lozančić, A.; Šipušić, J.; Munda, I.K.; Višić, L.; Blagojević, S.L.; et al. In-operando diffraction and spectroscopic evaluation of pure, zr-, and ce-doped vanadium dioxide thermochromic films derived via glycolate synthesis. *Nanomaterials* **2020**, *10*, 2537. [CrossRef] [PubMed]

19. Guan, H.; Zhang, D.; Yang, Y.; Liu, Y.; Zhong, A.; He, Q.; Qi, J.; Fan, P. A novel method for notable reducing phase transition temperature of VO_2 films for smart energy efficient windows. *Nanomaterials* **2020**, *10*, 58. [CrossRef]
20. Ji, Q. Design of Intelligent Infrared Control Nanomaterial System. Master's Thesis, Zhengzhou University, Zhengzhou, China, 2014.
21. Hao, Q.; Li, W.; Xu, H.; Wang, J.; Yin, Y.; Wang, H.; Ma, L.; Ma, F.; Jiang, X.; Schmidt, O.G.; et al. VO_2/TiN plasmonic thermochromic smart coatings for room-temperature applications. *Adv. Mater.* **2018**, *30*, 1705421. [CrossRef] [PubMed]
22. Kim, D.; Lee, H.; Yoon, J. Remote control of volume phase transition of hydrogels containing graphene oxide by visible light irradiation. *RSC Adv.* **2014**, *4*, 25379. [CrossRef]
23. Kim, D.; Lee, E.; Lee, H.; Yoon, J. Energy efficient glazing for adaptive solar control fabricated with photothermotropic hydrogels containing graphene oxide. *Sci. Rep.* **2015**, *5*, 7646. [CrossRef] [PubMed]
24. Lee, E.; Kim, D.; Yooh, J. Stepwise Activation of switchable glazing by compositional gradient of copolymers. *ACS Appl. Mater. Interfaces* **2016**, *8*, 26359–26364. [CrossRef] [PubMed]
25. Chou, H.; Chen, Y.; Lee, C.; Tai, N. Switchable transparency of dual-controlled smart glass prepared with hydrogel-containing graphene oxide for energy efficiency. *Sol. Energy Mater. Sol. Cells* **2017**, *166*, 45–51. [CrossRef]
26. Huang, H.; Ng, M.; Wu, Y.; Kong, L. Solvothermal Synthesis of Sb:SnO_2 nanoparticles and ir shielding coating for smart window. *Mater. Des.* **2015**, *88*, 384–389. [CrossRef]
27. Lee, Y.; Cai, Y.; Bi, S.; Liang, Y.N.; Song, Y.; Hu, X.M. A dual responsive nanocomposite towards climate adaptable solar modulation for energy saving smart windows. *ACS Appl. Mater. Interfaces* **2017**, *9*, 6054–6063. [CrossRef]
28. Xu, Z.; Wang, S.; Hu, X.; Jiang, J.; Wang, L. Sunlight-induced photo-thermochromic supramolecular nanocomposite hydrogel film for energy-saving smart window. *Sol. RRL* **2018**, *2*, 1800204. [CrossRef]
29. Wu, M.; Shi, Y.; Li, R.; Wang, P. Spectrally selective smart window with high near-infrared light shielding and controllable visible light transmittance. *ACS Appl. Mater. Interfaces* **2018**, *10*, 39819–39827. [CrossRef]
30. Brongersma, M.L.; Halas, N.J.; Nordlander, P. Plasmon-induced hot carrier science and technology. *Nat. Nanotechnol.* **2015**, *10*, 25–34. [CrossRef]
31. Cao, D.; Xu, C.; Lu, W.; Qin, C.; Chen, S. Sunlight-driven photo-thermochromic smart windows. *Sol. RRL* **2018**, *2*, 1700129. [CrossRef]
32. Zhu, H.; Wang, L. Smart window based on Cu_7S_4/hydrogel composites with fast photothermal response. *Sol. Energy Mater. Sol. Cells* **2019**, *202*, 110109. [CrossRef]
33. Liang, X.; Gou, C.; Chen, M.; Gou, S.; Zhang, L.; Li, F.; Guo, S.; Yang, H. A roll-to-roll process for multi-responsive soft-matter composite films containing Cs_xWO_3 Nanorods for energy-efficient smart window applications. *Nanoscale Horiz.* **2017**, *2*, 319–325. [CrossRef]
34. Ikeda, T.; Tsutsumi, O. Optical switching and image storage by means of azobenzene liquid-crystal films. *Science* **1995**, *268*, 1873–1875. [CrossRef]
35. Oh, S.; Baek, J.; Kim, S.; Yoon, T. optical and electrical switching of cholesteric liquid crystals containing AZO dye. *RSC Adv.* **2017**, *7*, 19497–19501. [CrossRef]
36. Oh, S.; Baek, J.; Kim, S.; Yoon, T. Self-shading with optically- and thermally-switchable liquid crystals. *SID Symp. Dig. Tech. Pap.* **2018**, *49*, 554–556. [CrossRef]
37. Kuang, Z.; Deng, Y.; Hu, J.; Tao, L.; Wang, P.; Chen, J.; Xie, H.L. Responsive smart windows enabled by the azobenzene copolymer brush with photothermal effect. *ACS Appl. Mater. Interfaces* **2019**, *11*, 37026–37034. [CrossRef]
38. Beydaghyan, G.; Basque, V.; Ashrit, P.V. High contrast thermochromic switching in vanadium dioxide (VO_2) thin films deposited on indium tin oxide substrates. *Thin Solid Films* **2012**, *522*, 204–207. [CrossRef]
39. Li, M.; Ji, S.; Pan, J.; Wu, H.; Zhong, L.; Wang, Q.; Li, F.; Li, G. Infrared response of self-heating VO_2 nanoparticles film based on Ag nanowires heater. *J. Mater. Chem. A* **2014**, *2*, 20470–20473. [CrossRef]
40. Li, M.; Wu, H.; Li, Z.; Wang, H.; Lou, Y.; Li, G. Active and dynamic infrared switching of VO_2(M) nanoparticle film on ITO glass. *J. Mater. Chem. C* **2016**, *4*, 1579–1583. [CrossRef]
41. Hao, R.; Li, Y.; Liu, F.; Sun, Y.; Tang, J.; Chen, P.; Jiang, W.; Wu, Z.; Xu, T.; Fang, B. Electric field induced metal-insulator transition in VO_2 thin film based on FTO/VO_2/FTO structure. *Infrared Phys. Technol.* **2016**, *75*, 82–86. [CrossRef]
42. Xu, Z.; Qin, G.; Bernussi, A.; Fan, Z. Electrothermally control of dynamic infrared switching of VO_2 thin film on FTO glass. *J. Alloys Compd.* **2020**, *858*, 157640. [CrossRef]
43. Skuza, J.; Scott, D.; Mundle, R.; Pradhan, A. Electro-thermal control of aluminum-doped zinc oxide/vanadium dioxide multilayered thin films for smart-device applications. *Sci. Rep.* **2016**, *6*, 21040. [CrossRef]
44. Behera, M.; Williams, L.; Pradhan, S.; Bahoura, M. Reduced transition temperature in Al:ZnO/VO_2 based multi-layered device for low powered smart window application. *Sci. Rep.* **2020**, *10*, 1842.
45. Chen, Y.; Fan, L.; Qi, F.; Xu, W.; Chen, S.; Zan, G.; Ren, H.; Song, L.; Zou, C. Free-standing SWNTs/VO_2/Mica hierarchical films for high-performance thermochromic devices. *Nano Energy* **2017**, *31*, 144–151. [CrossRef]
46. Chen, S.; Wang, Z.; Ren, H.; Chen, Y.; Yan, W.; Wang, C.; Li, B.; Jiang, J.; Zou, C. Gate-controlled VO_2 phase transition for high-performance smart windows. *Sci. Adv.* **2019**, *5*, eaav6815. [CrossRef]
47. Lee, S.J.; Choi, D.S.; Kang, S.H.; Yang, S.; Nahm, S.; Han, S.H.; Kim, T. VO_2/WO_3-based hybrid smart windows with thermochromic and electrochromic properties. *ACS Sustain. Chem. Eng.* **2019**, *7*, 7111–7117. [CrossRef]

48. Jia, H.; Cao, X.; Shao, Z.; Long, S.; Jin, L.; Ma, L.; Chang, T.; Xu, F.; Yang, Y.; Bao, S.; et al. Dual-response and Li+-insertion induced phase transition of VO_2-based smart windows for selective visible and near-infrared light transmittance modulation. *Sol. Energy Mater. Sol. Cells* **2019**, *200*, 110045. [CrossRef]
49. Zhou, Y.; Michael, L.; Wang, S.; Hu, P.; Ke, Y.; Magdassi, S.; Long, Y. Fully Printed Flexible Smart Hybrid Hydrogels. *Adv. Funct. Mater.* **2018**, *28*, 1705365. [CrossRef]
50. Gyenes, T.; Szilágyi, A.; Lohonyai, T.; Zrínyi, M. Electrically adjustable thermotropic windows based on polymer gels. *Polym. Adv. Technol.* **2003**, *14*, 757–762. [CrossRef]
51. Chen, F.; Ren, Y.; Guo, J.; Yan, F. Thermo- and electro-dual responsive Poly(ionic Liquid) electrolyte based smart windows. *Chem. Commun.* **2016**, *53*, 1595–1598. [CrossRef] [PubMed]
52. Zhou, Y.; Layani, M.; Boey, F.; Chiang, Y.; Sokolov, I.; Magdassiet, S.; Long, Y. Electro-thermochromic devices composed of self-assembled transparent electrodes and hydrogels. *Adv. Mater. Technol.* **2016**, *1*, 160069. [CrossRef]
53. Kiruthika, S.; Kulkarni, G. Energy efficient hydrogel based smart windows with low cost transparent conducting electrodes. *Sol. Energy Mater. Sol. Cells* **2017**, *163*, 231–236. [CrossRef]
54. Singh, A.K.; Shanmugam, K.; Mondal, I.; Kulkarni, G. Fabrication of solar and electrically adjustable large area smart windows for indoor light and heat modulation. *J. Mater. Chem. C* **2017**, *5*, 5917–5922. [CrossRef]
55. Garcia-Garcia, F.J.; Gil-Rostra, J.; Terriza, A.; González, J.C.; Cotrino, J.; Frutos, F. Low refractive index SiOF thin films prepared by reactive magnetron sputtering. *Thin Solid Films* **2013**, *542*, 332–337. [CrossRef]
56. Tao, X.; Liu, D.; Yu, J.; Cheng, H. Reversible metal electrodeposition devices: An emerging approach to effective light modulation and thermal management. *Adv. Opt. Mater.* **2021**, *8*, 2001847. [CrossRef]
57. Gil-Rostra, J.; García-García, F.; Yubero, F.; González-Elipe, A.R. Tuning the transmittance and the electrochromic behavior of $Co_xSi_yO_z$ thin films prepared by magnetron sputtering at glancing angle. *Sol. Energy Mater. Sol. Cells* **2014**, *123*, 130–138. [CrossRef]
58. Chen, C.; Brigeman, A.; Ho, T.; Khoo, I. Normally transparent smart window based on electrically induced instability in dielectrically negative cholesteric liquid crystal. *Opt. Mater. Express* **2018**, *8*, 691–697. [CrossRef]
59. Wang, L.; Bisoyi, H.; Zheng, Z.; Gutierrez-Cuevas, K.; Singh, G.; Kumar, S.; Bunning, T.J.; Li, Q. Stimuli-directed self-organized chiral superstructures for adaptive windows enabled by mesogen-functionalized graphene. *Mater. Today* **2017**, *20*, 230–237. [CrossRef]
60. Liang, X.; Guo, S.; Chen, M.; Li, C.; Wang, Q.; Zou, C.; Zhang, C.; Zhang, L.; Guo, S.; Yang, H. Temperature and electric field-responsive flexible smart film with full broadband optical modulation. *Mater. Horiz.* **2017**, *4*, 878–884. [CrossRef]

MDPI
St. Alban-Anlage 66
4052 Basel
Switzerland
Tel. +41 61 683 77 34
Fax +41 61 302 89 18
www.mdpi.com

Nanomaterials Editorial Office
E-mail: nanomaterials@mdpi.com
www.mdpi.com/journal/nanomaterials

www.ingramcontent.com/pod-product-compliance
Lightning Source LLC
Chambersburg PA
CBHW061605210926
51850CB00009B/2222
9783036582511